A Modern Approach to Chemistry

A Modern Approach to Chemistry

Edited by
Johana Meyer

WILLFORD PRESS

www.willfordpress.com

Published by Willford Press,
118-35 Queens Blvd., Suite 400,
Forest Hills, NY 11375, USA

ISBN: 978-1-68285-624-6

Cataloging-in-Publication Data

A modern approach to chemistry / edited by Johana Meyer.
   p. cm.
Includes bibliographical references and index.
ISBN 978-1-68285-624-6
1. Chemistry. I. Meyer, Johana.
QD31.3 .M63 2019
540--dc23

For information on all Willford Press publications
visit our website at www.willfordpress.com

WILLFORD PRESS

Contents

Preface

Chemistry is the scientific study of compounds, their composition, structure, properties and reactions that they undergo. It addresses varied questions on the interaction between atoms and molecules and their chemical bonds for the formation of compounds. It covers all types of bonding, such as ionic bonds, hydrogen bonds, Van der Waals force bonds, etc. Some of the applications of chemistry are in the extraction of metals from ores, fermentation of beer and wine, extraction of chemicals from plants, medicine, etc. Besides these, it also has applications in the production of various chemical products like soap, glass, perfume, etc. This book presents the complex subject of chemistry in the most comprehensible and easy to understand language. From theories to research to practical applications, case studies related to all contemporary topics of relevance to this field have been included in this book. It will serve as a reference to a broad spectrum of readers.

Various studies have approached the subject by analyzing it with a single perspective, but the present book provides diverse methodologies and techniques to address this field. This book contains theories and applications needed for understanding the subject from different perspectives. The aim is to keep the readers informed about the progresses in the field; therefore, the contributions were carefully examined to compile novel researches by specialists from across the globe.

Indeed, the job of the editor is the most crucial and challenging in compiling all chapters into a single book. In the end, I would extend my sincere thanks to the chapter authors for their profound work. I am also thankful for the support provided by my family and colleagues during the compilation of this book.

Editor

Biosorption of Copper Cu (II) in aqueous solution by chemically modified crushed marine algae (*Bifurcaria bifurcata*): Equilibrium and kinetic studies

Nawal Benzidia[1], Anas Salhi [2], Salem Bakkas [1] and Layachi Khamliche [1,*]

[1] Laboratory of Organic Chemistry, Bioorganic and Environment
[2] Laboratory of Water and Environment
Faculty of Science, University Chouaïb Doukkali, El Jadida, Morocco.

Abstract: The Moroccan coastlines are known to be rich in marine algae species. Among these seaweed species; *Bifurcaria bifurcata* with its important biomass has been employed as a new adsorbent for the removal of copper in aqueous solution. Indeed, adsorption tests showed that the equilibrium is established after 60 minutes. Our experimental results demonstrate that the adsorption of copper on the algae is dependent on the pH of the solution and the initial concentration of copper. The adsorption capacity was determined using the Langmuir and the Freundlich isotherms. The maximum adsorption capacity is 101,9 mg.g^{-1}. The kinetic studies carried out showed that the adsorption of copper by the algae follows pseudo-second order kinetic model with high R^2 values.

Keywords: *Bifurcaria bifurcata*; adsorption; biosorption; Copper, pollution; environment.

Introduction

Copper is an element found naturally in the earth's crust and diffuses into the environment through natural phenomena. This compound, essential for the development of all known forms of life, is widely used in metallurgy for making alloys and especially as a fungicide in viticulture [1,2]. The increase in its production in recent years is the origin of the increasing amounts found in nature.

Copper is one of the most toxic heavy metals. Therefore, when the soil contains high amounts of copper (II) ions, the effects are harmful on humans and other living beings. Acute poisoning due to inhalation following exposure to copper (II) ions is very common and can cause irritation, headaches, stomach upsets, dizziness, vomiting and diarrhea[3]. In contrast, oral poisoning is very rare, usually accidental and is accompanied by renal insufficiency, and in severe cases, loss of consciousness or other disorders leading to death [3].

Decontamination of heavy metals from wastewater has been a challenge for a long time. A number of methods have been developed for removal of copper from industrial effluents in order to protect the environment and to possibly re-use water, especially in arid countries [4,5].

However, these conventional technologies are expensive [6,7]. Hence, an alternative adsorption on biological materials (ie biosorption); available and relatively inexpensive; may give rise to a technically reliable and economically viable clearance process.

In this work, we proposed to study at first the biosorption capacity of copper by a marine biomass *Bifurcaria bifurcata,* from aqueous solutions prepared in the laboratory with different concentrations of copper. Thereafter, to optimize the influence of some key parameters on biosorption, we performed a modeling of isotherms and a kinetic study.

Materials and methods

Chemical pretreatment of biomass

Once collected from the Atlantic coast of the city of El Jadida (Morocco), algae *Bifurcaria bifurcata* were washed with water, dried in an oven at 60 ° C for 24 hours, then ground and sieved to obtain the fraction size of 0,50 mm. The crude material has been activated by successive immersions in solutions of NaOH (0,75 M), HCl (0,75 M) then NaCl (2 M) following a method developed in the laboratory.

Corresponding author: Layachi Khamliche
E-mail address: *Khamliche@yahoo.fr*

Preparation and dosage of Copper solutions

A copper stock solution was prepared by dissolving a given amount of copper sulphate in distilled water to obtain a concentration of 100 mg.L^{-1}, further concentrations are obtained if necessary by successive dilutions. The initial pH was adjusted with dilute solutions of HCl (0,1M) and NaOH (0,1M).

The residual copper concentration in the reaction mixture was determined by inductively coupled plasma atomic emission spectroscopy (ICP-AES) and the copper adsorption capacity was calculated using the following formula [8].

$$q_t = \frac{(C_0 - C_t) * V}{m}$$

Where q_t (mg.g^{-1}) is the amount adsorbed at time t (min), C_0 (mg.L^{-1}) is the initial concentration of copper, C_t (mg.L^{-1}) is the concentration at time t, V is the solution volume (mL) and m (g) is the amount of the adsorbent used.

Batch method

The experiments were conducted in "batch method". Basically, 0.1g of the algae was added to 50 ml of a copper solution of known concentration, in a 250 ml Erlenmeyer. This solution is left under constant stirring at room temperature. After 60 min, the solution is filtered and then stored away from light, at 3°C and a pH <2 in order to avoid any changes on the samples before analysis.

Adsorption kinetics

The kinetic study of copper biosorption was performed to determine the contact time required to reach equilibrium. During the contact time, the solution is kept under constant agitation, at temperature and pH both constant (pH = 5,6 and T = 298 K). Samples were taken every 10 min for 100 min.

Adsorption isotherm

In order to obtain the adsorption isotherms, we mixed 0,1g of the adsorbent with 50 ml of the copper solution at different initial concentrations ranging from 10 mg.L-1 to 1 g.L^{-1}, with constant stirring at room temperature.

Linear regression analysis

In this study, all the model parameters were evaluated by linear regression to determine the best fitting of the equation to the experimental data. The sole correlation coefficient (R^2) is not sufficient; to this purpose we have also used the residual root mean square error (RMSE) and the Chi-square test χ^2 to measure the goodness-of-fit. RMSE can be represented as follows [9]:

$$RMSE = \sqrt{\frac{1}{m-2}\sum_{i=1}^{m}(Y_i - y_i)^2}$$

Where:

Y_i: the experimental value;

y_i: the value obtained by calculating from the model;

m: the number of observations .

Smaller RMSE values indicate better curve fitting.

The Chi-square test is basically the sum of the squares of the differences between the experimental values and the values obtained by calculating from the model, with each squared difference divided by the corresponding values calculated from the model [9]. It can be defined as follows:

$$\chi^2 = \sum_{i=1}^{m}\frac{(Y_i - y_i)^2}{y_i}$$

If data from the model are similar to the experimental data, the χ^2 test is a small number.

Results and Discussion

Effect of pH

Copper biosorption by *Bifurcaria bifurcata* depends on the pH influencing the nature of the biosorbent surface, the degree of ionization and the nature of the adsorbate ionic species. In fact various adsorption studies at different pH values ranging from 1 to 10, show that copper adsorption is pH dependent as we can notice in Figure 1.

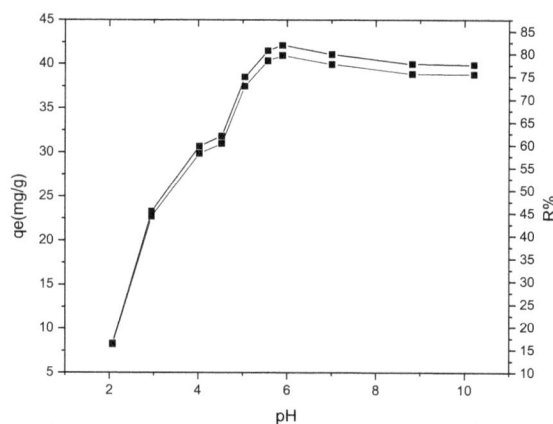

Figure 1. Effect of pH on copper adsorption by the algae. Adsorbent mass = 0.1 g; solution volume = 50 mL; concentration = 10 mg.L^{-1}; time = 60 min; temperature = 298 K.

Figure 1 clearly shows the influence of the pH on the adsorption rate. Based on the curve, we see that at low pH values, the adsorption rate is low, it reaches 15% at pH = 2.

When the pH value is between 2 and 6, the copper retention rate increases rapidly to a maximum of 82% at pH 5,6. Thereafter, the adsorption rate decreases to 75% between pH 6 and 10. These results allowed us to determine the pH corresponding to the best adsorption efficiency. Accordingly, all the copper adsorption studies were carried out at pH 5,6.

The main functional groups responsible for the absorption of metals, including copper on the algae are the carboxylic, hydroxylic and sulfonates groups [11] of the polysaccharide present within the surface of the biosorbent, the carboxylic groups being the most active.

Thus, at low pH, the adsorption of copper is low. The active groups are poorly ionized and the concentration of $[H_3O^+]$ is higher in solution favoring the protonation of the functional active sites on the surface of the algae [12,13].

Beyond pH = 2, the concentration of $[H3O^+]$ decreases compared to that of Cu^{2+} ions, which explains the increase in the rate of adsorption.

At high pH, the active sites are deprotonated enabling an interaction with the metallic ions instead of $H3O^+$ ions.

The reduction in the adsorption rate observed at pH=10 is probably due to the formation of soluble hydroxyl complexes displaying more affinity with the functional sites of algae at the expense of copper.

Effect of the initial concentration

To study the effect of the initial concentration of the pollutant on the adsorption process, we varied the concentration of copper from 10 to 1500 mg .L^{-1}. The results obtained are shown in Figure 2 below:

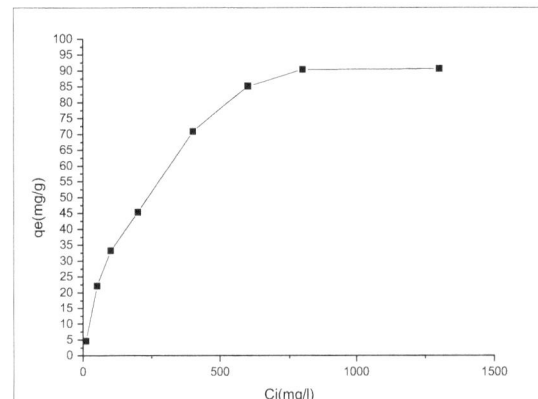

Figure 2. Effect of initial concentration on copper adsorption by the algae. Adsorbent mass = 0,1 g; solution volume = 50 ml; pH = 5,6; time = 60 min; temperature = 298 K

The curve indicates a rapid increase in storage capacity at low concentrations (10 to 200 mg.L^{-1}). Binding capacity continues to increase with the rise of the initial concentration, indicating the existence of attractions forces between the exchange surface and the metal cation. Then, an equilibrium plateau appears as the concentration reaches the value of 500 mg.L^{-1}. This plateau may reflect the saturation of active sites involved in the adsorption process.

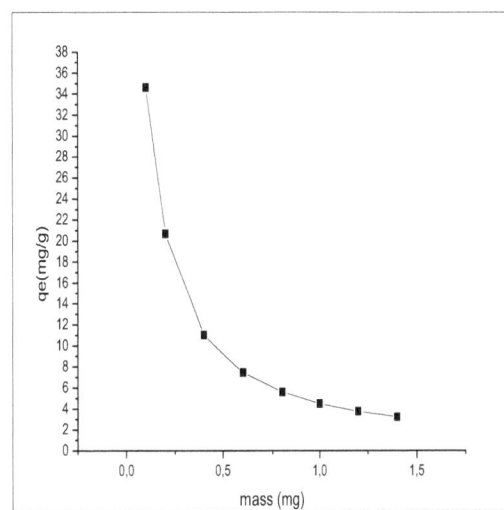

Effect of biosorbent dosage

To examine the effect of the biosorbent dosage, we varied the amount of biosorbent from 0,1 to 1,4 g using a fixed solution volume.

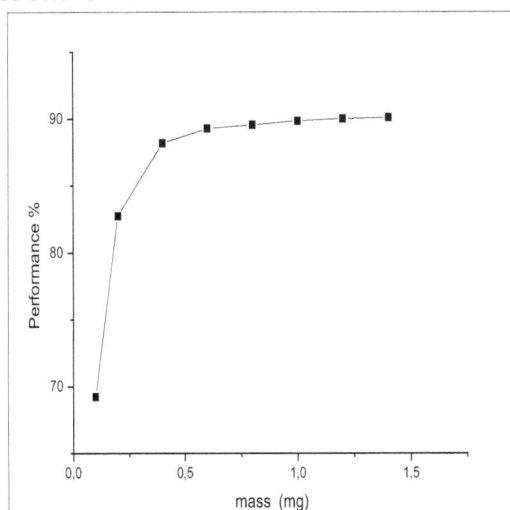

Figure 3a. Effect of the biosorbent dosage on copper adsorption efficiency by the algae. pH = 5,6; solution volume = 50 ml; time = 60 min; concentration = 100 mg.L^{-1}; temperature = 298K

Figure 3b. Effect of biosorbent dosage on the amount of copper absorbed by the algae. pH = 5,6; solution volume = 50 ml; time = 60 min; concentration = 100 mg.L^{-1}; temperature = 29

Figure 3a shows that the copper removal efficiency increases strongly with the increasing amount of the adsorbent till 0,5 mg. In contrast, Figure 3b shows that the amount of copper absorbed (q_e) in mg decreases rapidly. Starting from 0,5 mg of algae, a plateau appears on both studies (efficiency and q_e) [14, 15.16].

The decrease in biosorption efficacy with increasing doses of biomass can be explained as follows:
•   As the amount of adsorbent added to the copper solution is low, Copper ions can easily access the adsorption sites. The addition of adsorbent can increase the number of adsorption sites but copper ions have greater difficulty in approaching these sites because of congestion.
•   A large amount of adsorbent creates aggregates of particles, resulting in a reduction of the total surface area for adsorption, and therefore, a decrease in the amount of adsorbate per unit mass of adsorbent [17, 18].

According to both Figures 3a and 3b, we can deduce the following results:
For an amount of 1,4g of adsorbent, we have a copper removal efficiency of 90% and a q_e of 5 mg.g^{-1}. However, for an amount of 0,1 g of adsorbent, we have a 70% of efficiency and a q_e of 35 mg.g^{-1}. It is clear that the use of 0,1g of seaweed (14 times less) leads to the best results (35 mg of copper adsorbed per g of algae instead of 5 mg.g^{-1} for 1,4 g of algae). All these data, obtained with simple experiments may be of great interest to optimize the purification of industrial effluents. Therefore, in the following studies, we used the amount of 0,1g adsorbent with 50 ml of solution.

Adsorption kinetics
One of the most used parameters to estimate the purification performance of an adsorbent is the adsorption kinetics. Not only it estimates the amount of adsorbed pollutants over time but also it gives the contact time between the biosorbent and pollutants to reach equilibrium. Moreover, kinetics provides information on the adsorption mechanism and the mode of transfer of solutes from liquid to solid phase [19].

Contact time
Initially, we were interested in the effect of contact time on copper adsorption by the algae. We have studied the evolution of three solutions with concentrations of 10,100 and 400 mg.L^{-1}. The results obtained are summarized in Figure 4.

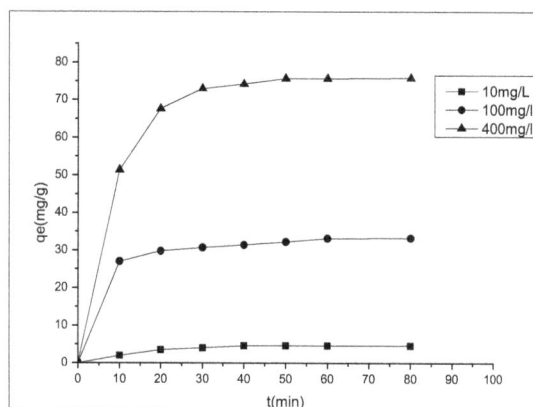

Figure 4. Effect of contact time on copper adsorption by the algae. Adsorbent mass = 0,1 g; solution volume = 50 ml; pH = 5,6; temperature = 298 K; time = 0-80 min

Figure 4 shows the effect of the initial concentration of copper on the retention rate at different contact times. For the three concentrations used, the retention rate increases with the increase in reaction time following two different slopes. The first is rapid, in the first 20 minutes, while the second is slow and may express the balance between retained and desorbed Copper fractions. The adsorption capacity increases with increasing concentration of the solution to reach the values of 76, 33 and 4,6 mg.g^{-1} respectively to the concentrations of 400, 100 and 10 mg.L^{-1}.

Kinetic study
Like the adsorption equilibrium, the kinetics of adsorption of a material can be modeled. In this regard, the literature reports a number of models, such as the model of Lagergren (first-order model), the kinetic model of second order and intra-particle diffusion model. The majority of works consulted evaluate the kinetic potential of biosorbents by second order kinetic model [20,21].

Then, to examine the mechanism of the adsorption processes (mass transfer or chemical reaction), we have used kinetic models to analyze our experimental results. Different models such as the external diffusion model on homogeneous area, the intra-particle diffusion model and the surface reaction model are applied in a batch system in order to describe the phenomenon of transport of the adsorbate through the pores of the adsorbent [22].

The external diffusion step:
The following kinetic expression is often cited and used to model the external diffusion of any solute transfer from liquid to any another phase, including solid one [23].

$$-\frac{dC_t}{dt} = k\left(\frac{a}{v}\right) \cdot (C_t - C_e)$$

C_e: concentration of solute in the solution at equilibrium;
a: area of the solid interface / liquid;
V: volume of solution.

The integrated form is the following:

$$\ln\left[\frac{(C_0 - C_e)}{(C_t - C_e)}\right] = k\left(\frac{a}{v}\right) \cdot t = Kt$$

A simple plot of ln [C_0- C_e) / (C_t - C_e)] in function of the reaction time should allow us to evaluate whether the external diffusion step is determining for the entire reaction [24].

Intra-particle diffusion step:
The model of the intra-particle diffusion is represented by the following equation [25, 26].

$$q_t = k_D t^{1/2} + C_D$$

K_D: rate constant of intra- particle diffusion (mg.g^{-1}.min$^{1/2}$);
C_D: constant connected to the trendline.

Surface reaction step:
Expression of pseudo first order
The expression of pseudo-first order, often cited, is the following [11, 27].

$$+\frac{dq_t}{dt} = k_1 \cdot (q_e - q_t)$$

The linear form is the following:

$$\ln(q_e - q_t) = \ln(q_e) - k_1 \cdot t$$

K_1: the first-order rate constant (min^{-1});
t: time (min).

Expression of pseudo- second order
The expression of pseudo- second order is often used [22, 28].

$$+\frac{dq_t}{dt} = k_2(q_e - q_t)^2$$

The linear form is the following:

$$\frac{t}{q_t} = \frac{1}{K_2 \cdot q_e^2} + \frac{1}{q_e}t$$

K_2: pseudo-second-order rate constant (g.mg^{-1}. min).

The set of kinetic parameters of biosorption is determined from the linearity straights according to the equations of the models described above.

In order to examine the reliability of the proposed kinetic models, we calculated the equations correlation coefficients, as well as the kinetic constants of each model which are summarized in Tables 1a and 1b.

Table 1a. Parameters of the external diffusion model and the intra-particle diffusion model of the kinetics of copper biosorption by the algae

C_0 (mg.l^{-1})	The external diffusion model				the intra-particle diffusion model				
	k	R^2	RMSE	χ^2	kd (mg.g^{-1}·min$^{-0,5}$)	C	R^2	RMSE	χ^2
10	0,0443	0,9784	0,1059	0,0095	0,6796	2,3456	0,9031	0,3380	0,1064
100	0,0717	0,9784	0,0992	0,0118	1,2713	23,460	0,9387	0,4958	0,0253
400	0,1218	0,9182	0,5684	0,2393	6,0025	36,577	0,8000	1,1245	0,9166

Table 1b. Parameters of the reaction surface model of kinetics of copper biosorption by the algae

C_0 (mg.l^{-1})	surface reaction models										
	Pseudo-first order					Pseudo-second order					
	k_1 (min^{-1})	q_e (mg .g^{-1})	R^2	RMSE	χ^2	k_2 (g.mg^{-1}.min^{-1})	q_e (mg.g^{-1})	R^2	RMSE	χ^2	
10	0,0027	27,66	0,9084	0,0331	0,0016	0,456	4,757	0,9994	0,0164	0,0014	
100	0,0020	28,31	0,8604	0,0393	0,0274	9,2.10^{-3}	34,55	0,9996	0,0159	0,0013	
400	0,0087	52,17	0,8573	0,0953	0,0407	4,59.10^{-3}	75	0,9976	0,0090	0,0008	

From these results, it appears that the adsorbed amount at equilibrium q_e increases with the increase of the initial concentration. Furthermore, the R^2 values obtained with the model of pseudo- second order are very high and greater than 0,99, the values of RMSE and χ^2 instead, are poor .They far exceed those obtained with the model of pseudo-first order, the intra- particle diffusion and the external diffusion. The amounts absorbed at equilibrium qe reach 4,757; 34.55 and 75 mg.g-1 corresponding respectively to the concentrations of 10; 100 and 400 mg.L^{-1} which are very close to the experimental values of 4.625, 33 and 76 mg.g-1. This finding leads us to confirm that the adsorption process follows the pseudo-second-order model.

Adsorption isotherm

The adsorption isotherms provide evaluation of copper adsorption capacity on the seaweed. These curves link the amount of copper adsorbed per unit mass of the algae (qe) to the concentration of copper remaining in solution (C_e).

Figure 5 shows the isotherm of copper adsorption by the algae.

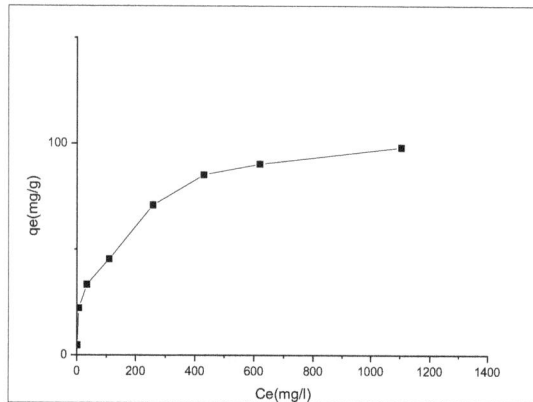

Figure 5. The isotherm of copper adsorption by the algae. Adsorbent mass = 0,1 g; solution volume = 50 ml; time = 60 min ; pH = 5,6 ; temperature = 298K

To determine the amount of copper adsorbed per unit mass of the algae (q_e), several authors have proposed theoretical or empirical models to describe the relationship between the amount adsorbed and the residual concentration both at equilibrium. The main models described in literature are [11,29].

• **Langmuir model:**

$$q_e = \frac{q_{m,L} * K_L * C_e}{1 + K_L * C_e}$$

$$\frac{C_e}{q_e} = \frac{1}{q_{m,L} * K_L} + \frac{1}{q_{m,L}} C_e$$

Where q_e: adsorption capacity at equilibrium (mg.g^{-1});

q_m: adsorption capacity of saturation (mg.g^{-1});

K_L: Langmuir constant.

The plot of C_e/q_e in function of C_e, allows to determine q_m and K_L.

Another parameter; R_L can explain the adsorption according to Langmuir as following:

$$R_L = \frac{1}{1 + K_L * C_0}$$

Based on the R_L value, we can say that the adsorption is favorable when R_L tends to zero ($R_L \rightarrow 0$) or unfavorable when R_L tends to one ($R_L \rightarrow 1$) [29].

• **Freundlich model:**
The simple and empirical model of Freundlich is the most commonly used[29].

$$q_e = K_F * C_e^n$$

The linear form is the following:

$$Ln q_e = Ln K_F + n Ln C_e$$

K_F: constant relative to the adsorption capacity (mg mg$^{(1-n)}$.L^n.g^{-1});

n: constant (dimensionless) giving an indication of the adsorption intensity.

It is generally accepted that low values of n (0,1 <n < 0,5) are characteristics of a good adsorption, whereas higher values indicate a moderate adsorption (0,5 <n<1) or low (n> 1). The constant « n » is often replaced by « 1/n » or heterogeneity factor [29].

The adsorption isotherm is an L type [30], suggesting a progressive saturation of the solid. When the equilibrium concentration (C_e) tends to zero, the slope of the isotherm is constant.

In order to explain the obtained equilibrium isotherms, experimental data were treated according to the isotherms models described above (Freundlich model and Langmuir model).

Table 2. Parameters of isotherms models of copper adsorption by the algae

Langmuir model						Freundlich model				
q_m (mg.g$^-$)	K_L (l.mg^{-1})	R_L	R^2	RMSE	χ^2	K_F mg$^{(1-n)}$.l^n.g^{-1}	n	R^2	RMSE	χ^2
101,9	70,466	0,124	0,9927	0,0276	0,0454	7,333	0,3974	0,9461	0,3660	0,1129

From the values of the correlation coefficients R^2, the residual root mean square error (RMSE) and the Chi-square test (Table.2) we can conclude that the Langmuir model is the closest to the experimental results. We also note that the R_L coefficient of this model tends to 0 which explains that the adsorption of copper on the algae is favorable.

Conclusion

In the present work, we studied copper biosorption capacity by a marine biomass; *Bifurcaria bifurcata*; very available along the Moroccan Atlantic coast. Experimental results show that the adsorption process depends on the solution pH, the initial concentration of copper and the amount of biomass. The kinetic study of copper adsorption on the algae showed that the adsorption process follows the pseudo-second-order model. The modeling of isotherms reveals that the Langmuir model best expresses this type of adsorption. The copper ions are adsorbed as monolayers, without interaction with adjacent ions, increasing the order of their distribution on the bioadsorbent surface. It is interesting to note however, that based on the Langmuir isotherm, the exceptional value of the maximum adsorption capacity reaches 101,9 mg.g^{-1}.

Taking together the results of this study, we suggest that *Bifurcaria bifurcata* algae may be considered a promising biological material to be used as an effective adsorbent for the removal of copper present in liquid effluents.

References

1- F. Edeline, Traitement des eaux industrielles chargées en métaux lourds, tribune de l'eau édition CEDEDOC. **1993**, N° 565,5.

2- VCH Verlags « Water » in Ullman's Encyclopedia of Industrial chemistry. **1995**, vol 8.

3- a) MG. Miquel, Les effets des métaux lourds sur l'environnement et santé- rapport de l'office parlementaire d'évaluation des choix scientifiques et technologiques, France. **2001**.
b) S. H. Abbas, I. M. Ismail, T. M. Mostafa, A. H. Sulaymon, Biosorption of Heavy Metals: A Review, Journal of Chemical Science and Technology, **2014**, 3(4), 74-102.

4- D. Hébert, R. Elwell, S. Travelos, J. Fitz, R. Bucher, Subchronic toxicity of cupric sulfate administered in drinking water and feed to ratsand mice, Fund. Appl. Toxical. **1993**, 21, 461.

5- E. Hosovski, A. Viakovic, J. Sunderic, Kidney injuries due to inhalation of copper dust and fumes- Abstracts 23 rd, International Congress on Occupational Health-Montréal.**1990**.

6- F. Berne, J. Cordonnier, Industrial water treatment, Edition Technip Paris. **1995**.

7- Deborah Chapman, Water quality assessments, second edition, UNESCO/WHO/UNEP. **1996**.

8- M. Montazer-Rahmatia, P. Rabbania, A. Atefeh, A. Keshtkarb, Kinetics and equilibrium studies on biosorption of cadmium, lead, and nickel ions from aqueous solutions by intact and chemically modified brown algae, Journal of Hazardous Materials. **2011**, 185, 401-407.

9- Y. Ho, W. Chiu, C. Wang, Regression analysis for the sorption isotherms of basic dyes on sugarcane dust, Bioresource Technology. **2005**, 96, 1285-1291.

10- J. Li, Q. Lin, X. Zhang, Y. Yan, Kinetic parameters and mechanisms of the batch biosorption of Cr(VI) and Cr(III) onto Leersia hexandra Swartz biomass, J. Colloid Interface Sci. **2009**, 333 , 71-77.

11- E. Fourest, B. Volesky, Contribution of sulfonate groups and alginate to heavy metal biosorption by the dry biomass of Sargassum fluitans, Environ. Sci. Technol. **1996**, 30, 277-282;

12- Y. Liu, Q. Cao, F. Luo, J. Chen , Biosorption of Cd^{2+}, Cu^{2+}, Ni^{2+} and Zn^{2+} ions from aqueous solutions by pretreated biomass of brown algae ,Journal of Hazardous Materials ,**2009**,16, 931-938.

13- Z. Aksu, Equilibrium and kinetic modelling of cadmium (II) biosorption by C. vulgaris in a batch system: effect of temperature, Sep. Purif. Technol. **2001**, 21, 285-294.

14- S. Karthikeyan, R. Balasubramanian, C. S. P. Iyer, Evaluation of the marine algae Ulva fasciata And Sargassumsp for the biosorption of Cu (II) from aqueous solutions, Bioresource Technology. **2007**, 98, 452-455.

15- V. S. Munagapati, V. Yarramuthi, S. K. Nadavala, S. R. Alla, K. Abburi, Biosorption of Cu(II), Cd(II) and Pb(II) by Acacia leucocephala bark powder: kinetics, equilibrium and thermodynamics, Chemical Engineering Journal. **2010**, 157, 357-365.

16- E. Fourest, J.C. Roux, Heavy metal biosorption by fungal mycelial by-products: mechanisms and influence of pH, Applied Microbiology and Biotechnology. **1992**, 37, 399-403.

17- V.K. Gupta, A. Mittal, V. Gajbe, Adsorption and desorption studies of water soluble dye, Quinoline Yellow, using waste materiels, J. Colloid. Interf. Sci. **2005**, 284, 89-98.

18- M. Arecoa, S. Hanelab, J. Duranb, M. Afonsoa, Biosorption of Cu(II), Zn(II), Cd(II) and Pb(II) by dead biomasses of green alga Ulvalactuca and the development of a sustainable matrix for adsorption, Journal of Hazardous Materials. **2012**, 213-214, 123-132.

19- P. Lodeiro, J.L. Barriada, R. Herrero, M.E. Sastre de Vicente, The marine macroalga Cystoseira baccata as biosorbent for cadmium (II) and lead (II) removal: Kinetic and equilibrium studies , Kinetic and equilibrium studies. Environmental Pollution. **2006**, 142, 264-273.

20-L. Liggieri, F. Ravera, A. Passerone, A diffusion-based approach to mixed adsorption kinetics, Colloids and Surfaces A: Physicochemical and Engineering Aspects. **1996**, 114, 351-359.

21-Y. Cheng et al., J. Environ. Eng., **2015**, 10.1061/(ACSE)EE.1943-7870.0000956, C4015001.

22-J.P. Vilar, Botelho, M.S. Boaventura, A.R. Boaventura , Equilibrium and kinetic modeling of Cd(II) biosorption by algae Gelidium and agar extraction algal waste , Water Res. **2006**, 40, 291–302.

23-Lian-Ming SUN, Techniques de l'ingénieur Adsorption Aspects théoriques. J 2 730.

24-P. Lodeiro, B. Cordero, J.L. Barriada, R. Herrero, M.E. Sastre de Vicente, Biosorption of cadmium by biomass of brown marine macroalgae, Bioresource Technology. **2005**, 96, 1796-1803.

25-O. Hamdaouia, E. Naffrechoux, Modeling of adsorption isotherms of phenol and chlorophenols onto granular activated carbon Part I. Two-parameter models and equations allowing determination of thermodynamic parameters ,Journal of Hazardous Materials. **2007**, 147, 381-394.

26-I. Langmuir, The constitution and fundamental properties of solids and liquids, J. Am. Chem. Soc. **1916**, 38, 2221-2295.

27-D. Kumar, J.P. Gaur, Chemical reaction- and particle diffusion-based kinetic modeling of metal biosorption by a Phormidium sp.-dominated cyanobacterial mat, Bioresource Technology. **2011**, 102, 633-640.

28-Y.S. Ho, Review of second-order models for adsorption systems, J. Hazard. Mater. **2006**, 136, 681-689.

29-A.M. Abdel -Aty, N.S. Ammar, H.H. Abdel Ghafar, R.K. Ali , Biosorption of cadmium and lead from aqueous solution by fresh water alga Anabaena sphaerica biomass, Journal of Advanced Research. **2013**, 4, 367-374.

30- J.P. Gaudet, L. Charlet, S. Szenknect, V. Barthès, M. Krimissa, Sorption isotherms: A review on physical bases, modeling and measurement, Applied Geochemistry. **2007**, 22, 249-275.

DNA binding and biological activity of mixed ligand complexes of Cu(II), Ni(II) and Co(II) with quinolones and N donor ligand

S.M.M. Akram[1,2], Aijaz Ahmad Tak[1], Peerzada, G. Mustafa[2] and Javid A. Parray[3]

[1]Department of Chemistry, Islamia College of Science & Commerce, Srinagar India
[2]Department of Chemistry, University of Kashmir, India
[3]Centre of Research for Development, University of Kashmir, India

Abstract: Mixed ligand complexes of Cu(II), Ni(II) and Co(II) have been synthesized by using levofloxacin and bipyridyl and characterized using spectral and analytical techniques. The binding behavior of the Ni(II) and Cu(II) complexes with hs-DNA were determined using electronic absorption titration, viscometric measurements and cyclic voltammetry measurements. The binding constants calculated for Cu(II) and Ni(II) complexes are 2.0 x 10^4 and 4.0 x 10^4 M^{-1} respectively. Detailed analysis reveals that the metal complexes interact with DNA through intercalative binding mode. The protective activity of Cu(II) and Ni(II) complexes with ct-DNA was carried out using agarose gel electrophoresis technique. The antioxidant activities for the synthesized complexes have been tested and the antibacterial activity for Ni(II) complex was also checked.

Key words: Intercalation, hypochromism, red shift, peak potential.

Introduction

Quinolones are large group of synthetic antibacterial agents containing 4-oxo-1,4-dihydroquinoline Skelton used in practice for treatment of a variety of bacterial infections[1-3]. Since the main targets of quinolones are DNA gyrase and topoisomerase IV participating in DNA replication[4], the interaction with DNA and the antibacterial activity are of great importance and thoroughly studied[5]. Levofloxacin (Figure 1) is used to treat infections including respiratory tract infections, cellulites, urinary tract infections, prostrates, anthrax, endocarditis, meningitis, pelvic inflammatory disease, traveler's diarrhea and Plague.

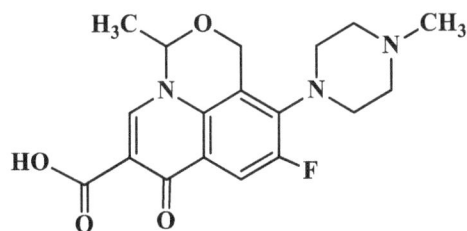

Figure 1. Structure of Levofloxacin

Based on the increased activity shown by metal-drugs complexes in comparison to their parent compounds[6], diverse complexes of metal and quinolone ligands have been synthesized and characterized. Their interaction with DNA[7-10] and serum albumin proteins[11-14], their antibacterial activity[15-17] and potential antitumor activity[18-20] have been evaluated in comparison to free quinolones. Mixed ligand of nickel and copper with quinolones using NN donor ligands have been synthesized and explored for their biological activities[21-23]. Nickel is one of the most essential elements to a healthy life for humans and higher animal species like chicken, rats, pigs, cows, sheep and goat[24-25]. It stabilizes DNA and RNA against thermal denaturation[24-25] and activates many enzymes like arginase, tyrosinase and phosphor-glucomutase. As for copper, the role of its compounds in the treatment of numerous chronic diseases is well established. Moreover, numerous metal compounds are able to act as anti-oxidants[26], antimicrobial[27], antiparasitic[28], anti-inflammatory, anticonvulsant29 and antitumor agents[30]. The chelation of metal (II) with ligand reduces the polarity of the metal ion and this, by the overlapping of the ligand orbital and the partial sharing of the metal ion positive charge with donor groups. The increase in the delocalization of π electrons over the whole ligand enhances the penetration of the complexes into the lipid membrane and the blocking of the metal binding sites on the enzymes of microorganisms[31]. The complexes may also disturb the respiration processes of the cell, block the protein synthesis and restrict further growth of the organism.

Corresponding author: Aijaz Ahmad Tak
E-mail address: mehroosh21@yahoo.in

Herein, we synthesized the mixed ligand complexes of Cu(II), Ni(II) and Co(II) with levofloxacin and bidentate bipyridyl ligands. The Cu(II) and Ni(II) complexes were tested for DNA interaction using UV-vis spectroscopy, cyclic voltammetry, viscosity and gel electrophoresis. The antioxidant activity for all the synthesized complexes was checked. The in vitro antimicrobial activity of Ni(II) was evaluated against different strains of bacteria.

Experimental studies:

Materials and Methods

All reagents, chemicals and solvents were of analytical grade and were used as such. Double distilled water was used throughout the experiment. Levofloxacin, NaCl, bipyridyl, DPPH, NBT, riboflavin, TBA, TCA, Tris and hs-DNA were purchased from Sigma Chemical Company. Cu(II), Ni(II) and Co(II) chlorides were purchased from E-Merck (India) Ltd.

DNA stock solution was prepared by diluting hs-DNA in tris buffer (containing 25 mM Tris HCl and 50 mM NaCl at pH 7.2) followed by the exhaustive stirring for three days, the solution was then kept at 4 °C for no longer than a week. The stock solution of DNA gave a ratio of UV absorbance at 260 and 280 nm (A_{260}/A_{280}) of 1.88, indicating that the DNA was sufficiently free from protein contamination. The DNA concentration was determined by UV absorbance at 260 nm after 1:20 dilution using the following absorption coefficient ε=6600 cm^{-1} [6].

IR spectra of the complexes were recorded on a FTIR (Fourier transform infrared) spectrometer with samples prepared in KBr pellets. Electronic spectra were recorded on Shamidzu-UV-3600-UV-VIS-NIR spectrophotometer. The NMR spectra were obtained on Bruker DRX 400 spectrometer operating at room temperature. Magnetic measurements were carried out on magnetic susceptibility balance of Sherwood Scientific (Cambridge U.K.) at room temperature. Elemental analysis was performed on Perkin Elmer 2408 elemental analyzer. Molar conductance was measured at room temperature on Systronic conductivity bridge. Cyclic voltammetry was performed on SAS SP 150 Biologic Science Instruments carried out in 30 ml three electrode electrolytic cell. The working electrode was platinum disk, a separate Pt single sheet electrode was used as counter electrode and Ag/AgCl electrode saturated with KCl was used as reference electrode. KNO$_3$ and tris buffer were used as supporting electrolyte. The Cyclic voltammogram of the complex was recorded in tris HCl buffer (pH= 7.2) at 100 mV/s.

All electrochemical measurements were performed at room temperature. Hydrodynamic measurements were carried out from the observed flow time of hs-DNA containing solution (t > 100 seconds) corrected for flow time of the tris buffer alone (t$_0$) using ostwalds viscometer at 25 ± 0.01 °C.

Flow time was measured with a digital stop watch with least count of 0.01s.

Hydroxyl radicals generated by Fenton reaction were used to induce oxidative damage to DNA. The reaction mixture (15 µL) contained 25 mg of DNA in 20 mM of phosphate buffer saline (pH 7.4), 500 µg of test compounds were added and incubated with DNA for 15 minutes at room temperature. The oxidation was induced by treating DNA with 1 µL of 30mM of H$_2$O$_2$, 1 µL of 20 mM ferric nitrate and 1 µL of 100 mM ascorbic acid and incubated for 1 h at 37 °C. The reaction was terminated by the addition of loading dye (40% sucrose and 0.25% bromophenol blue) and the mixture was subjected to gel electrophoresis[32] using Hi Media LA666 in 0.7% agarose/TAE buffer run at 100 Volt. DNA was visualized by Gel Doc system.

To evaluate the antibacterial activity of Ni(II) complex, seven standard bacterial strains i.e. *Staphylococcus. aureus, Proteus vulgaris, Escheria coli, Pseudomonas aeruginosa, Bacillus subtilis, Klebsiella pneumonia* were used. The antibacterial activity by agar disc diffusion assay was carried out as determined by Bauer et al[33]. Erythromycin was used as the reference antibacterial agent while 10% aqueous DMSO was used as vehicular control.

The antioxidant activity was carried out using different assays. In DPPH (2,2-Diphenyl-1-picrylhydrazyl Diphenyl-1-picrylhydrazyl) assay, quantitative measurement of radical scavenging properties of metal complexes were carried out according to the method by Blios et al[34]. Briefly 0.1 mM solution of DPPH was prepared in methanol and 1 mL of this solution was added to 3 mL of metal complex (100-300 µg/mL) and shikonin (300 µg/mL). α-tocopherol was used as a reference antioxidant. Discoloration of reaction mixture was measured at 517 nm after incubation for 30 minutes.

The Superoxide anion radical scavenging activity involves measurement of scavenging activity of all the metal complexes based on the method described by Liu et al[35] with slight modification. 100µL riboflavin solution (20µg), 200µL EDTA solution (12 mM), 200µL methanol and 100 µL nitrobluetetrazolium (NBT) solution (0.1 mg) were mixed in test tube and reaction mixture was diluted up to 3mL with phosphate buffer (50 mM). The absorbance of the solution was measured at 590 nm using phosphate buffer as blank after illumination for 5 minutes. Different concentrations (50 µl) i.e. 100 µg, 200 µg, 300 µg of complex solutions were used. Decreased absorbance of the reaction mixture indicates an increased super oxide anion scavenging activity.

In hydroxyl scavenging activity-deoxyribose assay[36], the colorimetric deoxyribose (TBARS) method was applied as the reference method of comparison for determining the hydroxyl radical scavenging activity of metal complexes. The reacting mixture for the deoxyribose assay contained in a final volume of 1mL the following reagents: 200 µL of KH$_2$PO$_4$-KOH (100 mM), 200 µL of deoxyribose (15 mM), 200 µL of Ferric Chloride (500 µM), 100 µL of EDTA (1 mM), 100 µL of ascorbic acid (1mM), 100 µL of Hydrogen peroxide (10mM) and 100 µL of complex (100-300µg/mL). Reaction mixtures were incubated at 37 °C for one hour. At the end of the incubation period, 1 mL of 1% (w/v) thiobarbituric acid (TBA) was added to each mixture followed by the 1 mL of 2.8% (w/v) trichloroacetic acid (TCA). The solutions were heated on a waterbath at 80 °C for 20 minutes to develop pink coloured malonaldehyde-thiobarbituric acid (MDA-TBA) adduct and the absorbance of the resulting solution was measured at 532 nm.

In Ferric thiocyanate method (FTC)[37] 2 ml of complex solution (100-300 µg/mL) was mixed with 2.88 ml of linoleic acid (2.51%, v/v in 4 ml of 99.9% ethanol), 8 ml of phosphate buffer 0,05M at pH 7 and 3.9ml of distilled water. The whole reaction mixture was incubated at 40 °C for 96 h. To 100, 300 and 400 µL of this solution, 9.7, 9.4 and 9.3 mL of 75% (v/v) ethanol were added respectively followed by 0.1 mL of 30% ammonium thiocyanate to each one. Precisely after three minutes, 0.1 ml of 3.5% v/v HCl was added to the reaction mixtures, the absorbance at 500 nm of the resulting solutions was measured and recorded again after 24 h, until the absorbance of the control has reached the maximum value. α- tocopherol was used as reference antioxidant substance.

In thiobarbituric acid assay, thiobarbituric acid was added to the reaction mixture where it interacts with malanoaldhyde and the TBARS produced was measured spectrophotometrically[38]. To 2 mL of reaction mixture of ferricthiocyanate assay, 2 mL of TCA (20%) and 2 mL TBA (0.67%) were added, kept in boiling water for 10 minutes and later on, cooled under tap water. The reaction mixture was centrifuged at 3000 rpm for 20 minutes and the supernatant was read at 500 nm. α- tocopherol was

used as reference antioxidant substance. The capacity to scavenge the radicals was calculated using the following equation:

% inhibition = $A_c - A_s/A_c$ x 100

Where 'A_c' is the absorbance of the control reaction (reaction mixture without any antioxidant substance) and 'A_s' is the absorbance of reaction mixture with reference substance or complex. The experiments were repeated thrice.

General synthesis of complexes

To a solution of Levofloxacin (3.61g, 10 mmol) and 2,2' bipyridyl (1.56g, 10 mmol) in 50 mL absolute ethanol was added Nickel(II) chloride in 1:1:1 molar ratio and refluxed for 10 minutes. A blue precipitate was obtained, isolated from the hot solution, washed with ether and dried in vacuo. Similar procedure was adopted for the preparation of Cu(II) and Co(II) complexes.

$C_{28}H_{27}N_5O_4FNiCl$: Yield 70%, m.p 350 °C, Anal. (%Calcd./Found) C; 55.06/55.18, H; 4.45/4.49, N; 11.46/11.50. FT-IR (KBr, cm^{-1}); 1632 v(C=O) pyridine, 1564 v_{asym}(COO), 1365 v_{sym}(COO), ^{1}H NMR (400MHZ, DMSO-d$_6$) δ(ppm) 7.7- 8.0(m), 2.4- 3.5(m), 4.3- 4.7(m), µ$_{eff}$ = 2.57, soluble in water.

$C_{28}H_{27}N_5O_4FCuCl$, yield 82%, m.p 437° C, C; 54.63/54.35, H; 4.42/4.36, N; 11.37/11.28. FT-IR (KBr, cm^{-1}); 1634 v(C=O) pyridine, 1580 v_{asym}(COO), 1382 v_{sym}(COO), µ$_{eff}$ = 1.86, soluble in ethanol.

$C_{28}H_{27}N_5O_4FCoCl$, yield 78%, m.p 393 °C, C; 55.04/54.09, H; 4.45/4.48, N; 11.46/11.49. FT-IR (KBr, cm^{-1}); 1630 v(C=O) pyridine, 1575 v_{asym}(COO), 1375 v_{sym}(COO), µ$_{eff}$ = 3.24, soluble in DMF and DMSO.

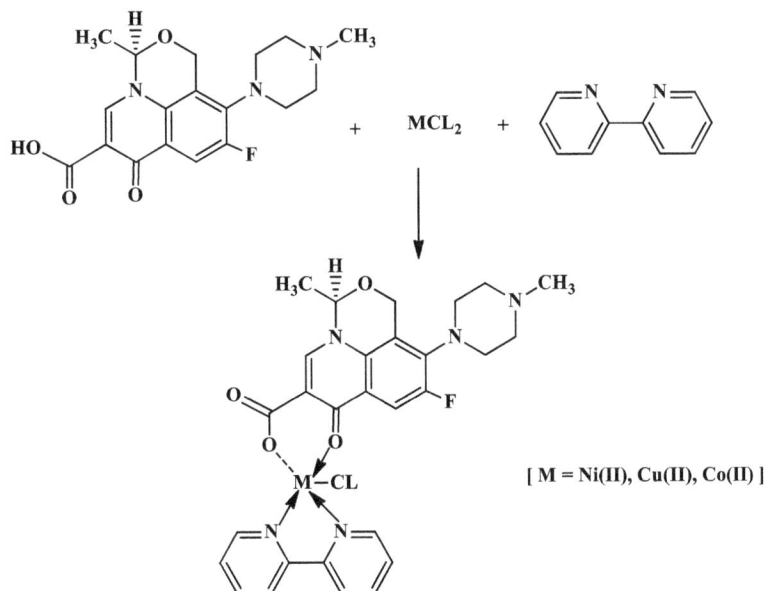

[M = Ni(II), Cu(II), Co(II)]

Scheme 1. General reaction sequence for the synthesis of the studied complexes

Results and Discussion

IR spectroscopy

IR spectroscopy confirmed the deprotonation and binding mode of levofloxacin. In the IR spectra of the complexes, the disappearance of ν(H-O) stretching vibration of free quinolones (3010, br.) is indicative of deprotonation of carboxylate group on binding to metal ion. The shifting of ν(C=O)$_p$ stretching vibration band in levofloxacin from 1708 to 1630-1634 cm^{-1} confirms the carbonyl oxygen of pyridine ring as the coordination site[31]. The characteristic ν(COO) asymmetric and symmetric vibrations as strong bands at 1624 and 1340 cm^{-1} also shift to 1564-1580 cm^{-1} and 1365-1382 cm^{-1} in the metal complexes. The unidentate nature of carboxylate group is confirmed by the frequency separation of 200 cm^{-1} (Δ= ν(COO$_{asym}$- COO$_{sym}$)[31,39]. The data are further supported by the appearance of ν(M-O) at 510-515 cm^{-1} and ν(M-N) at 535-542 cm^{-1} [40].

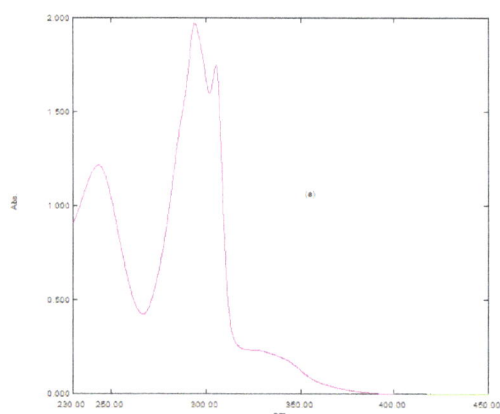

Electronic Absorption Spectra

The UV-vis spectrum of the Ni(II) complex was measured at room temperature in the region of 250-800 nm which exhibits bands at 294, 305, 325 nm respectively (Figure 2a) and a broad band at 557 nm assigned to $^3B_{1(F)} \rightarrow {}^3E_{(F)}$ transition (Figure 2b). The λ_{max} values are consistent with pentacoordinate geometry around Ni(II) ion[41]. The electronic spectrum of Cu(II) complex measured in the range of 240-800 nm displays bands in the region of 250-800 nm and 370 – 382 nm, assigned to intraligand $\pi – \pi^*$ and LMCT transitions. However, the d-d transitions observed in the region 600-675 nm are typical of penta-coordinated Cu(II) complexes having distorted square pyramidal geometry (d_{xy}, $d_{yz} \rightarrow d_x{}^2 – d_y{}^2$) [42]. The electronic spectrum of Co(II) complex displays band at 350 nm assigned to LMCT transition and another weak band due to d-d transitions at 625nm.

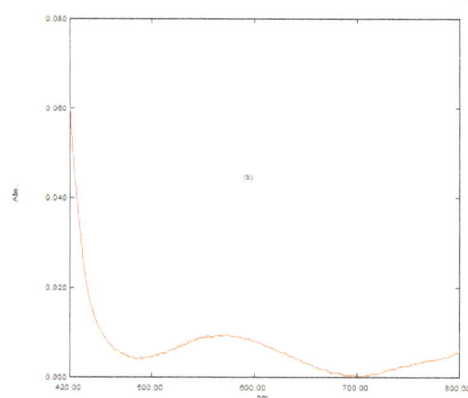

Figure 2. *Electronic spectra of Ni(II) complex(a,b)*

EPR Spectrum

The X–band EPR Spectrum of Cu(II) complex was recorded at liquid nitrogen temperature (LNT) in solid state using tetracyanoethylene (g = 2.00278) as field marker. The EPR spectrum exhibited broad band having 'g' isotropic values of 2.09 characteristic of square pyramidal geometry. The spectral lines usually result from intermolecular spin exchange which broadens the spectral lines. The intermolecular spin exchange is caused by strong spin coupling, which occurs during collision of paramagnetic centers.

Magnetic susceptibility measurements

The magnetic susceptibility measurements for the Cu(II) complexes at room temperature lies in the range of 1.8-1.9 BM corresponding to one unpaired electron and are consistent with d^9 configuration around Cu(II) ion[43]. The magnetic measurements for Ni(II) complex show the magnetic value of 2.57 BM consistent with the pentacoordinated geometry.

¹H NMR

The ^{1}H NMR spectra of the Ni(II) complex were recorded in DMSO-d$_6$ medium. The complex does not show any signal attributable to the carboxylic acid suggesting its involvement in the coordination of metal ion. The complex shows a broad multiplet at δ 7.7-8.0 ppm corresponding to aromatic protons and the peaks at δ 2.4-3.5 ppm are due to methyl protons. The methylene protons show peaks at δ 4.3-4.7 ppm.

DNA Binding Studies

UV-vis titration

Electronic absorption is an effective tool to examine the binding mode of complex with DNA[44-46]. Metal complexes bind to DNA via both covalent and non-covalent interactions[47]. Non-covalent interaction includes intercalation, binding to minor and major groove, sugar phosphate backbone and electrostatic binding mode[48]. Drugs binding with DNA via intercalation usually results in hypochromism and bathochromism of absorption band due to strong interaction between the aromatic chromophore of the molecule and base pairs of the DNA. On the other hand, absorption intensities of drugs are increased (hyperchromism) upon increase of the DNA concentration due to damage of the DNA double helix structure. The extent of hyperchromism is indicative of the partial or non-intercalative binding modes such as electrostatic and van der Waals interaction, hydrogen bonds and hydrophobic interaction.

The interaction of Ni(II) complex in water with hs-DNA was investigated through the change of absorbance at 256 nm with increasing concentration of hs-DNA. The spectra clearly show that bands at 256 exhibit hypochromism with red shift of 16 nm (Figure 3a). The bands at 294 nm and 305 nm completely disappear but the band at 325 nm displays a hyperchromism with blue shift of 20 nm to 305 nm (Figure 3b). These changes are typical of complexes bound to DNA through non-covalent interaction [49]. The red shift in absorbance are accompanied by increase in molar absorptivity so that isobestic points are formed at 285 nm indicating single mode of binding. The spectral features suggest that the Ni(II) complex bind to hs-DNA by intercalative binding. The observed hypochromism could be attributed to the contraction of hs-DNA helix axes as well as its conformational changes. The hyperchromism could be the result of secondary damage of hs-DNA double helix structure[44,50]. The

carbonyl group in the Ni(II) complex could form hydrogen bonds with suitable donors like N_7 and O_6 of adjacent guanine bases of hs-DNA supported by hydrophobic interaction of pyridyl ring on surface of hs-DNA contributing to hyperchromism. To quantify the extent of binding of the complex with hs-DNA, the intrinsic binding constant K_b was calculated using the following equation

$$[DNA]/[\varepsilon_a-\varepsilon_f) = [DNA]/\varepsilon_b-\varepsilon_f+1/k_b(\varepsilon_b-\varepsilon_f)$$

Where εa, εb and ε_f are apparent, bound and free extinction coefficient respectively. Fit the plot of $[DNA]/[\varepsilon_a-\varepsilon_f)$ vs $[DNA]$, the K_b was obtained from the ratio of slope to intercept. The binding constant of the complex is $4.0 \times 10^4 \, M^{-1}$(Figure 4). From the binding constant, it is clear that the complex binds strongly to hs-DNA, however the binding constant is lower than the classical intercalators[47,48].

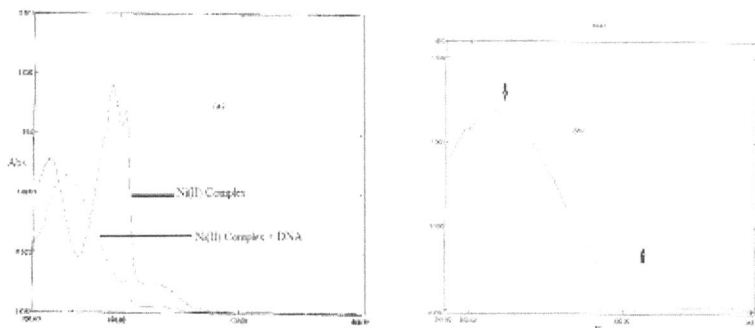

Figure 3. (a). Shift in absorption band of Ni(II) complex after addition of hs-DNA. (b). Electronic spectra of Ni(II) complex on addition of increasing amounts of hs-DNA

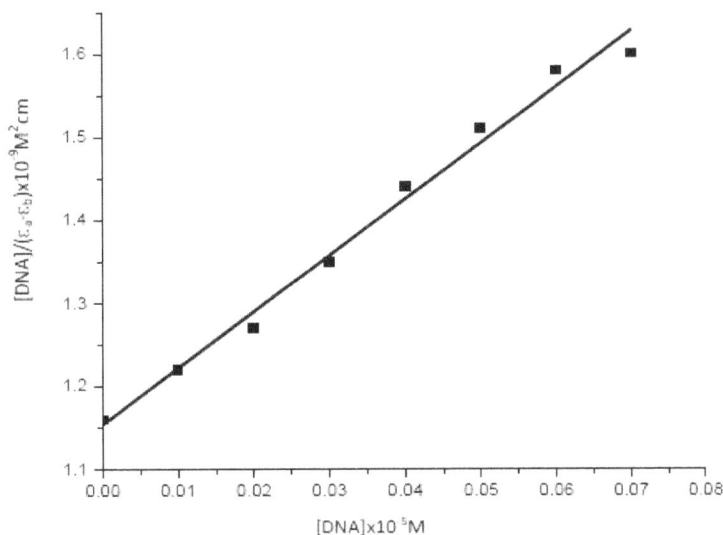

Figure 4. Plot of [DNA]/(ε_a-ε_b) x10^{-9} M^2 cm vs [DNA] x 10^{-5}M for titration of hs-DNA with Ni(II) complex.

The interaction of Cu(II) complex with hs-DNA was also carried out. The absorption spectra of the Cu(II) complex in the presence and absence of hs-DNA shows red shift of 4 nm and hypochromism (Figure 5) revealing that the complex binds to hs-DNA by intercalation[51]. This may be due to the fact that π orbital of the base pairs of hs-DNA couples with the π orbitals of the ligand leading to bathochromism. Further, the π orbital is partially occupied by electrons thereby decreasing the transitional probabilities and resulting in hypochromism. The binding constant calculated for Cu(II) complex is 2.0×10^4 M^{-1}.

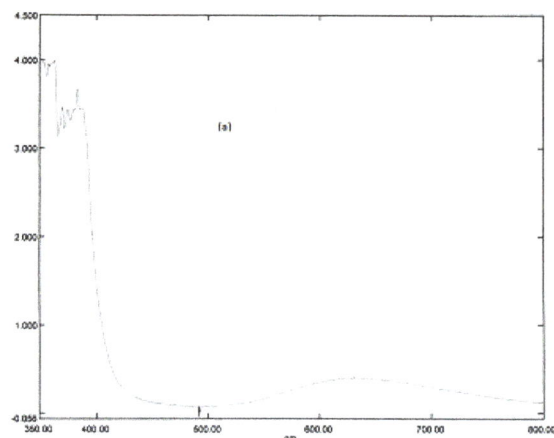

Figure 5(a). Electronic spectra of copper(II) complex

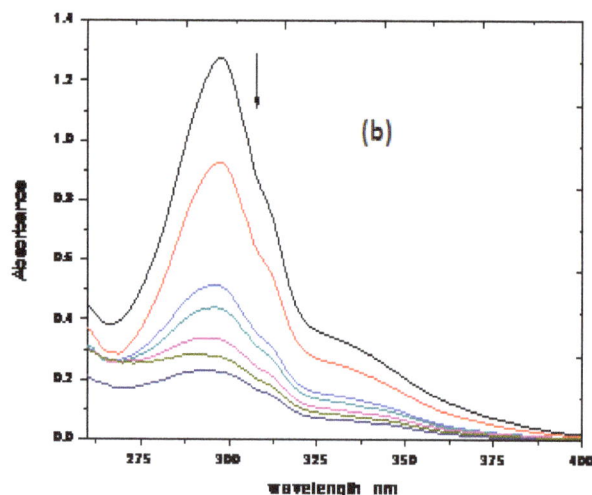

Figure 5(b). Electronic spectra of Cu(II) complex in presence of increasing amounts of hs-DNA.

Electrochemical studies

Cyclic voltammetry is widely used as simple and rapid method to study DNA interaction with metal complex. The cyclic voltammetry of Cu(II) complex alone and with hs-DNA were recorded in the range of -0.8 to 0.5 V with a scan rate of 100 mv/s in tris buffer (pH= 7.2). The cyclic voltammogram for Cu(II) complex exhibits one electron redox process involving Cu(II)/Cu(I) couple. The cathodic peak potential (E_{pc}) appears at -0.558V with I_{pc} at 4.466 mA. The addition of 10^{-3} M hs-DNA to Cu(II) complex causes a decrease in cathodic current (0.296 mA). Further, the peak potential E_{pc} shifts to a more positive potential (-0.450V). The observed decrease in current may be attributed to slow diffusion of an equilibrium mixture of free and DNA bound Cu(II) complex to the electrode surface[52]. Bard and co-workers[53] have discussed the binding modes between small molecules and DNA, If the interaction is through electrostatic binding mode, the formal potential shifts to a more negative potential while intercalative binding mode results in a more positive potential. In presence of hs-DNA, the Cu(II) complex shows positive shift revealing an intercalative binding mode between Cu(II) complex and hs-DNA base pairs (Figure 6)

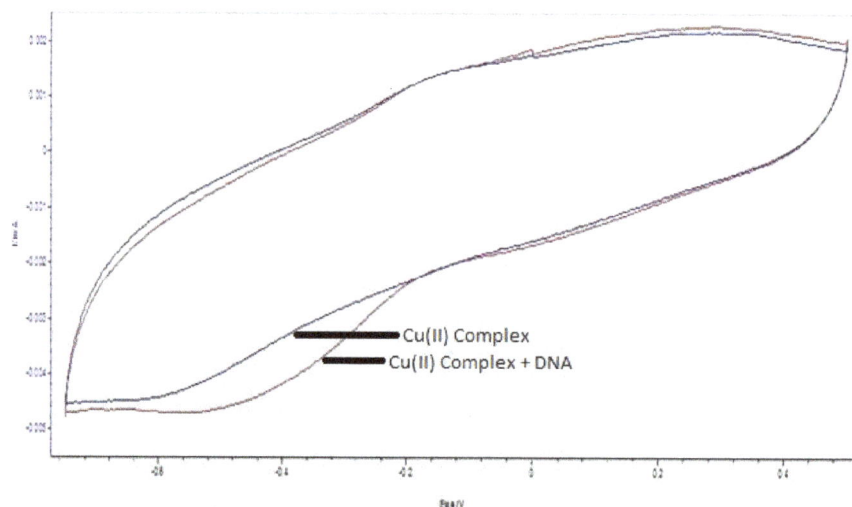

Cu(II) Complex
Cu(II) Complex + DNA

Figure 6. *Cyclic voltammogram of Cu(II) complex at scan rate of 100 mV/s in the presence and absence of hs-DNA.*

Hydrodynamic measurements

A hydrodynamic method such as viscosity which is sensitive to changes in DNA length offers a least ambiguous and definitive method to determine the binding mode of DNA binding agents[54]. A classical intercalation binding demands that DNA helix must lengthen as base pairs are separated to accommodate the binding ligand leading to an increase in DNA viscosity[55]. There is little effect on viscosity if electrostatic or groove binding occurs[56]. Intercalating agents are expected to destabilize base pairs, causing lengthening of the double helix resulting in the increase of viscosity of DNA, while non-classical intercalators or groove binding of the complex could bend or kink the DNA helix reducing its effective length and concomitantly its viscosity[57]. The relative viscosity was calculated using the equation $t-t_0/t_0$, where t_0 is the flow time for the used buffer and t is the observed flow time for DNA in absence and presence of the complex. The results were presented as η/η_0 vs r (binding ratio), where r = [complex]/[DNA], η is the viscosity of hs-DNA in presence of the complex and η_0 is the viscosity of hs-DNA alone. The effect of Ni(II) complex on relative viscosity of hs-DNA by varying the concentration of complex shows pronounced increase. The Cu(II) complex also experiences an increase in viscosity of hs-DNA. The results reveal that metal complexes bind to hs-DNA *via* intercalation. The results obtained from the viscosity studies validate data obtained from uv-vis and electrochemical titration.

Gel electrophoresis

The ability of the complex to interact with DNA for protection from hydroxyl radicals is depicted in figure 7. The principle of the method is that molecules migrate in the gel as a function of their mass, charge and shape. We observed that Cu(II) and Ni(II) complexes show considerable DNA damage protective effect. The reference substance BHT used show protective effect. BHT is a known antioxidant substance, as it has the property of scavenging the free radicals. In Lane 2 reaction mixture was added to ct-DNA and was kept as negative control as no protective/antioxidant substance was used in this lane and as seen in Lane 2, radicals generated completely inhibit the DNA. However in Lane 3 only DNA was used as acting as positive control.

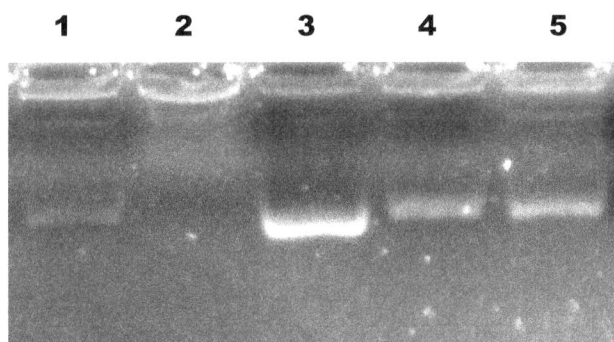

Figure 7: *Protective DNA damage by levofloxacin- bipyridyl complexes.*

Lane 1. BHT+ Reaction mixture +ct-DNA Lane 2. ct-DNA + Reaction mixture
Lane 3. ct-DNA alone Lane 4. Ni(II) complex (500 µg) + Reaction mixture + ct-DNA Lane 5. Cu(II) complex (500 µg) + Reaction mixture + ct-DNA .

Biological Activity
Antibacterial activity

The Ni(II) complex exhibits significant inhibition on the tested pathogenic strains. 10% aq. DMSO was used as negative control and showed no activity against all strains. The highest inhibition (36 mm diameter) was observed at high concentration of Ni(II) complex (200 µg) on *B. subtilis* with respect to the other strains and to erythromycin (25 mm) used as a positive control. The activity of Ni(II) complex was concentration dependent and the inhibition diameter of strains increases with the increase in Ni(II) complex concentration with an inhibition zone between 28-36 mm. In another experimental setup, the minimum inhibitory concentration (MIC) of the Ni(II) complex (Table 2) against the tested strains was determined. The MIC of Ni(II) complex was reported against *Staphylococcus aureus* (10 µg/ml) followed by *E.coli* (12 µg/ml) as compared to the studied strains.

Table 1 Antibacterial activity of Ni(II) complex against pathogenic bacterial strains

Bacterial Strain	Zone of inhibition mm				
	200µg	100µg	50µg	25µg	Erythromycin (25mcg/disc)
Proteus vulgaris	29 ± 1.32	26 ± 2.15	24 ± 3.45	20 ± 2.12	25 ± 1.17
Bacillus subtilis	36 ± 1.69	29 ± 1.56	26 ± 2.56	25 ± 2.72	29 ± 3.09
Staphylococcus aureus	30 ± 1.09	23 ± 1.97	21 ± 0.32	21 ± 3.42	22 ± 1.05
Escherichia coli	31 ± 2.17	26 ± 3.32	25 ± 1.71	22 ± 4.12	28 ± 1.91
Klebsiella pneumonia	28 ± 1.08	24 ± 2.12	23 ± 0.32	21 ± 3.12	ND
Psuedomonas aeruginosa	30 ± 1.0	26 ± 1.12	24 ± 0.39	22 ± 1.03	25 ± 2.03

Values are represented as mean ± Sd (*n* = 3), 10% aqueous DMSO used as negative control showed no activity

Table 2: MIC determination of Ni(II) complex against tested pathogenic bacterial strains

Bacterial Strain	MIC (µg/mL)	
	Ni(II) complex	Erythromycin
Proteus vulgaris	20 ± 2.41	18 ± 2.08
Bacillus subtilis	18 ± 2.327	24 ± 4.02
Staphylococcus aureus	10 ± 1.43	16 ± 2.17
Escherichia coli	12 ± 1.06	14 ± 3.95
Klebsiella pneumonia	24 ± 3.49	-
Psuedomonas aeruginosa	-	16 ± 1.49

Values are represented as mean ± SD (*n* = 3). The concentrations of the tested substances were in the range (24, 22, 20, 18, 16, 14, 12, 10, 7.5 µg/ml). 10% aqueous DMSO (was) used as negative control showed no inhibitory activity.

Antioxidant activity

The antioxidant activity of the metal complexes was compared with a positive control (BHT) and Ascorbic acid, which is known to protect tissues and cells against various oxidative stresses[58]. The mechanism of scavenging of radicals cannot be evaluated by a single method; therefore, five different antioxidant models were used in this study[59]. The antioxidant activity of metal complexes in DPPH method was determined through the decrease in absorption strength by radical scavenging activities of Cu(II), Ni(II), Co(II) complexes, Ascorbic acid, and BHT at 500 µg/mL and were found to be 20%, 70%, 53%, 66% and 80% respectively (Figure 8a). We found that the free radical scavenging activity of Ni(II) complex was stronger than Cu(II) and Co(II) complexes. However BHT was found to have the highest scavenging activity. The activity of all the complexes were found to be concentration dependent.

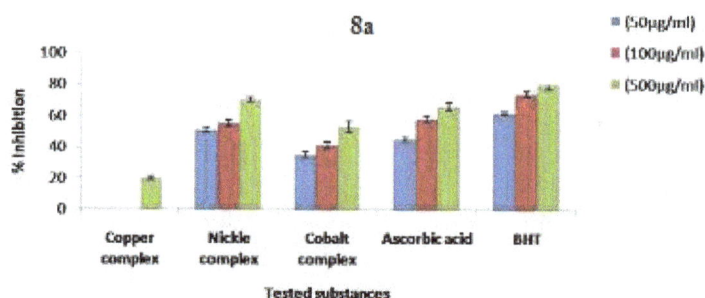

8b

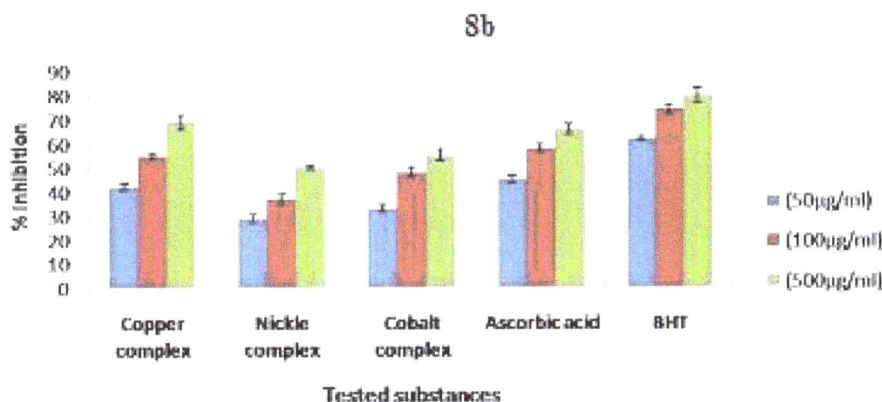

8c

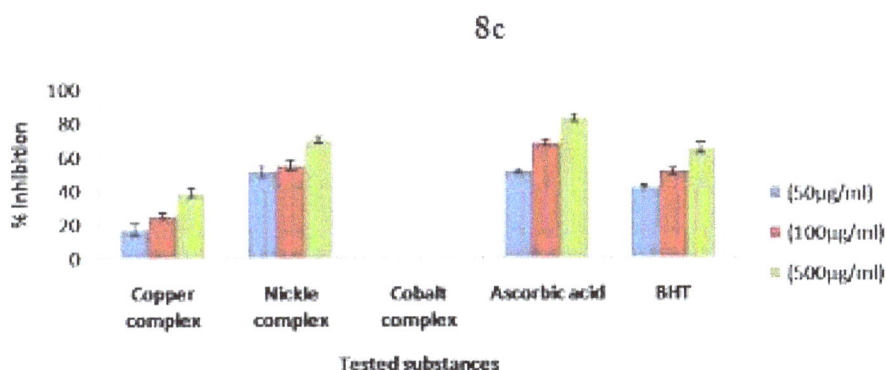

All the complexes exhibited moderate to high superoxide dismutase activity at variable concentrations. The Cu(II) complex exhibited higher scavenging activity (69%) followed by Co(II) and Ni(II) complexes at 500 μg/mL(Figure 8b).

The hydroxyl radicals are known to cause DNA damage by degradation of deoxyribose moiety which contributes to carcinogenesis, mutagenesis and cytotoxicity[60], however, the scavenging or chelation of radicals by any substance is due to the antioxidant capacity of that particular substance[61]. In our study, Ni (II) complex exhibited the highest chelating activity of hydroxyl radicals at 500 μg/ml. The scavenging activity decreased in the following order: Ascorbic acid (83%) > Ni(II) (70%) > BHT (65%)> Cu(II) (39%) at 500 μg/ml (Figure 8c).

FTC evaluates the effect of a reference antioxidant and metal complexes on preventing peroxidation of polyunsaturated fatty acids and linoleic acid. The % of inhibition was recorded after every 24 h and results were calculated for three consecutive days. The percentage of protective effect of linoleic acid peroxidation was 80% for Cu(II) complex, 70% for Ni(II) complex and 40% for Co(II) complex at 300 μg/ml concentration (Figure 8d).

In the TBA method, formation of malonaldehyde is the basis for evaluating the extent of lipid peroxidation. At low pH and high temperature malonaldehyde, the end product of lipid peroxidation, binds TBA to form a red colored complex. The concentrations used were 100– 300 μg/mL. The FTC method measures the amount of peroxide produced during the initial stage of lipid peroxidation. Subsequently at later stages of oxidation, peroxides decompose to form carbonyl compounds that are measured by the TBA method. Thiobarbituric acid assay is determined in the reaction mixture from the assay of FTC while in TBA assay, inhibitory activity of peroxide radicals is determined from the residual mixture earlier used in FTC method. At 300 μg/mL, the Cu(II) complex exhibited 61% inhibition of radicals, followed by Ni(II) complex (56%) (Figure 8e). Some of the reports mentioned the ferric reducing power of bioactive compounds such as phenolic substances and flavonoids[62-64]. The findings of this work confirm that the metal complexes can be used as an alternative therapy to combat various indigenous as well as exogenous stresses.

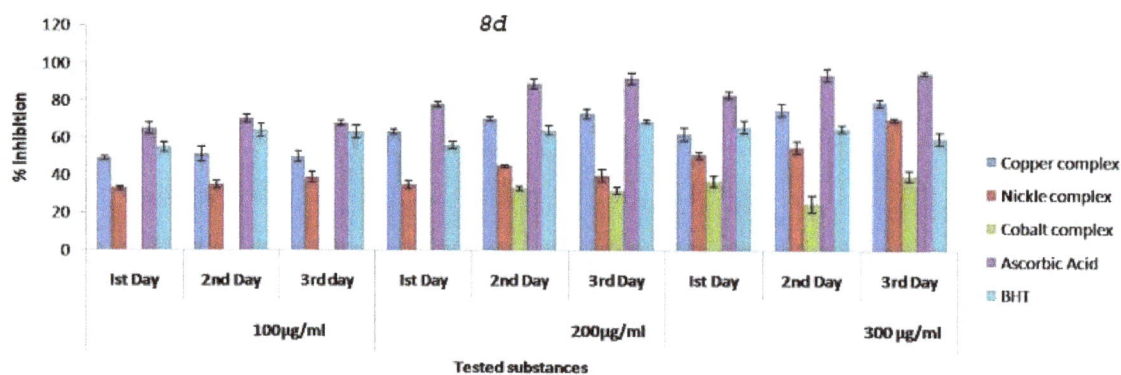

8d

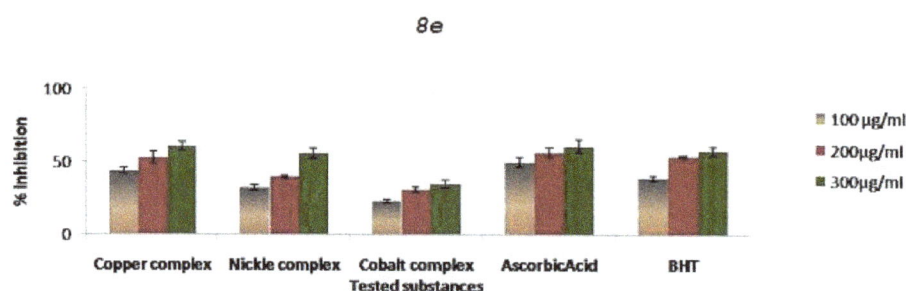

8e

The inhibitor concentration for scavenging of radicals (IC_{50} µg/mL) was determined for all the antioxidant methods. From the results, it was observed that IC_{50} values varied with the complexes as well as with the type of method. The Cu(II)

complex showed lowest value of 72 µg/mL in scavenging of superoxide radicals followed by Ni(II) complex with 95 µg/mL (Table 3). The Co(II) complex showed the comparatively higher value of IC_{50} in all the methods employed.

Table 3. IC_{50} determination of metal complexes

Complexes	IC_{50} µg/mL				
	DPPH assay	SOD	Hyroxyl scavenging assay	FTC	TBA
$C_{28}H_{27}N_5O_4FNiCl$	140	150	95	240	260
$C_{28}H_{27}N_5O_4FCoCl$	320	120	-	>300	-
$C_{28}H_{27}N_5O_4FCuCl$	>350	72	>300	100	140
Ascorbic acid	175	60	97	70	100
BHT	117	40	193	70	160

Conclusion

In this work, mixed ligand complexes of Cu(II), Ni(II) and Co(II) have been synthesized and characterized by various physicochemical studies with an aim to develop robust therapeutic agents. Complexes possess pentacoordinate geometry. The results reveal that both Ni(II) and Cu(II) complexes interact with DNA showing strong intercalation. The results of gel electrophoresis prove the cu(II) and Ni(II) complexes show considerable DNA damage protective effect. The MIC shows lower values for most of the tested strains respect to control confirming antimicrobial activity. All synthesized complexes show considerable antioxidant activity. Overall the mixed ligand complexes derived from

levofloxacin and bipyridyl possess strong potential to be used as possible therapeutic agents.

Acknowledgement

We are grateful to University Grants Commission New Delhi for financial support vide grant reference number F. No 8- 3 (32)/2011(MRP/NRCB).

References

1- In: V. T. Andriole (Ed.), The quinolones, Academic Press, **2000**.

2- I. Turel, Coordination Chem Rev., **2002**, 232, 27-47.

3- D. E. King, R. Malone, S. H. Lilley, Am. Fam. Physician., **2002**, 612741-2748.

4- J. Tuma, W. H. Conors, D. H. Stitelman, C. Richert, J. Am. Chem. Soc., **2002**, 124, 4236-4246.

5- M. Zampakou, M. Akrivou, E. G. Andreadou, C. R. Raptopoulou, V. Psycharis, A. A. Pantazaki,
   G. Psomas, J. Inorg. Biochem. **2013**, 121, 88-99

6- F. Dimiza, A. N. Papadopoulos, V. Tangoulis, V. Psycharis, C. P. Raptopoulou,
   D. P. Kessissoglou, G. Psomas, Dalton Trans., **2010**, 39, 4517-4528.

7- N. Jimmenez-Garrido, L. Perello, R. Ortiz, G. Olzuet, M. Gonzalez-Alvarez, E.
   Canton, M. Liu-Gonzalez, S. Garcia-Granda, M. Priede, J. Inorg. Biochem. **2005**, 99, 677-689.

8- M. N. Patel, C. R. Patel, H. N. Joshi, Iorganic Chem. Commun.**2013**, 27, 51-55.

9- P.Drevensek, T.zupancic, B.Pihlar, R.Jerala, U.Kolitsh, A.Plaper, I.Turel, J. Inorg. Biochem. **2005**, 99, 432-442.

10- I.Turel, J. Kljun, F. Perdih, E. Morozova, V. Bakulev, N. Kasyanenko, J.A.W. Byl, N. Oshroff,
   Inorg. Chem., **2010**, 49, 10750-10752.

11- Y. Wang, R. Hu, D. Jiang, P. Zhang, Q. Lin, W.Yang, J. Flouresc., **2011**, 21, 813-832.

12- K. C. Skyrianou, F. Perdih, I. Turel, D. P. Kessissoglou, G. Psomas, J. Inorg. Biochem., **2010**, 104, 161-170.

13- K. C. Skyrianou, V. Psycharis, C. P. Raptopoulou, D. P. Kessissoglou, G. Psomas, J. Inorg.
   Biochem., **2011**, 105, 63-74.

14- A. Tarushi, C. P. Raptopoulou, V. Psycharis, A. Terzis, G. Psomas, D. P. Kessissglou, Bioorg. Med. Chem., **2010**, 18, 2678-2685.

15- I. Turel, A. Golobic, A. Klavzar, B. Pihlar, P. Buglyo, E. Tolis, D. Rehder, K. Sepcic, J. Inorg.
   Biochem., **2003**, 95, 199-207.

16- A.Tarushi, E. K. Efthimiadou, P. Christofis, G. Psomas, Inorg. Chim. Acta, **2007**, 360, 3978-3986.

17- K. C. Skyrianou, E. K. Effthimaidou, V. Psycahris, A. Terzis, D. P. Kessissglou, G. Psomas,
   J. Inorg. Biochem., **2009**, 103, 1617-1625.

18- E. K. Efftimidou, M. E. Katsarou, A. Karaliota, G. Psomas. J. Inorg. Biochem., **2008**, 102, 910-920.

19- E. K. Efthimiadou, H. Thomadaki, Y. Sanakis, C. P. Raptopoulou, N. Katsaros,
   A. Scorilas, A. Karaliota, G. Psomas, J. Inorg. Biochem. **2007**, 101, 64-73.

20- M. Katsarou, E. K. Eftimaidou, G. Psomas, A. Karaliota, D.Vourloumis, J. Med. Chem., **2008**, 51, 470-478.

21- S. Niroomand, M. Khorasani-Motlagh, M. Norrizafar, A. Moodi, J. PhotoChem. Photobiol. B.,

2012, 117, 132-139.

22- E. K. Efthimiadou, N. Katsaros, A. karaliota, G. Psomas, Inorg. Chim Acta, **2007**, 360 4093- 4102.

23- M. N. Patel, C. R. Patel, H.N. Joshi, Inorg. Chem. commn., **2013**, 27, 148-155.

24- S. Shobana, J. Dharamaraja, S. Selvaraj, SPectrochem. Acta Part A. **2013**, 107, 117-132.

25- T. Philips, S. L. Tank, J. Wirtz, L. Brewer, A. Coyner, L. S. Ortego, A. Fairbrother, Environ. Rev., **2002**, 10, 209-261.

26- Y. J. liu, Z. H. Liang, X. L. Hong, Z. Z. Li, J. H. Yao, H. L. Hang, Inorg. Chem. Acta, **2012**, 387, 117-124.

27- D. L. Klayman, G. P. Sconill, G. F. Bafosevich, J. Bruce, J. Med. Chem., **1983**, 26, 35-59.

28- K. C. Agarwal. A. C. Santorelli. J. Med Chem., **1978**, 21, 218-221.

29- D. H. Jones, R. Slack, S. Squires, K. R. H. Wooldrige, J. Med. Chem., **1965**, 8, 676-680.

30- S. Tabassum, S Amir, F. Arjmand, C. Petinarri, F. Marchetti, N. Masciocchi, G. Lupidi,, R. Pettinari, Eur. J. Med. Chem. 2013, 60, 216-232.

31- Z. H. Chohan, C.T. Suparan, A. Scozzafava, J. Enz. Inhi. Med. Chem., **2005**, 21, 303- 307.

32- L. Q. Sun, D. W. Lee, Q. Zhang, Genes Dev., 2004, 18, 1035–1046.

33- A.W. Bauer, B. Kirby, E. M. Sherris, M. Turk, American Journal of Clinical Pathology, **1966**, 45, 493-496.

34- M. S. Blois, Nature, **1958**, 181, 1199-1200.

35- S.Y. Liu, F. Sporer, M. Wink, J. Jourdane, R. Henning, Y. L. Li, A. Ruppel, Trop. Med. and Int.
   Health, **1997**, 2(2), 179-188.

36- M. Padmaja, M. Sravanthi, K. P. J. Hemalatha, J. Phytology, **2011**, 3(3), 86-91.

37- H. Kikuzaki, N. J. Nakatani, Food Sci., **1993**, 58, 1407-1410.

38- E. Kishida, S. Tokumaru, Y. Ishitania, M. Yamamoto, M. Oribe, H. Iguchi, S. Kojo. Journal
   Agri. Food chem., **1993**, 41, 1598-1600.

39- K. Nakamoto, Infrared and Raman Spectra of Inorganic and coordination compounds 4th edition Wiley Interscience Publication New York, **1986**.

40- H.H. Freedman, J. Am. Chem. Soc., **1961**, 83, 2900-2905.

41- M. Duo, Y. M. Guo, X. H. Bu, J. Ribas, M.Monfort New journal chemistry, **2002**, 26, 939-945.

42- A. M. Herrera, R. J. Staples, S.V. Krjatov, A.Y. Nazarenko, E.V.R. Akimova, Dalton Trans, **2003**, 5, 864-856.

43- (a). M. Devereux, D. O. Shea, A. Kellett, M. McCann, M. Walsh, D. Egain, K. Kedziora, G. Rosair, H. M. Bunz, J. Inorg. BioChem., **2007**, 101, 881.

   (b). R. Haung, A. W. Wallquist, D. G. Covell, BioChem. Pharmacol., **2006**, 100, 1389.

(c) C. S. Dhar, M. Nethaji, A. R. Chakarvarty, Inorganic Chem., **2005**, 44, 8876.

44- J. K. Barton, A. T. Danishhefsky, J. M. Goldberg, J. Am. Chem. Soc., **1984**, 106, 2172-2176.

45- J. M. Kelly, A. B. Tossi, D. J. McConnell, Nucleic Acid Res., **1985**, 13, 6017-6034.

46- A. M. Pyle, J. P. Rehman, J. P. Meshoyrer, C.V. Kumar, N. J. Turro, J. K. Barton,
J. Am. Chem. Soc., **1989**, 111, 3051-3058.

47- M. Cory. D. D. McKee, J. Kagan, D. W. Henry, J. A. Miller, J. Am. Chem. Soc., **1985**, 107, 2528-2536.

48- M. J. Waring, J. Mol. Biol., **1965**, 13, 269-282.

49- T. Hirohama, Y. Kuranuki, E. Ebina, T. Sugizaki, H. Arii, M. Chikira, P. T. Selvi, M. Palaniandavar, J. Inorg. Biochem., **2005**, 99, 1205-1219.

50- C. S. Chow, J. K. Barton Methods Enzymol., **1992**, 212, 219-242.

51- G. Pratviel, J. Bernadou, B. Meunier, Adv Inorg. Chem., **1998**, 45, 251-262.

52- K. Abdi, H. Hadazadeh, M.Well, M. Salimi, Polyedron, **2012**, 31, 638.

53- M. T. Carter, M. Rodrigues, A. J. Bard, J. Am. Chem. Soc., **1989**, 111, 8901.

54- L. F. Tan, H. Chao, H. Li, Y. J. Liu, B. Sun, W. Wei, L. N. Ji, Inorganic Biochem., **2005**, 99, 513-520.

55- Z. C. Liu, B. D. Wang, Z. Y. Yang, D.D. Li, T. R. Qin, Eur. J. Med. Chem., **2009**, 44, 4477-4484.

56- G. Zhang, X. Hu, N. Zhao, W. Li, L. He, Biochem. Physiol., **2010**, 98, 206-212.

57- R.Vijayalakshmi, M. Kantimathi, R. Parthasarthi, B.U. Nair, Bioinorganic and medicinal
Chemistry, **2006**, 14, 3300-3306.

58- P. Duangporn, P. Siripong, American-Eurasian J Agric Environ Sci., **2009**, 5, 258–263.

59- J. A. Parray, A. N. Kamili, R. Hamid, Z. A. Reshi, R. A. Qadri. Frontiers in life Science **2014**;
doi: 10.1080/21553769.2014.951774.

60- P. Hochestein, A. S. Atallah. Mutant Res., **1988**, 202, 363-375.

61- J. A. Parray, A. N. Kamili, R. Hamid, B. A. Ganai, K. G. Mustafa, R.A. Qadri, J. Pharmacy Res.,
2011, 4(7), 2170-2174.

62- P. Siddhuraju, K. Becker, J. Agric Food Chem., **2003**, 51, 2144–2155.

63- E. Karimi, R. Oskoueian, H. Z. Hendra, E. Jaafar, Molecules, **2010**, 15, 6244–6256.

64- P. L. Ting, L. Lusk, J. Refling, S. Kay, D. Ryder, Journal of American Society of Brewing Chemists, 2008, 66(2), 116-126.

3

Click chemistry approach to ionic liquids (ILs) supported organic synthesis

Alioune Fall [1], Insa Seck [2], Massène Sène [1], Mohamed Gaye [1], Matar Seck [2], Generosa Gómez [3] and Yagamare Fall [3,*]

[1] Laboratoire de Chimie de Coordination Organique (LCCO), Département de Chimie, Faculté des Sciences et Techniques: Université Cheikh Anta Diop de Dakar, Sénégal.
[2] Département de Chimie, Faculté de Medecine, de Pharmacie et d'Odonto-stomatologie, Université Cheikh Anta Diop, Dakar, Sénégal.
[3] Departamento de Química Orgánica, Facultad de Química and Instituto de Investigación Biomedica (IBI), University of Vigo, Campus Lagoas de Marcosende, 36310 Vigo, Spain

Abstract: A furan substrate anchored to an ionic liquid bound *tert*-butyldiphenylsiloxane was synthesized using a "click" chemistry approach. The ionic liquid supported furan moiety underwent singlet oxygen oxidation to afford a butenolide intermediate which after removal of the siloxane group gave a bicyclic lactone.

Keywords: "Click" chemistry; Ionic liquids; *tert*-butyldiphenylsiloxane; Singlet oxygen; Butenolide; Lactones.

Introduction

In the last years, ionic liquids (ILs) have attracted considerable interest as environmentally benign reaction media because of their unique properties such as high thermal and chemical stability, negligible vapour pressure, tunable polarity, nonflammability, friction reduction, antiwear performance, high loading capacity and easy recyclability [1].

An attractive feature of ionic liquids is that their solubility can be tuned readily. Therefore, phase separation from organic solvent or aqueous phase is allowed depending on the choice of cations and anions. This suggests the possibility of using ILs as soluble support for organic synthesis. Substrates anchored on ionic liquids are expected to retain their reactivity, as in reactions in solution. One advantage of ILs supported synthesis over solid phase synthesis is that conventional spectroscopic analysis can be carried out during the synthetic process. The feasibility of ionic liquid supported organic synthesis has been demonstrated by many research groups [2].

Silicon protecting groups are of upmost importance in organic synthesis and silicon-containing linkers are valuable for the attachment of substrates to solid support. Brown and co-workers described the synthesis and applications of *tert*-alkoxysiloxane linkers in solid-phase chemistry [3]. This prompted us to design the synthesis of a *tert*-butyldiphenylsiloxane linked to an ionic liquid support (Figure 1).

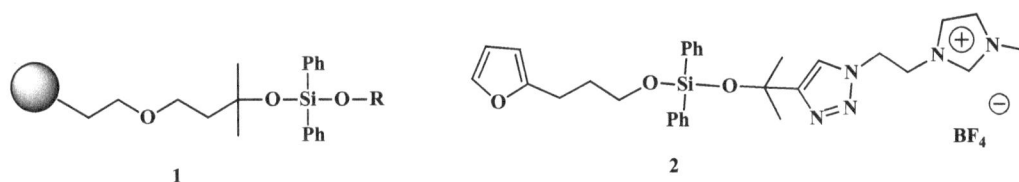

Figure 1. Structures of Brown's resin bound to a *tert*-butyldiphenylsiloxane linker (**1**) and the targeted ionic liquid supported *tert*-butyldiphenylsiloxane linker (**2**).

Results and Discussion

As part of our ongoing programme focused on the synthesis of ionic liquids and their application in organic reactions [4]. We now wish to report the synthesis of IL-supported furan **2**. Our retrosynthetic basis is outlined in Scheme 1.

Corresponding author: Yagamare Fall
E-mail address: yagamare@uvigo.es

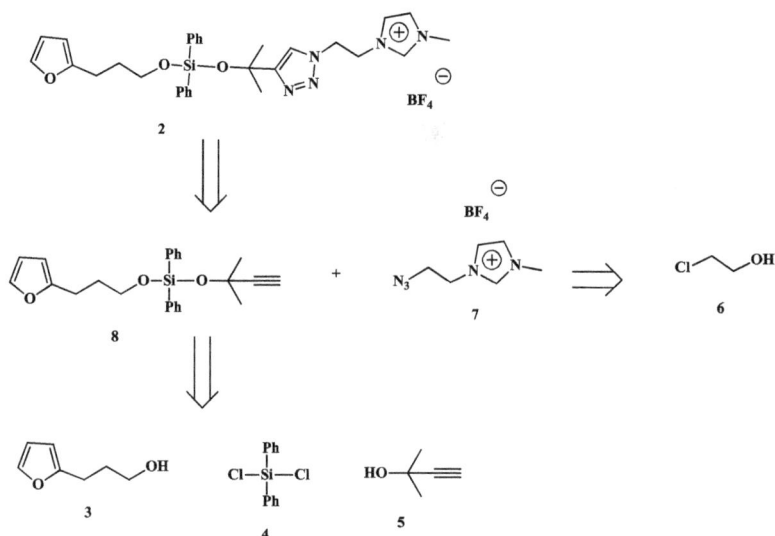

Scheme 1. Retrosynthetic analysis of IL supported furan **2**

We anticipated that IL supported furan **2** could be obtained using a "Click" chemistry approach [5] between alkyne **8** and azide **7** bearing the IL moiety.

Accordingly compounds **7** and **8** were prepared as outlined in Scheme 2.

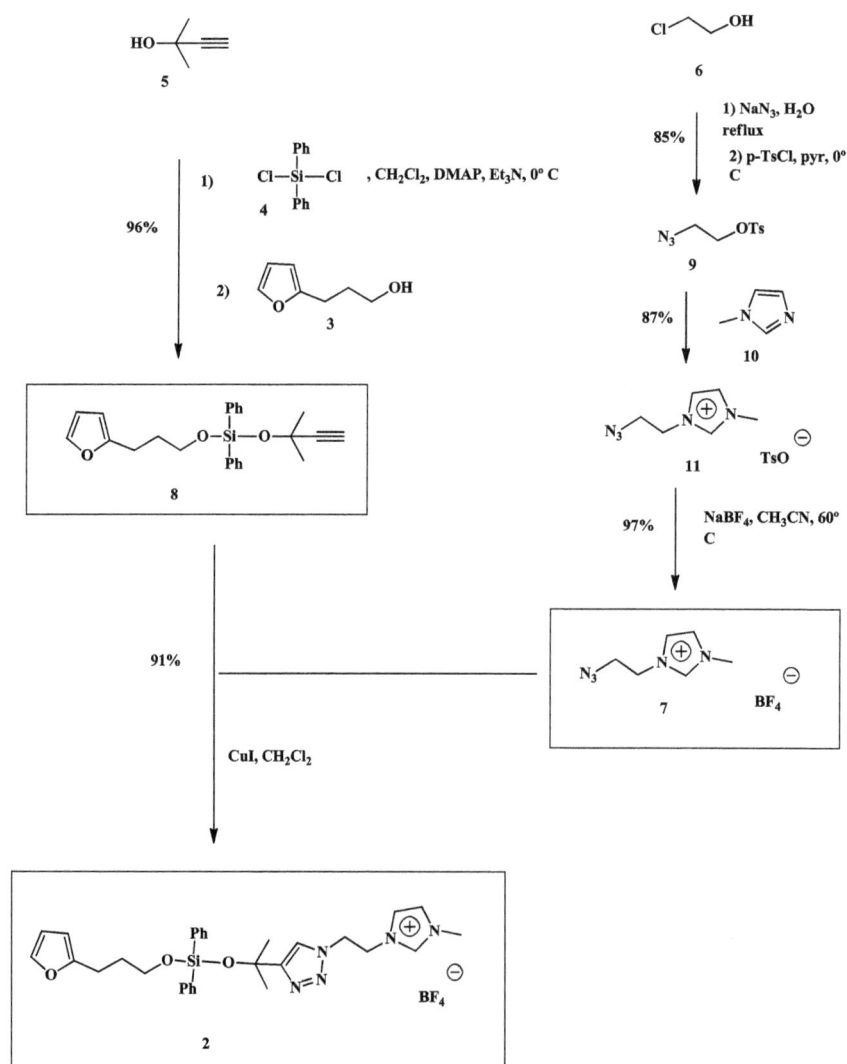

Scheme 2. Synthesis of IL supported furan **2**

Tosylate **9** was easily prepared from commercially available chloride **6** in 85% overall yield. Quaternization of **9** with methyl imidazole **10** gave **11** in a 87% yield. The latter underwent a metathesis reaction to afford the target azide **7** (97% yield). Compound **8** was obtained in a one pot reaction by sequential reaction of propargylic alcohol **5** with dichlorodiphenylsilane **4** followed by furan **3** to afford the target alkyne **8** in an overall yield of 96%. With alkyne **8** and azide **7** in hand, the stage was set for the "click" chemistry reaction which occurred uneventfully giving IL-supported furan **2** in 91% yield.

Our research group developed some years ago a new and original methodology for the synthesis of oxacyclic systems based on the oxidation of a furan ring with singlet oxygen, which we coined "the furan approach[6]. We anticipated that IL-supported furan **2**, could undergo a singlet oxygen oxidation to afford methoxy butenolide **12**, which after protecting group removal would give bicyclic lactone **13**, via an oxa Michael addition (Scheme 3).

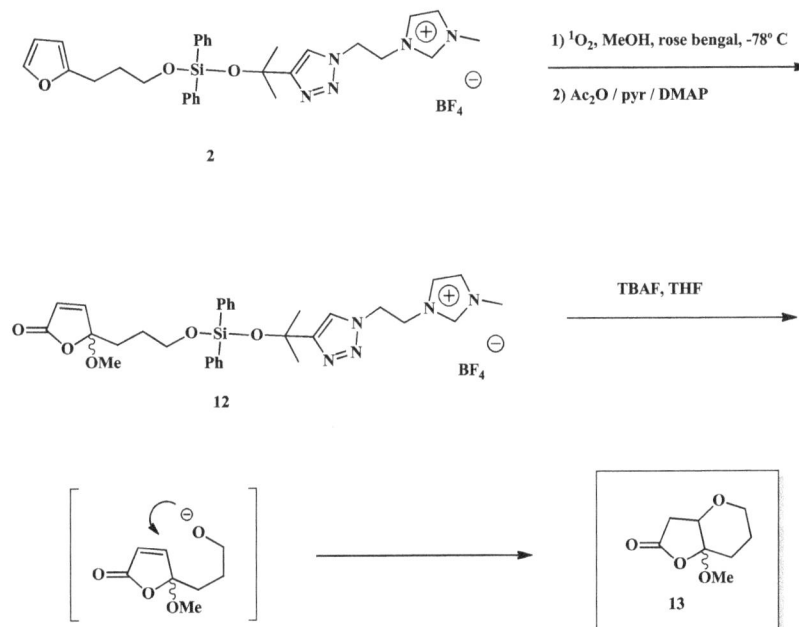

Scheme 3. Synthesis of bicyclic lactone **13** from IL-supported furan **2**

Accordingly, singlet oxygen oxidation of compound **2** gave butenolide **12** which was not isolated but treated with TBAF to give the target lactone **13** in an overall yield of 60%.

Conclusion

In conclusion we have described a straightforward synthesis of ionic liquid bound *tert*-butyldiphenylsiloxane. The ionic liquid supported furan moiety underwent singlet oxygen oxidation to afford a butenolide intermediate which after removal of the siloxane group gave a bicyclic lactone. This procedure could be a good alternative to solid-phase synthesis and use of this methodology for ionic liquid supported synthesis of natural products is now under way in our laboratories.

Acknowledgements

This work was supported financially by the Xunta de Galicia (projects CN 2012/184 and the Galician Network on Ionic Liquids (REGALIs2 (Axuda R2014/015)). The work of the NMR, SC-XRD and MS divisions of the research support services of the University of Vigo (CACTI) is also gratefully acknowledged. A.F; M.G and M.S. thank the University Cheikh Anta Diop (Dakar) for financial support for a research stay at the University of Vigo. I.S and M.S. thank the Senegalese Ministry of Scientific Research: FIRST (fonds d'impulsion de la recherche scientifique et technique) for a research grant

Experimental Section

General Procedures

Solvents were purified and dried by standard procedures. Flash chromatography was performed on silicagel (Merck 60, 230–400 mesh). Analytical TLC was performed on plates precoated with silica gel (Merck 60 F254, 0.25 mm). Melting points were obtained using a Gallenkamp apparatus and are uncorrected. Optical rotations were obtained using a Jasco P-2000 polarimeter. IR spectra were obtained using a Jasco FT/IR-6100 Type A spectrometer. ^1H NMR (400 MHz) and ^{13}C NMR (100 MHz) spectra were recorded on a Bruker ARX-400 spectrometer using TMS as the internal standard;

chemical shifts (δ) are quoted in ppm and coupling constants (J) in Hz. Mass spectrometry (MS and HRMS) was carried out using a Hewlett-Packard 5988A spectrometer. Electrospray mass spectra (ESI-MS) were measured on a Bruker APEXQe FT-ICR MS

2-azidoethyl 4-methylbenzenesulfonate (9).

To a solution of NaN$_3$ 5.8 g (89.6 mmol, 2equiv) in water (70 mL) was added a portionwise 3.00 mL (44.8 mmol, 1equiv) of 3-chloroethanol (6). The mixture was heated under reflux for 16 h, then cooled to room temperature and extracted with CH$_2$Cl$_2$ (60 mL x 3). The combined organic phases were dried over Na$_2$SO$_4$, filtered and evaporated under reduced pressure, affording 3.6 g of 3-azidoethanol as a colourless liquid (98%) used in the next step without any further purification.

2-azidoethanol : R$_f$ = 0.46 (EtOAc/Hexane 3:7); ^{1}H-NMR (CDCl$_3$, δ): 3.67 (2H, t, J = 4.84 Hz); 3.66 (1H, s); 3.33 (2H, t, J = 4.87 Hz); ^{13}C-NMR (CDCl$_3$, δ): 60.86; 53.04; MS (EI$^+$) (m/z, %): 83.96 (100); 85.95 (22); 87.04 (11); 87.95 (8).

To a solution of 2-azidoethanol (7.8 g, 90 mmol) in pyridine (40 mL) was added a portionwise of Tscl (21 g 108 mmol) and the mixture was stirred at 0° C for 3 h. Water (60 mL) was added and the product extracted with Et$_2$O (4 x 80 mL). The combined organic phases were washed with a 10% aqueous solution of HCl and brine. After a drying with Na$_2$SO$_4$, the filtration and solvent evaporation afforded 20 g of tosylate 9 (93%).

R$_f$ = 0.53 (EtOAc/Hexane 3:7); ^{1}H-NMR (CDCl$_3$, δ): 7.83 (2H, d, J = 8.26 Hz**)**; 7.38(2H, d, J = 8.08 Hz); 4.18 (2H, t, J = 5.16 Hz); 3.50 (2H, t, J = 5.04 Hz), 2.48 (3H, s); ^{13}C-NMR (CDCl$_3$, δ): 145.26; 132.61; 129.99; 127.98 ; 68.04; 49.59; 21.67.

Preparation of imidazolium salt 11

A mixture of tosylate 9 (7 g, 29 mmol) and methyl imidazole 10 (2.3 mL, 29 mmol) was stirred at room temperature for 24 h. The residue was washed with EtOAc (3 x 25 mL) to afford 2 g (87%) of salt 11; ^{1}H-NMR (CDCl$_3$, δ): 9.63 (1H, s); 7.73 (2H, d, J = 8.10 Hz); 7.56 (1H, s); 7.38 (1H, s); 7.15 (2H, d, J = 8.00 Hz); 4.39 (2H, t, J = 5.51Hz); 3.89 (3H, s); 3.77 (2H, t, J = 5.34 Hz), 2.34 (3H, s); ^{13}C-NMR (CDCl$_3$, δ): 143.68, 169.32, 137.59, 128.78, 125.59,123.55,122.94, 50.30, 48.48, 36.14,21.21; IR-(CDCl$_3$, v(cm^{-1})): 2924; 2103; 1736; 1575; 1450; 1349; 1289; 1192; 1122; 1034; 1011; 818; 683; 567; MS (ESI) (m/z, %]: ESI$^+$ 475.16 (2cations + 1anion, 10); 152.09 (cation, 100). ESI$^-$ 494.12 (1cation + 2anions, 100); 171.01 (anion, 99).

Preparation of imidazolium salt 7

A mixture of imidazolium salt 11 (1 g, 3.4 mmol) and NaBF$_4$ (0.37 g, 3.4 mmoL) in CH$_3$CN (20 mL) was stirred at 60° C for 24 h. The precipitate was filtered and the organic phase was concentrated to give 0.7 g (99%) of salt 7.
^{1}H-NMR (D$_2$O, δ): 8.72 (1H, s),7.48 (1H, s), 7.41 (1H, s), 4.32 (2H, t, J = 5.4 Hz), 3.85 (3H, s), 3.71 (2H, t, J = 6.4 Hz); ^{13}C-RMN (D$_2$O, δ): 136.62,

123.92, 122.56, 50.30, 48.62, 36.02, ;IR-(CDCl$_3$, v(cm^{-1})): 2924; 2853; 2360; 2105; 1576; 1455; 1351; 1292; 1168; 1055; 770; 649; 622; 521; MS (ESI) (m/z, %]: ESI$^+$ 391.18 (2cations + 1anion, 18); 152.09 (cation, 100). ESI$^-$ 649.20 (2cations + 4anions, 39); 565.19 (2cations + 3anions, 100); 326.10 (1cation + 2anions, 70); 171.01 (2anions, 47).

Preparation of compound 8

To a solution of propargylic alcohol 5 (0.5 mL, 5.2 mmol) in CH$_2$Cl$_2$ (10 mL) at 0° C were added Et$_3$N (1.12 mL, 15.5 mmol), dichlorodiphenylsilane 4 (1.1 mL, 5.2 mmol) and a catalytic amount of DMAP (monitoring the course of the reaction by tlc). The organic solvent was concentrated and water (20 mL) was added. The product was extracted with Et$_2$O (2 x 15 mL). The combined organic phases were washed with brine (2 x 15 mL), dried over Na$_2$SO$_4$. Filtration and solvent evaporation afforded a residue which was chromatographed on silica using 30% EtOAc / Hexane as eluent, affording 3.7 g (96% yield) of alkyne 8 as a colourless oil. Rf: 0,82 (30% EtOAc/Hexane); ^{1}H-NMR (CDCl$_3$, δ): 7.69 (4H, dd, J = 2.8 Hz, J = 1.6 Hz); 7.39 (6H, m); 7.29 (1H, m); 6.22 (1H, dd, J = 2.9 Hz, J = 1.7 Hz); 5.91 (1H, dd, J = 2.6 Hz, J = 2.00 Hz); 3.83 (2H, m,); 2.75 (2H, t, J = 7.7 Hz); 2.29 (1H, s); 1.95 (2H, m); 1.53 (6H, s); ^{13}C-NMR (CDCl$_3$, δ): 155.86, 140.70, 135.36, 134.14, 129.74, 127.57, 110.02, 104.82, 88.37, 70.69, 67.91, 62.29, 32.53, 30.53, 24.35; IR-(CDCl$_3$, v(cm^{-1})): 2984, 2935, 2874, 1591, 1507, 1429, 1380, 1361, 1226, 1151, 1116, 1095, 1046, 1007, 799, 739, 718, 700, 524; MS (EI$^+$) (m/z, %) 413.15 ([M + Na]$^+$, 38); 391.17 ([M + H]$^+$, 100); 325.13 ([C$_{19}$H$_{20}$O$_3$Si]$^+$, 72); 265.09 ([C$_{17}$H$_{17}$OSi]$^+$, 21); 136.06 ([C$_6$H$_4$O$_2$Si]+, 37); HRMS (EI$^+$): 390.1621 calcd for C$_{24}$H$_{26}$O$_3$Si; found: 307.1651.

Preparation of IL supported furan 2

To a solution of azide 7 (423.4 mg, 1.44 mmol) in CH$_2$Cl$_2$ (3 mL) were added alkyne 8 (609 mg, 1.56 mmol) and CuI (27.5 mg, 0.14 mmol). The mixture was stirred at room temperature for 48 h and the solvent concentrated. Ether was added to the residue and the organic phase was decanted, in order to remove excess of starting alkyne. CH$_2$Cl$_2$ (100 mL) was added and the solution was filtered through a short pad of silicagel. Solvent evaporation afforded 899 mg of 2 (91%) as a solid. M.p: 77-79° C;
^{1}H-NMR (CDCl$_3$, δ): 8.70 (1H, s), 7.58-7.56 (4H, m), 7.56 (1H, s), 7.36-7.28 (6H, m), 6.96 (2H, d, J = 1.25 Hz), 6.23 (1H, dd, J = 1.88 Hz, J = 1.92 Hz); 5.89 (1H, dd, J = 1.72 Hz, J = 2.00 Hz); 4.72 (4H, s), 3.77 (3H, s), 3.74 (2H, t, J = 6.20 Hz), 2.69 (2H, t, J = 7.52 Hz), 1.85 (1H, m), 1.65 (6H, s); ^{13}C-NMR (CDCl$_3$, δ): 156.10,155.72,140.77,137.35,134.86, 134.81,134.56, 134.32,134.28,134.01,130.25,130.06, 127.91.127.83,127.76,127.73,122.91, 122.73,121.75,110.05,104.87,72.49,62.19,48.97,48.88, 36.31, 30.91, 30.54, 24.31; MS (ESI) (m/z, %): ESI$^+$ 542.26 (cation, 100). ESI$^-$ 1061.82 (33); 884.29

(cation, 2anions, 88); 748.25 (cation + 2 anions); 524.52 (47); 357.72 (28); 131.54 (100).

7a-methoxyhexahydro-2H-furo[3,2-b]pyran-2-one (13).

To a solution of furan 2 (599 mg, 0.84 mmol) in dry methanol (12 mL) was added Rose bengal (17 mg). The mixture was purged several times with O_2 (balloon), cooled to −78 °C and irradiated with a 200 W lamp for 3 h, stirring under oxygen atmosphere. The mixture was allowed to reach room temperature and the solvent was evaporated. After solvent evaporation the residue was dissolved in pyridine (5 mL), acetic anhydride (1.5 mL) then DMAP (catalytic) were added and the mixture stirred at room temperature overnight. MeOH (5 mL) was added and stirring continued for 30 min. The methanol was rotatory evaporated. The residue was treated with H_2O (20 mL) and the product extracted with CH_2Cl_2 (3 x 15 mL). The combined organic phases were washed with a 10% aqueous solution of $CuSO_4$ (2 × 15 mL), dried over Na_2SO_4 and the solvent concentrated giving a residue which was dissolved in THF (15 mL). Tetrabutylammonium fluoride (0.84 mL of 1.0 M solution in THF, 0.84 mmol) was added and the mixture was stirred at room temperature for 4 h. Aqueous saturated solution of $NaHCO_3$ (30 mL) was added and the product extracted with ethyl acetate (3 x 20 mL). The combined organic phases were washed with brine (2 x 15 mL), dried over Na_2SO_4, filtered and the solvent evaporated giving a residue which was chromatographed on silica using 20% EtOAc / Hexane as eluent, affording 87 mg (60% overall yield) of bicyclic lactone 13; Rf: 0,45 (30% EtOAc/Hexane); ^{1}H-NMR (CDCl_3, δ): 3,88 (1H, d, J=4,4 Hz, CH-1); 3,.85 (1H, m, CH-3); 3,37 (1H, dd, J=11,7 Hz, J=1,7 Hz, CH-3); 3,33 (3H, s, OCH_3); 2,88 (1H, dd, J=17,2 Hz, J=4,4 Hz, CH-9); 2,52 (1H, m, CH-5); 2,33 (1H, d, J=17,2 Hz, CH-9); 1,75 (1H, m, CH-5); 1,66 (2H, m, CH-4); ^{13}C-NMR (CDCl_3, δ): 176,00 (CO); 104,61 (C-6); 76,48 (CH-1); 65,39 (CH_2-3); 49,.35 (CH_3O); 37,02 (CH_2-9); 27,16 (CH_2-5); 21,52 (CH_2-4); MS (EI[+]) [m/z, (%)]: 141,05 (M[+]-OMe, 41); 123,04 (22); 113,06 (8); 102,06 (8); 101,05 (100); 100,05 (10); 99,04 (29); 97,06 (36); 72,06 (8); 71,04 (10); 69,03 (13); HRMS (EI[+]): calcd for $C_8H_{12}O_4$, 172,0736; found: 172,0743.

References

1 - For recent reviews, see (a) T. Welton Chem. Rev. **1999**, 99, 2071-2083. (b) P. Wasserscheid, W. Keim, Angew. Chem., Int. Ed. **2000**, 39, 3772-3789; (c) J. S. Wilkes Green Chem. **2002**, 4, 73-80;

(d) J. Dupont, R. F. de Souza, P. A. Z. Suarez, Chem. Rev. **2002**, 102, 3667-3691;
(e) P. Wasserscheid, T. Welton Ionic Liquids in Synthesis; Wiley-VCH: Weinheim, Germany, 2003.

2 - (a) J-Y. Huang, M. Lei, Y-G. Wang, Tetrahedron Lett. **2006**, 47, 3047-3050 and referencies therein. (b) W. Miao, T. H. Chan, Acc. Chem. Res. **2006**, 39, 897-908.

3 - M. M. Meloni, P. D. White, D. Armour, R. C. D. Brown Tetrahedron **2007**, 63, 299-311.

4 - (a) A. Fall, M. Sène, M. Gaye, G. Gómez, Y. Fall, Tetrahedron Lett. **2010**, 51, 4501-4504; (b) A. Fall, M. Sène, E. Tojo, G. Gómez, Y. Fall, Synthesis **2010**, 20, 3415-3417; (c) A. Fall, M. Sène, O. Diouf, M. Gaye, G. Gómez, Y. Fall, Open Org. Chem.J. **2012**, 6, 21-26.

5 - (a) H. C. Kolb, M. G. Finn, K. B. Sharpless Angew. Chem., Int. Ed. **2001**, 40, 2004-2021; (b) V. V. Rostovtsev, L. G. Green, V. V. Fokin, K. B. Sharpless, Angew. Chem., Int. Ed. **2002**, 41, 2596-2599; (c) C. W. Toroe, M. Christensen, M. Meldal J. Org. Chem. **2002**, 67, 3057-3064; (d) A. Krasiñski, V. V. Forkin, K. B. Sharpless Org. Lett. **2004**, 6, 1237-1240; (e) T. Thirumurugan, D. Matosiuk, K. Jozwiak Chem. Rev. **2013**, 113, 4905-4979.

6 - (a) Y. Fall, B. Vidal, D. Alonso, G. Gómez Tetrahedron Lett. **2003**, 44, 4467-4469; (b) M. Pérez, P. Canoa, G. Gómez, C. Terán, Y. Fall Tetrahedron Lett. **2004**, 45, 5207-5209; (c) D. Alonso, M. Pérez, G. Gómez, B. Covelo, Y. Fall Tetrahedron. **2005**, 61, 2021-2026; (d) M. Teijeira, P. L. Suárez, G. Gómez, C. Terán, Y. Fall Tetrahedron Lett. **2005**, 46, 5889-5892; (e) I. García, G. Gómez, M. Teijeira, C. Terán, Y. Fall Tetrahedron Lett. **2006**, 47, 1333-1335; (f) P. Canoa, M. Pérez, B. Covelo, G. Gómez, Y. Fall Tetrahedron Lett. **2007**, 48, 3441-3443; (g) C. Álvarez, M. Pérez, A. Zúñiga, G. Gómez, Y. Fall Synthesis **2010**, 22, 3883-3890; (h) P. Canoa, Z. Gándara, M. Pérez, R. Gago, G. Gómez, Y. Fall Synthesis **2011**, 3, 431-436; (h) M. Gónzalez, Z. Gándara, B. Covelo, G. Gómez, Y. Fall Tetrahedron Lett. **2011**, 52, 5983-5986; (i)A. Zúñiga, G. Pazos, P. Besada, Y. Fall Tetrahedron Lett. **2012**, 53, 4293-4295; (j) M. González, Z. Gándara, G. Pazos, G. Gómez, Y. Fall Synthesis **2013**, 45, 625-632; (k) M. Gónzalez, Z. Gándara, A. Martínez, G. Gómez, Y. Fall Tetrahedron Lett. **2013**, 54, 3647-3650; (l) M. Gónzalez, Z. Gándara, A. Martínez, G. Gómez, Y. Fall Synthesis **2013**, 45, 1693-1700; M. González, Z. Gándara, M. Seck, G. Gómez, Y. Fall Mediterr. J. Chem. **2015**, 4(3), 18-29.

4

Effect of temperatures and residence time of calcinations on substitution process during enrichment of Tunisian phosphorites

Sarra Elgharbi[1,*], Karima Horchani-Naifer[1], Nabil Fetteh[2] and Mokhtar Férid[1]

[1] Laboratory of Physical Chemistry of Mineral Materials and their Applications, National Research Center in Materials Sciences, Technopole Borj Cedria B.P. 73- 8027 Soliman, Tunisia.
[2] Research Center, CPG Métlaoui 2134 Gafsa, Tunisia

Abstract: The present study deals with the effects of temperature and residence time of calcinations on substitution process. The upgrading of sedimentary phosphorites coming from the Ras-Dhraa deposit which belongs to the Gafsa-Metlaoui basin (Tunisia) was conducted. The characterization of raw and calcined phosphorites and the evolution of rare earth elements and yttrium (REEs+Y) content and associated gangue were evaluated with different techniques (X-ray powder diffraction (XRD), scanning electron microscopy (SEM/EDX), Fourier transform infrared (FT-IR) and inductively coupled plasma mass spectrometry (ICP-MS)). The effect of increasing residence time and temperature of calcinations on phosphate structure were determined. FT-IR spectroscopy indicates that PO_4^{3-} was substituted by SiO_4^{4-} and SO_4^{2-} and these substitutions cause a worsening of phosphorites properties for sulfuric acid production. A temperature of calcination approximately of 900°C and a period of 30 minutes is sufficient, because higher temperature and longer residence time accentuate these substitutions. Indeed, in these conditions we obtain a phosphate with a better grade of P_2O_5. It permits to increase the phosphate content from 29.58% to 33.07%. Moreover, REEs+Y contents increased by 5%.

Keywords: Tunisian phosphorite; Calcinations; Substitution; Infrared spectroscopy; REEs+Y.

Introduction

Tunisian phosphorite is an essential element used in the process of phosphoric acid production. It consists mostly of carbonate fluorapatite and the major impurities are calcite and dolomite, while quartz and feldspar are present like traces[1]. Apatite with the formula $Ca_{10}(PO_4)_6F_2$ may have several substitutions in the structure[2,3]. The most common substitutions in natural phosphates are Mg^{2+} and Na^+ for Ca^{2+}, CO_3^{2-} for PO_4^{3-} and OH^- for F^-. Substitutions modify the structure and often show marked effects on crystallinity, solubility and thermal stability[4]. The presence of carbonate in mineral phosphates requires a supplementary adding of sulfuric acid in the plants of production of phosphoric acid; and causes the formation of foams in the reactors during the addition of sulfuric acid. Also the presence of carbonates at high contents reduces the price value of phosphate rock. More studies have been carried out to reduce the calcium carbonate content of the phosphate rocks by flotation and/or calcination[5,6].

Enrichment of the ores by calcination is one of the well-known processes. It is based on the dissociation of the calcium carbonate by thermal treatment[7-9]. Calcination is the process of ore heating to a high temperatures ranging from 800° to 1000° C to decompose the $CaCO_3$ and $MgCO_3$ to CaO, MgO and gaseous CO_2. Then, after calcination, the formed CaO and MgO may be separated by slaking and washing. Several authors[8,10] have investigated the mineralogical effects of thermal treatment on sedimentary apatites. They showed that the main changes in the apatite structure are the structural CO_2 expulsion, cell parameter a increase and crystallite size variation. These transformations are accompanied by the formation of free CaO. There are several parameters that determine calcinations such as the residence time, the temperature and the physico-chemical properties of the phosphates[9,11]. Not enough investigations have been carried out to study the effect of calcinations parameters on substitutions. In the present study, the upgrading of Tunisian phosphorites by thermal treatment was conducted. The effects of temperature and residence time of calcinations on the substitution process were investigated. Calcinations tests as well as measurement techniques such as XRD, FT-IR, scanning electron microscopy (SEM), and ICP analysis were used to characterize and analyze the raw and calcined phosphorites.

Experimental Section

Samples selection and preparation

Phosphorite used in this investigation was provided from phosphate deposit in Ras-Dhraa

Corresponding author: Sarra Elgharbi
E-mail address: elgharbisarra@gmail.com

(Tozeur). The 70-2000 µm grain size fraction was investigated. As published elsewhere in our paper[12], the chemical composition was determined as: 29.58% P_2O_5, 49.45% CaO, 6.48% CO_2, 3.58% SO_3, 2.84% F, 3.08% SiO_2, 1.39% Na_2O, 0.25% F_2O_3, 0.86 Al_2O_3, 0.054% K_2O and 0.7% MgO. The thermal behavior was studied to predict the optimal temperature for the phosphate rock enrichment. It was found[12] that calcination at 900°C removes humidity, organic matter and decomposes carbonates.

Equipment and instruments

XRD analyses were performed using an X'PERT Pro PAN Analytical diffractometer with CuKa radiation (λ=1.5418 Å). The samples were scanned in the diffraction angle range (2θ) varying from 5° to 80° with steps of 0.02° for 5 min. The crystalline phases have been determined by comparison of the registered patterns with the International Center for Diffraction Data (ICDD)-Powder Diffraction Files (PDF).

Infrared spectra were measured from 400 to 4000 cm^{-1} (wavenumber) by the standard KBr pellet method using a Fourier transform infrared spectrometer (Perkin–Elmer). The resolution of 4 cm^{-1} was used collecting 20 scans for each sample. The micromorphology of the phosphate ore samples before and after calcination was evaluated using scanning electron microscopy (QUANTA 200 microscope). In addition, SEM–EDX technique was used to identify the chemical composition of the products.

The rare earth elements concentrations were carried out at the Department of Earth Sciences, Università della Calabria, and were determined by Inductively Coupled Plasma-Mass Spectrometry (ICP-MS) using an Elan DRCe (Perkin Elmer/SCIEX).

Phosphorous was determined by spectrophotometric method using "Perkin-Elmer, model Lambda 20" Spectrophotometer, by measuring the color density of yellow complex phospho-vanado-molybdate at a wavelength of 430 nm.

Digestion procedure

For the ICP-MS analysis phosphorites samples were powdered in an agate mortar and then dissolved by microwave digestion using a Mars5 microwave apparatus (CEM technologies). About 100 mg (±0.01 mg) of powder was placed in a microwave vessel with a mixture of Merck "suprapur" quality hydrofluoric acid (6 ml HF), nitric acid (5 ml HNO_3) and perchloric acid (3 ml $HClO_4$), covered and sealed with a cap and subjected to an oven method that consisted of a 15min ramp to 200 °C and a pressure of 600 PSI, then held at temperature for 15 min and then a 15 min cool down. An unclear solution containing some residue was obtained. The content of the vessel was allowed to heat up to 200 °C. Before the acid complete evaporation, 3ml of pure $HClO_4$ were added and the mixture was heated up to 200°C. Before the complete evaporation, 5 ml of HNO_3 (5%) was then added. A clear, colorless solution was finally obtained. This latter, was left to cool down gently and made up to a standard volume in a 100 ml volumetric flask with Millipore water in order to prepare the mother solution. External calibration curves were prepared using Perkin Elmer "multi-element Calibration Standard 2 solution" to analyze rare earth elements. Standard reference materials Micaschist (SDC1) were prepared in the same way and were used as unknown samples during the analytical sequence. The elements concentrations were compared with certified values to evaluate accuracy and precision of analytical data.

Calcination experiments

Calcination was carried out in electric furnace. 2 g of phosphorite were putted in platinum crucible, then the temperature was raised gradually (10°C/min) from room temperature until reaching the required level and maintained for the predetermined time. A series of experiments were carried on the phosphorite covering the temperature 900, 1000 and 1050 °C for 30, 40, 60 and 90 minutes.

Results and Discussion

SEM and EDX of raw and calcined phosphorites

The scanning electron microscopy (SEM) observation of raw phosphorites reveals the irregular particle morphology, as shown in Fig.1.The dispersive analysis in energy of X-rays (EDX) (Fig.2) indicates high contents of Ca, P, O and impurity levels of Si, Mg, Na and S. This observation confirmed the results found by XRD, FT-IR and chemical analysis.

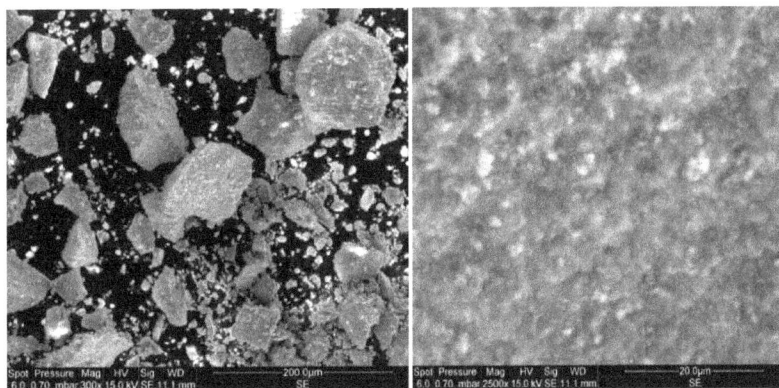

Figure 1. SEM images of raw phosphorites with two different resolutions

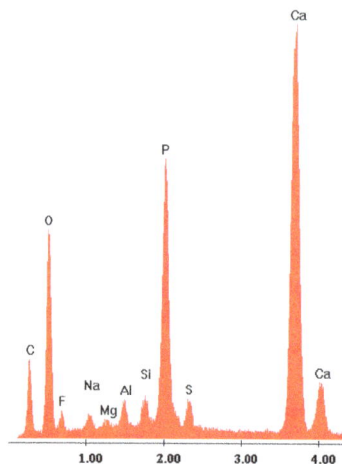

Fig. 2. EDX pattern of raw phosphorite

The SEM micrographs of phosphorites particles after calcinations at 900°C for 30 min (Fig.3) indicates that the calcined particles appear more compacted as well as the surface of particles becomes relatively smooth due to the sintering of the particles at high temperature. Also, SEM micrographs showed macropores occurrence in the phosphate after calcinations. In addition, the EDX results of calcined phosphorites are shown in Fig. 4. As it is clear, peaks of C, Al and O elements are present in addition to Ca, P, and Si. The high intensity of carbon and aluminum peaks can be explained by two reasons: In one hand, carbon and aluminum peaks are associated with the agents used to immobilize the particles for analysis. On the other hand, the carbon peak may depict the calcium carbonate presence after annealing for 30 min. Thus, it may imply the incomplete decomposition of $CaCO_3$.

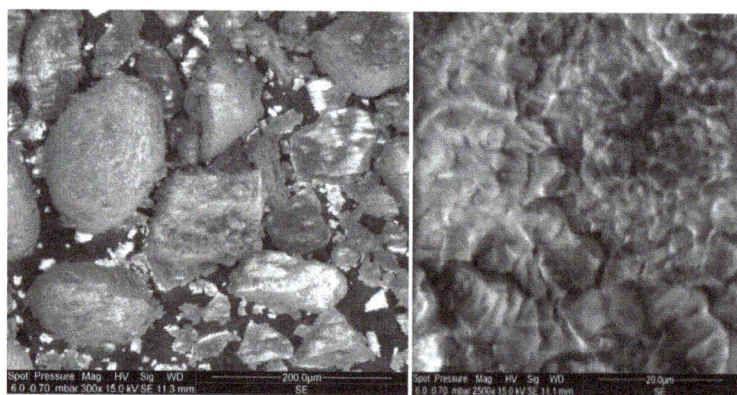

Figure 3. SEM images of phosphorites particles calcined for 30 min at 900°C with two different resolutions

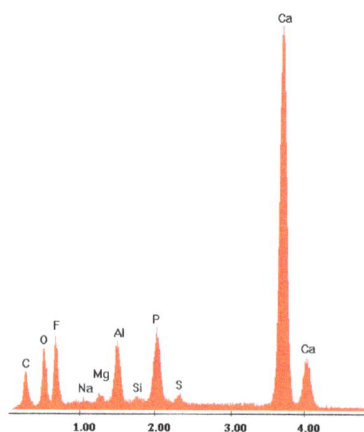

Figure 4. EDX pattern of phosphorites particles calcined for 30 min at 900°C.

XRD analysis of raw and calcined phosphorites

The XRD patterns of raw and calcined phosphorites are shown in Fig.5. XRD pattern of raw

phosphorites (Fig.5.a) presents various reflections attributed to carbonate fluorapatite, fluorapatite, dolomite, calcite and quartz[12]. After calcination at 900°C, a slight modification of the reflection positions is noted. During calcinations carbonate fluorapatite changes to fluorapatite and carbonates decompose to form CaO. The relative intensities of the calcite, dolomite, and quartz reflections in the XRD patterns disappears and the lime reflections were formed and became clearly distinguishable (Fig.5.b).

As a result of CO_3^{2-} removement from phosphorite at temperatures above 900°C the unit cell parameter achieves almost the value of pure fluorapatite and the crystallinity improves again[13]. It is important to note that regardless of the residence time of calcinations; the obtained X-ray patterns of calcined phosphorites are similar suggesting that there is no significant structural difference between the phosphorites calcined 30 min or 90 min. As it is supposed, carbonate decomposition is the main process during calcination, and 30 min is long enough to complete the transformation.

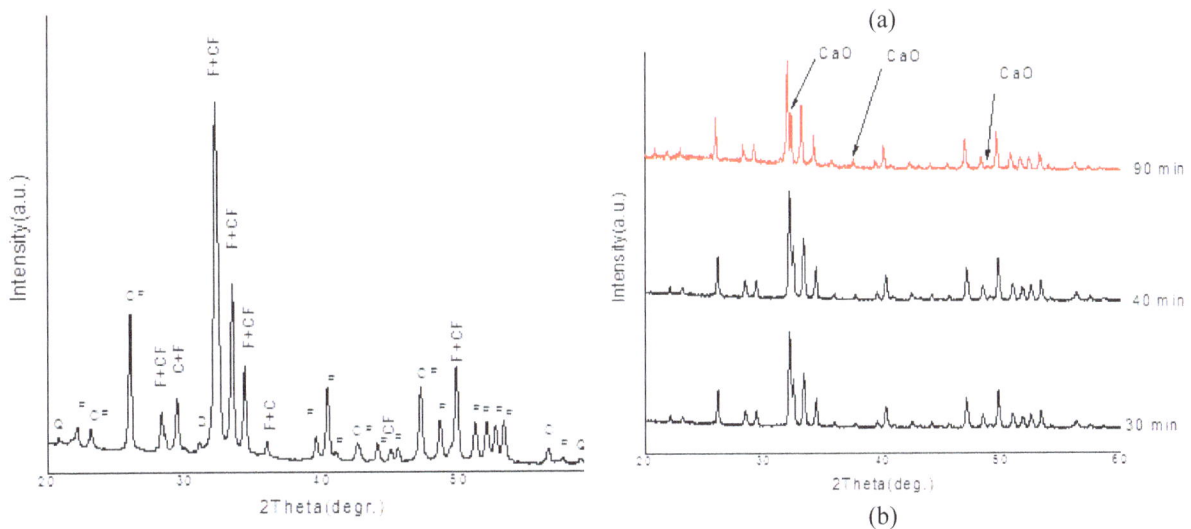

Fig. 5. XRD patterns of (a) raw and (b) calcined phosphorites at 900° for 30, 40 and 90 min
F:fluorapatite ; C:calcite ; D:dolomite ; Q:quartz ; CF: carbonate fluorapatite

FT-IR spectroscopy

The FT-IR spectroscopic study was conducted in order to check the following complex ions: PO_4^{3-}, CO_3^{2-}, SiO_4^{4-}, SO_4^{2-} and SiO_3^{2-}. Fig.6 shows infrared absorption spectrum of phosphorites before and after thermal treatment. The IR spectral analyses of Tunisian phosphorite examined in details in our previous publication[12] have evidenced the bands of PO_4^{3-}, SiO_2, CO_3^{2-}, OH^- referring to the apatite, quartz and calcite.

The effect of residence time of calcination on phosphorites was determined by calcinations of phosphate particles at 900°C for 30 min, 40 min, 60 min and 90 min. FT-IR spectra of calcined samples are shown in Fig. 6. It is quite worthy to note that no appreciable changes in phosphate matrix have been happening at rising residence time of calcinations. Bands characteristic of fluorapatite are not affected. The absorption bands assigned to the PO_4^{3-} group was detected at 1096, 1040, 962, 606, 577 and 474 cm^{-1} [14]. Most of the bands due to the phosphate vibrations of fluorapatite have largely increased in intensity after increasing time of calcination. This behavior has been described by other authors[15,16] as a consequence of the thermal treatment of apatites.

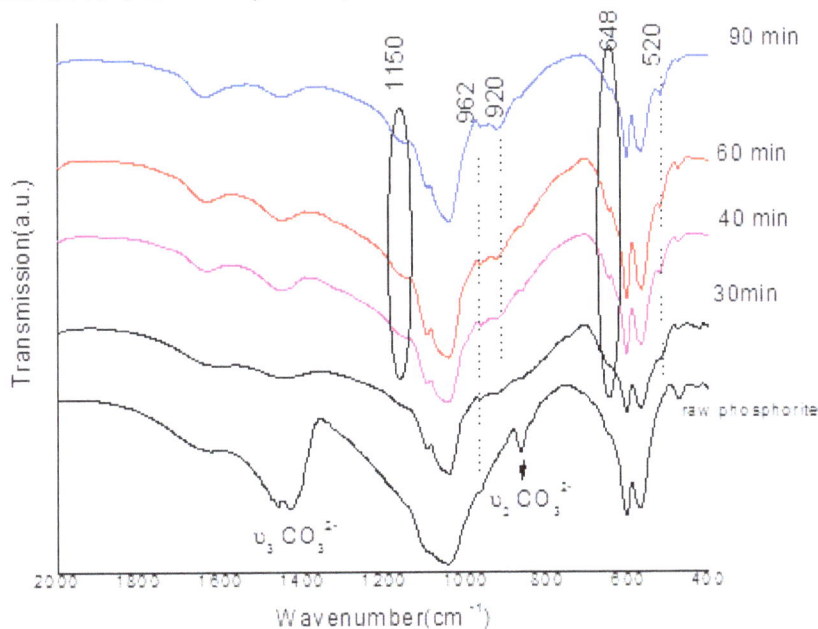

Figure 6. FT-IR spectra of raw and calcined phosphorites at 900°C for 30, 40, 60 and 90 min

FT-IR Spectra showed additional absorption bands at 520 cm^{-1}, 926 cm^{-1} and 647 cm^{-1}. These results resemble those found by Petkova[17] during thermal treatment of nanosized carbonate apatite from Syria. These bands are attributed to the bending and stretching modes of the SiO_4^{4-} and SiO_3^{2-}. The established change gives ground to assume that during calcinations quartz mineral present in raw phosphorites transform into structure of higher symmetry. Bands around 520 cm^{-1} and 926 cm^{-1}

indicate substitution of PO_4^{3-} by (SiO_4^{4-})[12]. The peak observed around 647 cm^{-1} was attributed to (SiO_3^{2-}) [17]. The intensity of this peak increases with increasing residence time of calcinations.

Previous study[18] on apatite-type lanthanum silicates showed an increase in the intensity of stretching and bending mode of the SiO_4 tetrahedra with firing temperature. These increases indicate increase of cation vacancies at high temperature. These vacancies were produced by the ousting of the carbonate.

Small band was observed at 1150 cm^{-1} in the FT-IR spectra of phosphorite calcined up to 40 min at 900°C. This band indicates substitution of PO_4^{3-} by SO_4^{2-}. Thermal oxidation and decomposition of organic matter occurs, sulfur is liberated, mainly as SO_2 and some of it enters the apatite structure as SO_4. The SO_4^{2-} and SiO_4^{4-} enter the apatite structure in the vacant orthophosphate positions produced by the ousting of the carbonate. Other workers [19] in this field argued that the presence of carbonate groups in the raw apatite played an important role in the incorporation of silicates during heating and observed that at low silicate content, the latter would substitute for carbonate during the heat treatment. Baumer et

al.[20] argue for the occurrence of a coupled substitution ($2PO_4^{3-} \rightarrow SO_4^{2-} + SiO_4^{4-}$).

We note that increasing residence time increases the intensity of bands corresponding to the SO_4^{2-}, SiO_3^{2-} and SiO_4^{4-} substitutions. It is obvious that these bands have been intensified with increasing the degree of substitution.

On the other hand, we have raised temperature of calcinations (900; 950; 1050°C) to investigate the effect of temperature on substitutions. As it can be seen from Fig.7 the intensity of bands at 520, 920, 648 and 1150 cm^{-1} increases with increasing temperature of calcinations. An increase in the operating temperature was accompanied by an increase in the extent of calcination and sulfation[21].

Additionally, the typical bands of carbonate substituting for phosphate site (type B) in the apatite lattice (peak at 871 cm^{-1} and double band at 1410/1445 cm^{-1}) [12] are not observed at 1050°C. The calcination at higher temperatures (900-1200 °C) results in the complete removal of carbonate from the mineral[22]. Higher temperature is not desirable in spite of almost complete ousting of structural carbonate, because the reactivity of the calcined product was lowered by raising the calcination temperature[12].

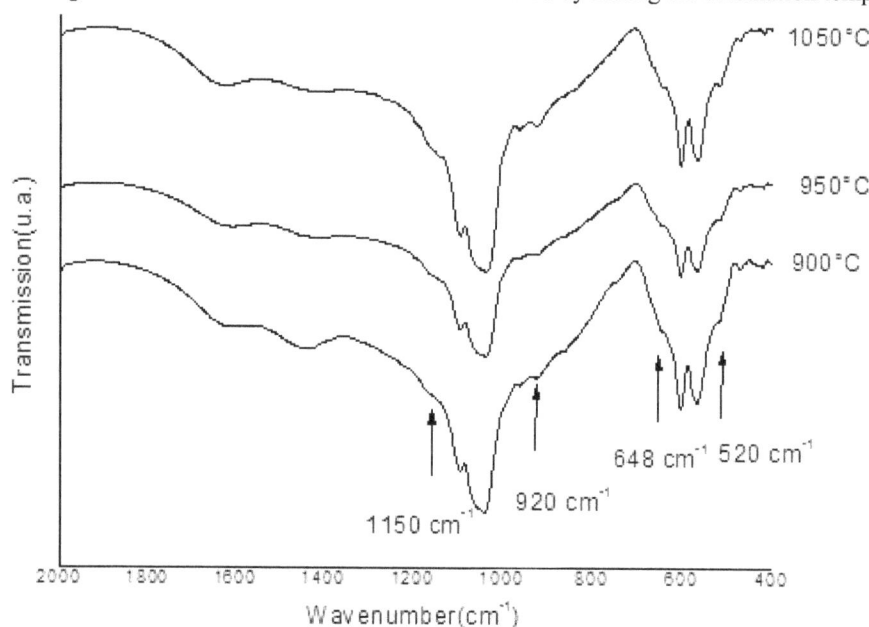

Figure 7. FT-IR spectra of phosphorites calcined at 900, 1000 and 1050°C for 30 min

Higher temperatures and longer residence times can accentuate the substitution of PO_4^{3-} by SO_4^{2-} and (SiO_4^{4-}) [23] and these substitutions cause a worsening of phosphorite properties for sulfuric acid production[24].

The obtained results show that calcination at 900°C for 30 min is sufficient to enrich phosphorites without accentuation of the substitution rate. In addition, these heating time and temperature were more economical. It was shown in another work[6] that increasing temperature and time of calcinations may result in the loss of phosphate grains but has not affected the P_2O_5 content significantly. Flash calcinations reduce the

residence times at the high temperature[23]. In this calcination the entering of sulfur and silicon in the apatite structure practically takes place only at 950°C. Knubovets et al.[23] showed that flash calcination at 870° (residence time a few seconds) is enough for the beneficiation of phosphorites from the Nahal Zin deposit. This shows that use of flash for upgrading phosphorite is desirable because the samples are heated at a high temperature for a very short time (a few seconds).

REEs + Y contents in raw and calcined phosphorites

Rare earth elements are required for many different applications such as metal and glass additives, catalysts for petroleum refining and phosphors used in electronic displays [25-27].
Increased commercial demand for REEs has resulted in a research of new rare earth elements bearing minerals, as well as process required to separate lanthanides. Phosphorites ores are potential resource of REEs and it is important to determine the concentration of these elements for their recovering. Esmaeil et al.[28] have studied the extraction of REEs from apatite and they show that it is possible to extract REEs with a percentage of 80% for neodymium, cerium, lanthanum and yttrium. These elements are of significant industrial interest and even a small quantity could be extracted from phosphate.

The REEs+Y contents of raw phosphorites from Rhas-Dhraa deposit were presented in Table 1.

Table 1 Rare earth elements concentration (ppm) of raw and calcined phosphorites (30min/900°C) from Ras-Dhraa deposit.

Element	Raw phosphorites	Calcined phosphorites
La	54.63	58.56
Ce	67.63	62.64
Pr	10.53	10.57
Nd	56.44	60.41
Sm	7.69	7.43
ΣLREEs	**196.92**	**199.61**
Eu	1.91	1.87
Gd	8.7	8.34
Tb	1.38	1.32
Dy	8.99	8.67
Ho	2.02	1.97
Er	6.5	6.42
Tm	0.91	0.88
Yb	7.45	7.34
Lu	0.97	0.96
Y	92.17	107.99
ΣHREEs	**38.83**	**37.77**
ΣRREs+Y	**327.92**	**345.37**

Light REEs (LREEs: La to Sm), Heavy REEs (HREEs: Eu to Lu)

The sum of rare-earth elements concentration (ΣREEs+Y= 327.9 ppm) is lower than the average content in the phosphorite from Sra Ouertane phosphorites (ΣREE=570,56 ppm) and Jebel Jebs phosphorites (ΣREEs=547.79 ppm)[29] but Ras Dhraa phosphorites have higher REEs content compared to the Gafsa Metlaoui particles (ΣREEs=162.3 ppm)[29]. A significant concentration of yttrium, lanthanum, cerium and neodymium was observed in Rhas-Dhraa phosphorites samples. The concentration of the light rare elements is higher than the concentration of the heavy rare elements; the ionic radii of the light elements are very close to that of Ca^{2+}, which facilitates their substitutions in the crystal lattice of the francolite[30].

Phosphorites are composed of francolite as the main mineral and the asseccory minerals like silicates, carbonates and organic matter, which incorporate REEs+Y disproportionately and unpredictably. Such accessory phase must be ruled out in order to interpret properly REEs+Y content of phosphorites. After calcinations of phosphorites at 900°C for 30 min we note that the REEs content increases from 327.92 to rich 349.77 ppm. REEs+Y contents increased by 5% .Dolomite and quartz contain insignificant amount of REEs as compared to carbonate fluorapatite[31].The effect is not strong and may be attributed to total mass decrease due to CO_2 loss.

Effect of calcination on P_2O_5 content of phosphorites

Chemical analysis of calcined phosphorite show higher P_2O_5 grade in the phosphate product after calcinations 30 min at 900°C. The decomposition of adsorbed and combined water, burning of organic matter and most of the combined carbon dioxide increase the phosphate grade in the calcined product. It permits to increase the phosphate content from 29.58% to 33.07%. Many investigators[31] recommended that further increase in calcinations temperature has no detectable effect on P_2O_5 content.

Conclusion

In this study, the calcination of Ras-Dhraa phosphorite was conducted in terms of calcinations temperature and time .Chemical analysis; FT-IR spectroscopy and XRD analysis were used to investigate changes in the raw and calcined phosphorites. The elemental analysis of raw phosphorites verified the presence of many impurities such as S, Al, Si and Mg and the morphology of calcined phosphorite shows formation of macropores after calcinations. No significant effect of substitution in apatite structure was detected by XRD analysis. In addition, FT-IR spectroscopy results highlighted that increasing residence time of calcinations can accentuate SiO_4^{4-} and SO_4^{2-} substitutions and these substitutions cause a worsening of phosphorites properties for sulfuric acid production. It was also found that calcination at 900°C for a period of 30 min increase the phosphate grade by 10.5% and increase REE+Y content by 5%.

As a result, in the light of these observations, it can be concluded that the calcination of phosphorite at 900°C for 30 min minimize the rate of substitution and permit to obtain phosphorite with better phosphate grade and REEs +Y content and minimize calcite. This phosphorite will be suitable for the manufacture of clear green phosphoric acid with the possibility of recovering REEs.

Acknowledgments

The authors gratefully thank Professor Donatella Barca (Department of Earth Sciences, University of Calabria) for expert technical assistance in ICP-MS investigations.
This work is supported by the Ministry of Higher Education and Scientific Research of Tunisia.

References

1- V. Kolevaa, V. Petkova, IR spectroscopic study of high energy activated Tunisian phosphorite, *Journal of Vibrational Spectroscopy*, **2012**, 58, 125-132.

2- Y. Pan, M.E. Fleet, Compositions of the apatite group minerals: substitution mechanisms and controlling factors, *Reviews in Mineralogy and Geochemistry,* **2002**, 48, 13-50.

3- M. Pasero, T.J. White et al, Nomenclature of the apatite super group minerals, *European Journal of Mineralogy*, **2010**, 22, 163-179.

4- M. Veiderma , Studies on thermochemistry and thermal processing of apatite. *Proceeding of the Estonian Academy of Science. Chemistry*, **2000**, 49, 5-18.

5- N.A. Abdel-Khalek, E. El-Shall, M.A.Abdel-Khalek, A.M.El-Mahdy, S.E. El-Mofty, Carbonate separation from sedimentary phosphates through bioflotation, *Minerals and Metallurgical Processing*, **2009**, 26, 85-93.

6- I.Z. Zafar, M.M. Anwar, D.W. Pritchard, Optimization of thermal beneficiation of a low grade dolomitic phosphate rock, *International Journal of Mineral Processing*, **1995**, 43, 123-134

7- M. Ashraf, Z.I. Zafar, T.M. Ansari, Selective Leaching Kinetics and Upgrading of Low-Grade Calcareous Phosphate Rock in Succinic Acid, *Hydrometallurgy*, **2005**, 80, 286-92.

8- S. El Asri , A. Hakam et al, Structure and thermal behaviors of Moroccan phosphate rock (Bengurir), *Journal of Thermal Analysis and Calorimetry*, **2009**, 95, 15-19.

9- M. Gharabaghi, M. Irannajad, M. Noaparast, A review of the Beneficiation of Calcareous Phosphate Ores Using Organic Acid Leaching, *Hydrometallurgy*, **2010**, 103, 96-07.

10- A. K. Özer, M. S. Gulaboglu, S. Bayrakceken, W. Weisweiler, Changes in physical structure and chemical composition of phosphate rock during calcination in fluidized and fixed beds, *Advanced Powder Technology*, **2006**, 17, 481-494.

11- A.Z.M. Abouzeid, Physical and thermal treatment of phosphate ores - An overview, International Journal of Mineral Processing, 2008, 85, 59-84.

12- S. Elgharbi, K.Horchani-Naifer, M. Férid, Investigation of the structural and mineralogical changes of Tunisian phosphorite during calcinations, *Journal of Thermal Analysis and Calorimetry*, **2015**, 119, 265-271

13- K. Tonsuaadu, M.Peld , V. Bender V, Thermal analysis of apatite structure, *Journal of Thermal Analysis and Calorimetry* , **2003**, 72, 363-371.

14- V.C. Farmer, The layer silicates: The Infrared Spectra of Minerals, Monograph 4; ed. by V.C. Farmer; Mineralogical Society: London, **1974**, pp. 331-361.

15- C.Y. Ooi, M. Hamdi,S. Ramesh, Properties of hydroxyapatite produced by annealing of bovine bone, *Ceramics Internationals*, **2007**, 33, 1171-1177.

16- S.E. Etok, S.L. Woodgate et al, Structural and chemical changes of thermally treated bone apatite, *Journal of Materials Science,.* **2007**, 42, 9807-9816.

17- V. Petkova, V. Yaneva, Thermal behavior and phase transformations of nanosized carbonate apatite (Syria), *Journal of Thermal Analysis and Calorimetry*, **2010**, 99, 179-189.

18- E. Rodríguez-Reyna, A.F. Fuentes, M. Maczka, J. Hanuza, U. Amador, Structural, microstructural and vibrational characterization of apatite-type lanthanum silicates prepared by mechanical milling, *Journal of Solid State Chemistry*, **2006**, 179, 522-531.

19- M. Palard, E. Champion, S. Foucaud, Synthesis of silicated hydroxyapatite $Ca_{10}(PO4)_{6-x}(SiO_4)_x(OH)_{2-x}$, *Journal of Solid State Chemistry*, **2008**, 181, 1950-1960.

20- A. Baumer, R. Caruba, M. Ganteaume, Carbonate-fluorapatite: mise en evidence des substitutions couplees 2 PO4 → SiO4 + SO4 par spectrometrie infrarouge, *European Journal of Mineralogy*. **1990**, 2, 297-304.

21- M. Sınırkaya, H. Bayrakçeken, A.K. Özer, M.S. Gülaboglu, The effect of carbon dioxide during the desulfurization of flue gas with Mardin-Mazıdagı phosphate rock, *Fuel*, **2008**, 87, 3200–3206.

22- M. Figueiredo, H. Figueiredo et al, Effect of the calcination temperature on the composition and microstructure of hydroxyapatite derived from human and animal bone, *Ceramics Internationals*, **2010**, 36, 2383-2393.

23- R. Knubovets, Y. Nathan, S. Shoval, J. Rabinowitz, Thermal transformation in phosphorite, *Journal of Thermal Analysis and Calorimetry*, **1997**, 50, 229-239.

24- R. Knubovets, Structural mineralogy and properties of natural phosphate, *Reviews in Chemical Engineering*, **1993**, 9, 161-216.

25- L. Meyer, B. Bras, Rare earth metal recycling. International Symposium on Sustainable Systems and Technology, 2011, 1-6.

26- C. Preinfalk, G. Morteani, The industrial applications of rare earth elements, Special Publication of the Society for Geology Applied to Mineral Deposits. **1989**, 7, 359-70.

27- M. Peter, C. Peter, S. Francis, Lanthanides, Tantalum and Niobium. Society for Geology Applied to Mineral Deposits. **1989**, 7, 359-70.

28- J. Esmaeil, S. Malek, The production of rare earth elements group via tributyl phosphate extraction and precipitation stripping using oxalic acid. *Arabian Journal of Chemistry*, **2012**; doi:10.1016/j.arabjc.2012.04.002.

29- G. Hechmi, B. Salah, B. Donatella, C. Chaker, Application of LA-ICP-MS to sedimentary phosphatic particles from Tunisian phosphorite deposits: Insights from trace elements and REE intopaleo-depositional environments, *Chemie der Erde-Geochemistry*, **2012**, 72, 127-139.

30- Y. Nathan, Phosphate Minerals: the mineralogy and geochemistry of phosphorite; eds. Heidelberg; Springer: Berlin, **1984**, pp. 275-291.

31- A.A. El-Midany, F.A. Abd El-Aleem, T.F. Al-Farissa, Why do relatively coarse calcareous phosphate particles perform better in astatic-bed calciner, *Powder Technology*, **2013**, 237, 180-185.

5

Crystal Structure and DFT Calculation Studies of Ni (II) Cinnamaldehyde Thiosemicarbazone Complex

Karima Benhamed[1], Leila Boukli-Hacene[1,*] and Yahia Harek[2]

[1] Laboratoire de Chimie Inorganique et Environnement, BP 119, Université de Tlemcen, Algeria
[2] Laboratoire de Chimie Analytique et Electrochimie, BP 119, Université de Tlemcen, Algeria

Abstract: Slow evaporation of a dilute DMSO solution of the title compound at room temperature, provided a brown crystal of $Ni(CMTSC)_2DMSO$ suitable for X-Rays study (space group: P-1, a(Å)=8.1750 (3), b(Å) =11.3400 (4), c(Å) =15.1940 (5), α(°)=68.581 (3), β(°)=78.894(4), γ (°)=79.265(5)). Two Ni atoms were located on special positions providing two molecules of $Ni(CMTSC)_2$ different for their torsion angles and intermolecular interactions. In both molecules, the thiosemicarbazone coordinates as an anionic ligand via the thiosemicarbazone moiety's azomethine nitrogen and thiolatosulphur in a square-planar geometry. In the aim of investigating structural features, Density Functional Theory calculations of both ligand and complex were fully optimised with respect to the energy using B3LYP level. The predicted geometry parameters are compared with their corresponding X-ray crystallographic data.

Keywords: Ni(II) complex; thiosemicarbazone; cinnamaldehyde; DFT calculations; crystal structure.

Introduction

Thiosemicarbazones (TSCs) constantly attract the interest of chemist and pharmacist because of their well-known and remarkable biological and pharmacological properties (antiviral, antibacterial or antitumoral activities); that is why the literature concerning these subjects is steadily increasing [1–5]. Such pharmacological activities are due to the strong chelating ability of these ligands with biologically important metal ions such as Fe^{2+}, Cu^{2+}, Ni^{2+} and their reductive capacities[6,7].

However, not only the bioinorganic relevance of the complexes but also the chemistry of transition metal complexes of the thiosemicarbazones is receiving significant current attention because of the variable binding modes displayed by these ligands in their complexes[8–14]. Thiosemicarbazones usually bind to a metal ion as bidentate N, S donor ligands via dissociation of the hydrazinic proton, forming five-membered chelate rings. Because the ligands can bind either as the neutral species or monodeprotonated, crystallographically the difference can generally be monitored by the length of the C–S bond. The neutral ligand contains a formal C–S double bond with bond lengths of the order of 1.67–1.72 Å, while the deprotonated ligand undergoes tautomerization to produce a formal C–S single bond with bond lengths of the order of 1.71–1.80 Å[15].

In some cases TSC can act as a C, N, S donor, forming cyclometallated complexes[16,17]. An overview of thiosemicarbazone complexes was published by Lobana and al.[18]. The presence of multiple donor atoms within the same ligand multiplying coordination modes affects the properties of ligands and complexes[19,20] and in the same time thiosemicarbazones exist in equilibrium of various tautomers or conformers which greatly affects their chelating ability[21,22]. On the other hand Nickel can take up different coordination environments (such as octahedral, square-planar and tetrahedral). These special features represent a real challenge to explore the coordination mode of the present complex further.

Till now, no experimental and theoretical data on the main geometric and structural characteristics of $Ni(CMTSC)_2$, which are necessary for a deep understanding of their chemical and physical behavior, have been reported. In this perspective, we report here synthesis, single X-ray structure of $[Ni(CMTSC)_2]$ DMSO and DFT calculations of both ligand and complex (scheme 1). They were fully optimised with respect to the energy using B3LYP/6-311G(d,p) level.

Corresponding author: Leila Boukli-Hacene
E-mail address: *l_bouklihacene@mail.univ-tlemcen.dz

Experimental and theoretical methods:

Synthesis

The complex was obtained in the same way as previously described[23]. It was prepared by the addition of Ni(OAc)$_2$ 4H$_2$O (0.249 g, 1 mmol) in absolute EtOH to a hot absolute EtOH solution of the ligand (0.371 g, 2 mmol). The mixture was heated under reflux with stirring for 1h. The precipitate was filtered off hot, washed successively with EtOH and dried in a vacuum desiccator over silica gel. Additional details will be reported in further report.

X-ray Crystallography and Data Collection.

Crystals suitable for an X-ray structure determination were obtained by slowly evaporating a DMSO solution of the compound in air at room temperature. X-ray data were recorded on a BrukerAPEX2 diffractometer using Mo Kα radiation (0.71073 A). Data collection and reduction were performed using SAINT softwares[24]. The structure was solved bySIR97[25], and refined by SHELXL97 [26]. ORTEP-3 for Windows [27]and MERCURY [28] were used for molecular graphics. To prepare material for publication, we used WinGX publication routines [29].

All non-hydrogen atoms were subjected to anisotropic refinement by full-matrix least squares on F^2. The hydrogen atom positions were fixed geometrically at calculated distances and allowed to ride on the parent atoms, except hydrogen atoms bonded to C51, C61, C71, C52, C62, and C72 which were located in a difference Fourier map and refined independently.

A summary of the crystal data, experimental details and refinement results are listed in Table 1. Selected bond lengths and angles are in Table 2, torsion angles in Table 3 and the relevant hydrogen bond parameters are provided in Table 4.

Table 1. Crystal data, data collection and structure-refinement parameters for **[Ni(CMTSC)$_2$] DMSO**

Formula	C$_{22}$H$_{26}$N$_6$NiOS$_3$
Color	Brown
Wavelength (Å)	0.71073
T (K)	293(2)
Crystal size (mm)	0.3 ×·0.2 ×·0.2
Crystal system	Triclinic
Space group	P-1
a (Å)	8.1750 (5)
b (Å)	11.3400 (4)
c (Å)	15.1940 (4)
α (°)	68.581 (3)
β (°)	78.894(4)
γ (°)	79.265(5)
V(A^3)	1276.31(10)
Z	2
D calc. (g cm^{-1})	1.403
F w	539.33
Absorption coefficient. (mm^{-1})	1.031
F(0 0 0)	556
Index ranges	0 ≤ h ≤ 5 -14 ≤ k ≤ 15 -19 ≤ l ≤ 20
θ range (°)	1.45-28.31
Number of data measured	3979
Number of data with I >2σ(I)	3053
Number of variables	296
R_1($R1$ all data)	0.066(0.089)
wR_2	0.20
Max Δe/Å^3	1.070
Max shift	0.003
Goodness-of-fit on F^2	1.1350

Theoretical and computational details

A theoretical quantum-chemical method most widely used today is density functional theory (DFT). DFT method combines accuracy with computational speed and ease of use. This is particularly true for hybrid DFT methods, which consistently have been shown to be highly reliable. Of all hybrid DFT methods, the B3LYP functional is the most widely used [30] and yields accurate results for many systems containing transition metal atoms [31].

For the complex NiL_2, initial molecular geometry was taken from crystal structure while initial molecular geometry of ligand HL was taken from crystal structure of similar compound (4-Phenyl-1-(3-phenylallylidene) thiosemicarbazide)[32]. Then, Full geometry optimizations were carried out using semi-empirical PM3 method[33] included in hyperchem 8.0.10[34]. Therefore, DFT[35] calculations with the B3LYP hybrid functional, in which Becke's three parameter exchange functional[36,37] and the Lee-Yang-Parr non-local correlation functional[38] at basis set 6-311G (d,p)[39] were performed with the Gaussian 09 software package[40] and Gauss view 5.08 program[41]. The harmonic vibrational frequencies were calculated for the optimized geometries and no imaginary frequencies were found indicating that the optimized structures were stable.

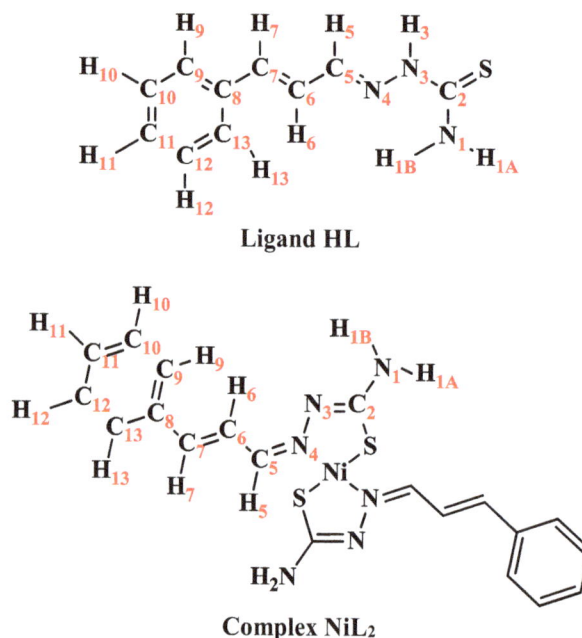

Ligand HL

Complex NiL₂

Scheme 1. Chemical structures of the ligand HL and the complex NiL₂ used in this study.

Some electronic properties such as energy of the highest occupied molecular orbital (E_{HOMO}), energy of the lowest unoccupied molecular orbital (E_{LUMO}), energy gap (E_{GAP}) between HOMO and LUMO, dipole moment (μ), hardness (η), Softness (σ) and Mulliken charges of all studied compounds were also obtained from the DFT calculations. These quantum chemical parameters (η, σ) were calculated using the following equations:

$$\eta = 1/2(I - A) \qquad (1)$$

$$\sigma = 1/\eta \qquad (2)$$

where I and A are the ionization potential and electron affinity of the molecule, respectively. According to Koopman's theorem[35,42-44] using the frontier orbital energies, I and A are given as follows:

$$I = -E_{HOMO} \quad \text{and} \quad A = -E_{LUMO} \qquad (3)$$

hence,

$$\eta = -1/2(E_{HOMO} - E_{LUMO}) = -E_{GAP}/2 \qquad (4)$$

where E_{GAP} is the HOMO-LUMO energy gap.

Results and discussion:

Structural description

A perspective view of the title compound with the atomic numbering scheme is shown in Figure 1. The selected bond distances and angles are listed in Table 2 and Table 3. Selected torsion angles are reported in Table 4. Hydrogen bonding interactions are listed in Table 5.

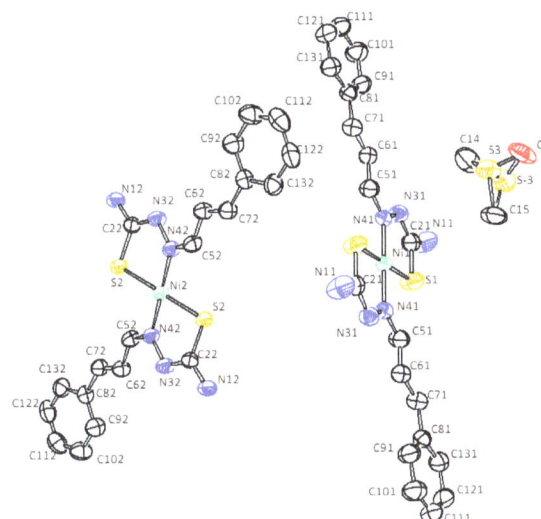

Figure 1. ORTEP drawing with the numbering scheme of the non-hydrogen atoms. Displacement ellipsoids are drawn at the 50% probability level.

Table 2: Selected bond lengths (Å) for [Ni(CMTSC)$_2$] DMSO

Ni(1)-S(1)	2.162(2)	Ni(2)-S(2)	2.171(1)
Ni(1)-N(41)	1.932(4)	Ni(2)-N(42)	1.914(5)
S(1)-C(21)	1.722(5)	S(2)-C(22)	1.737(5)
N(11)-C(21)	1.333(7)	N(12)-C(22)	1.345(5)
N(31)-C(21)	1.310(7)	N(32)-C(22)	1.302(6)
N(31)-N(41)	1.387(6)	N(32)-N(42)	1.403(5)
N(41)-C(51)	1.301(6)	N(42)-C(52)	1.312(7)
C(61)-C(51)	1.427(6)	C(62)-C(52)	1.432(7)
C(71)-C(61)	1.342(7)	C(62)-C(72)	1.332(7)
C(81)-C(71)	1.447(6)	C(72)-C(82)	1.455(7)

Table 3: Selected angles (°) for [Ni(CMTSC)$_2$] DMSO

S(1)-Ni(1)-N(41)	94.1(2)	S(2)-Ni(2)-N(42)	94.6(2)
Ni(1)-S(1)-C(21)	96.9(2)	Ni(2)-S(2)-C(22)	95.8(2)
Ni(1)-N(41)-C(51)	126.7(3)	Ni(2)-N(42)-C(52)	125.4(3)
S(1)-C(21)-N(31)	123.3(4)	S(2)-C(22)-N(32)	123.1(3)
S(1)-C(21)-N(11)	118.0(4)	S(2)-C(22)-N(12)	117.1(4)
Ni(1)-N(41)-N(31)	121.1(3)	Ni(2)-N(42)-N(32)	120.5(2)
N(31)-N(41)-C(51)	112.1(4)	Ni(2)- N(42)- C(52)	125.4(3)
N(41)-C(51)-C(61)	127.3(5)	N(42)-C(52)-C(62)	128.5(4)

Table 4: Selected (of) torsion angles (°) for [Ni(CMTSC)$_2$] DMSO

N(41)-Ni(1)-S(1)-C(21)	-2.3(2)	N(42)-Ni(2)-S(2)-C(22)	13.49(17)
S(1)-Ni(1)-N(41) -C(51)	178.9(4)	S(2)-Ni(2)-N(42)-C(52)	15.5(3)
S(1)-Ni(1)-N(41)-N(31)	1.1(3)	S(2)-Ni(2)-N(42)-N(32)	-162.2(3)
Ni(2)-S(2)-C(22)-N(12)	169.3(3)	Ni(1)-S(1)-C(21)-N(11)	-177.2(4)

Table 5. Selected hydrogen bonding interactions for [Ni(CMTSC)$_2$] DMSO

D-H····A	d(D-H) (Å)	d(H····A) (Å)	d(D····A) (Å)	∠(DHA) (°)
N12-H12A····O #1	0.86	2.121	2.930	156.55
N12-H12B ····O #2	0.86	2.134	2.902	148.50
N11-H11B ····S2 #3	0.860	2.768	3.461	138.71
C51 -H51····S1 # 4	0.937	2.460	3.073	123.50

Symmetry transformations used to generate equivalent atoms: #1 -x+1,-y,-z+1; #2 x-1,+y,+z-1; #3x,+y,+z+1; #4-x+1,-y+1,-z+1

Asymmetric unit includes two semi-molecules (1 and 2) and one DMSO solvent molecule. Both Ni(II) ions are situated on special positions which leads to two different molecules Ni(CMTSC)$_2$. In both molecules, the equivalent anionic CMTSC ligands coordinate to the central metal atom in a bidentate manner via the azomethine nitrogen N41(N42) and the thiolato sulfur atoms S1(S2), in square-planar geometry (figure1) forming five membered chelate rings. The sulfur atoms S1(S2) and nitrogen atoms N41(N42) of the two ligands are in a *trans*-configuration.

The molecules themselves are very close to planar, as shown by the dihedral angles of 4.0 (3) and 6.3 (2) between the two end groups for molecules 1 and 2, respectively.

The negative charge of the deprotonated ligand of the nickel complex is delocalized over the thiosemicarbazone moiety and the C-S bond distances are consistent with the increased single bond character, while the imine C-N distances and both thioamide C-N distances indicate considerable double bond character. This is explained by the difference noted in the level of the bond distances in the deprotonated ligand of the nickel complex which is for S1-C21:1.722(5) Å (S2-C22= 1.737(5) Å); C21-N31 = 1.310(7) Å (C22-N32=1.302(6) Å) (thiolato form) and the bond distances in the case of similar non-deprotonated free ligand, for S-C1 =1.690(2) Å, C1-N2 =1.338(2) Å (thione form)[45].

Ni-N and Ni-S bond distance are similar to those found in other square-planar nickel thiosemicarbazone complexes [23,46]. Although having similar geometries, the main molecules have significant differences in geometrical parameters around Ni(II) ions (see torsion angles Table 4). Whereas Ni1 ion is located in a perfect square plane environment (± 0.028 Å), Ni2 is in a distorted square plane environment (± 1.308 Å). This can be explained by the presence of solvent molecule (DMSO) implicated in Hydrogen bonds with molecule 2 and causing a puckering effect (see Figure 2, Figure 3 and Table 4).

DMSO molecule presents a high thermal agitation on sulfur atom (S3). This latter was set on two different positions with different multiplicities (see   deposit).

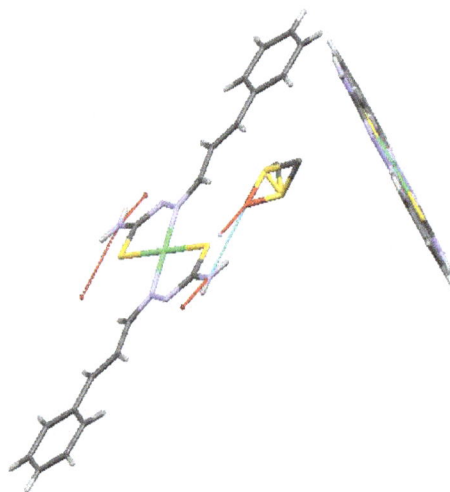

Figure 2. Figure showing molecule 2 implicated in H-bonds.

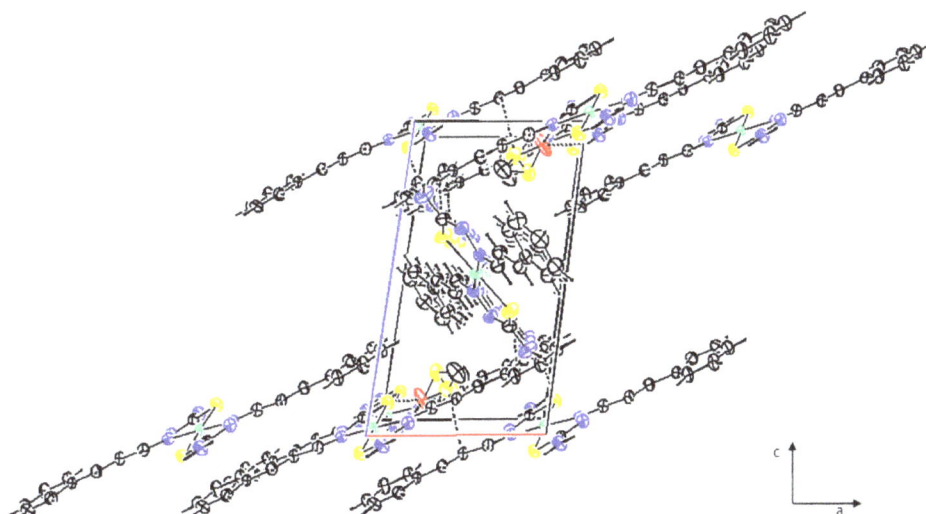

Figure 3. Packing arrangement of the title compound where molecules are packed to form infinite layers parallel along b axis.

Theoretical results
Geometric structures

For the two compounds, geometry optimizations shown in Fig. 4 have been performed at B3LYP/6-311G(d,p) level of theory and the calculated results are also listed in Table 6. Thus, the comparisons between the experiments and the calculations can be more straightforward.

As seen in Table 6, for the two compounds, most of the predicted geometric parameters have higher values than those determined experimentally. This is most likely due to the fact that the experimental data describe the compounds in the solid state, whereas the calculated data correspond to the molecules in the gas phase. For the ligand HL, comparing the predicted values with the experimental ones, it can be found that the biggest difference in bond length occurs at bond C(6)-C(7) with the difference being 0.0213 Å. Considering the bond angles, the biggest variation between the experimental and predicted values can be found at bond angle C(5)-N(4)-N(3)

with the difference being 1.60°. For the complex NiL$_2$, the biggest differences of bond length and bond angle between the experiments and calculations are at the Ni(1)-S(1) and N(31)-N(41)-C(51), respectively, with the values being 0.053 Å and 3.46°. Since in the calculations, no intermolecular interactions are considered, which results in the theoretical molecules being more spreading themselves and having longer bond lengths and bigger bond angles. In addition, comparing all the parameters of the free ligand HL and the coordinated L$^-$ ion in complex NiL$_2$, there are also some differences between them since the L$^-$ is coordinated with Ni(II) ion (see Table 1). Aforementioned comparisons indicate that, although there exist some differences on the geometrical parameters between the experiments and the predictions, the optimized geometries of the ligand HL and complex NiL$_2$ resemble closely to their crystal structures and B3LYP/6-311G(d,p) level of theory can provide satisfying calculational precision for the system studied here.

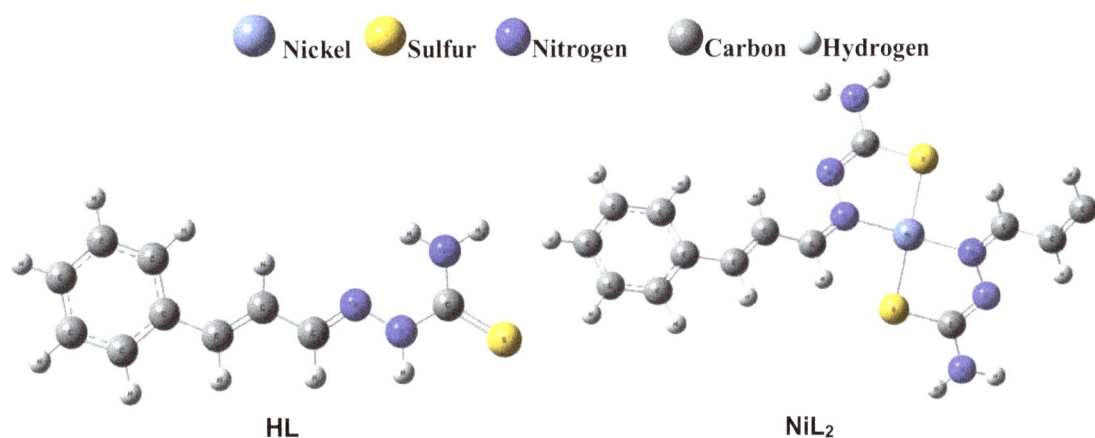

Figure 4. B3LYP/6-311G(d,p) optimized geometry of ligand HL and complex NiL$_2$.

Table 6. Selected geometric parameters by X-ray and theoretical calculations at B3LYP/6-311(d,p) level of theory for the ligand HL and complex NiL_2.

Compound		Bond length (Å)			Bond angle (°)	
		Exp.	Calc.		Exp.	Calc.
HL	C2-S	1.670(1)	1.6787	N1-C2-N3	114.4(2)	114.96
	N4-N3	1.369(2)	1.3524	C2-N3-N4	120.2(2)	121.72
	C2-N3	1.354(2)	1.3724	C5-N4-N3	116.1(6)	117.70
	C2-N1	1.340(2)	1.3431	N4-C5-C6	121.7(4)	120.95
	C5-N4	1.278(2)	1.2894	C5-C6-C7	123.9(2)	122.73
	C5-C6	1.423(3)	1.4386	C6-C7-C8	128.2(2)	127.61
	C6-C7	1.328(3)	1.3493			
	C7-C8	1.455(3)	1.4605			
NiL2	Ni(1)-S(1)	2.162(2)	2.2156	S(1)-Ni(1)-N(41)	94.6(2)	94.93
	Ni(1)-N(41)	1.932(4)	1.9147	Ni(1)-N(41)-C(51)	126.7(3)	123.33
	S(1)-C(21)	1.722(5)	1.7513	C(71)-C(61)-C(51)	120.4(4)	120.99
	N(11)-C(21)	1.334(7)	1.3618	C(81)-C(71)-C(61)	128.7(5)	127.35
	N(41)-C(51)	1.301(6)	1.3068	N(41)-C(51)-C(61)	127.3(5)	125.92
	C(81)-C(71)	1.447(6)	1.4589	S(1)-C(21)-N(11)	118.0(4)	118.07
	N(31)-C(21)	1.310(6)	1.3044	S(1)-C(21)-N(31)	123.3(4)	123.81
	C(61)-C(71)	1.342(7)	1.3519	N(11)-C(21)-N(31)	118.6(4)	118.06
	C(61)-C(51)	1.427(6)	1.4351	N(41)-N(31)-C(21)	112.6(3)	113.22
	N(41)-N(31)	1.387(4)	1.3764	Ni(1)-N(41)-N(31)	121.1(3)	120.96
				N(31)-N(41)-C(51)	112.1(4)	115.56

HOMO–LUMO energy gap and related molecular properties

Electron distributions in the frontier molecular orbitals (the highest occupied molecular orbital (HOMO) and the lowest unoccupied molecular orbital (LUMO)) of ligand and Ni (II) complex were calculated and shown in Fig. 5 and Fig. 6 respectively. While the corresponding orbital energy, energy gap, hardness, softness and dipole moment values are presented in Table 7.

As seen in Fig. 5 the HOMO is mainly localized on double bonds, whereas LUMO is generally localized on simple bonds. For the complex (Fig. 6), the HOMO is mainly located around the Ni atom, and the LUMO is localized along the ligand.

HOMO   **LUMO**

Fig. 5. Calculated HOMO and LUMO molecular orbitals of the studied Ligand HL using the B3LYP/6-311G(d,p) method

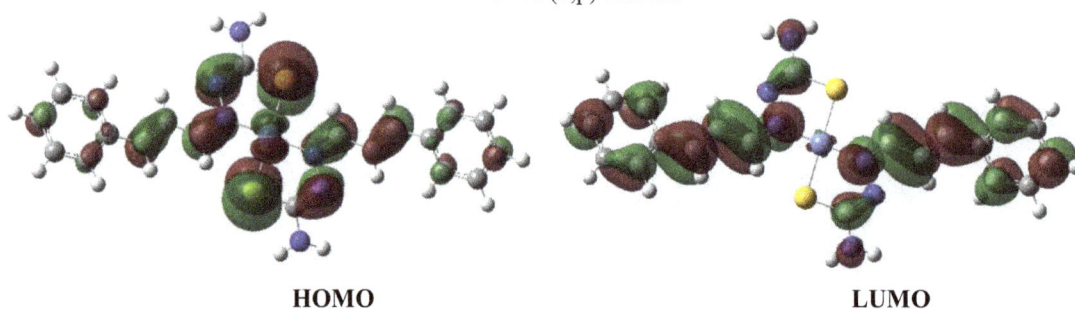

HOMO   **LUMO**

Figure 6. Calculated HOMO and LUMO molecular orbitals of the studied complex NiL_2 using the B3LYP/6-311G(d,p) method

Table 7. Calculated HOMO and LUMO energies, HOMO-LUMO gap energies, hardness (η) in eV, softness (σ) in eV^{-1} and dipole moment (μ) values in Debye.

Compounds	HOMO	LUMO	E_{GAP}	η	σ	μ
HL	-5.7658	-2.2751	3.4907	1.7453	0.5729	6.1398
NiL$_2$	-5.2425	-2.2841	2.9584	1.4792	0.6760	0.3983

The energy of HOMO is often associated with the electron-donating ability of a molecule; such that high energy values of HOMO are likely to indicate a tendency of the molecule to donate electrons. The energy of LUMO is related to the electron affinity. The binding ability of the ligand to the metal increases with increasing HOMO energy values[47,48]. The calculations show that HL ligand has the highest HOMO level at -5.7658 eV and the lowest LUMO level at-2.2751 eV when compared to the obtained parameters for the NiH$_2$ complex.

It has been revealed that the HOMO-LUMO energy gap is an important stability index[49,50]. A large energy gap indicates high stability for the molecule in the chemical reaction. The energy gap for the complex is lower than that of the ligand. This result is consistent with the similar complexes[51].

The effect of the molecular structure on the chemical reactivity has been object of great interest in several disciplines of chemistry. The hardness and softness are other important parameters, there are commonly used as a criterion of chemical reactivity and stability.

The hardness (η) and softness (σ) can be estimated from the calculated HOMO and LUMO energies using Equations (1)-(4)[35,52-54].

The smaller values of hardness imply higher reactivity, which means that a molecule with a small HOMO-LUMO gap is more reactive and is a softer

molecule. The soft molecules undergo changes in electron density more easily and are more reactive than hard molecules. The hardness value for complex is lower than the ligand and hence its reactivity increases (Table 7).

The dipole moment (μ) is another parameter of the electronic distribution in a molecule which can be related to the dipole-dipole interaction of molecules and metal surface[55]. Ligand with larger dipole moment forms more stable complex[56]. In our case, HL ligand has higher dipole moment than the NiL$_2$ complex (Table 7).

Mulliken atomic charges

Mulliken atomic charge calculation has an important role in the application of quantum chemical calculation to molecular system because of atomic charges effect dipole moment, molecular polarizability, electronic structure and a lot of properties of molecular systems[57]. The charge distribution over the atoms suggests the formation of donor and acceptor pairs involving the charge transfer in the molecule. Atomic charge has been used to describe the processes of electronegativity equalization and charge transfer in chemical reactions[58,59]. The Mulliken charges of the atoms were calculated for each compounds and shown in Table 8.

Table 8. Mulliken atomic charges calculated by B3LYP/6-311G (d,p) method.

Atom	NiL$_2$	HL
Ni	0.916	
S	-0.348	-0.264
C2	0.174	0.199
N3	-0.264	-0.265
N4	-0.434	-0.210
N1	-0.412	-0.426
C5	0.167	0.106
C6	-0.129	-0.123
C7	-0.056	-0.062
C8	-0.078	-0.073
H1A	0.243	0.263
H1B	0.223	0.240
H5	0.142	0.252
H6	0.119	0.082
H7	0.098	0.101

The Mulliken atomic charges calculated by B3LYP/6-311G (d,p) method are collected in Table 3. It is worthy to mention that C2 and C5 atoms of the ligand and complex exhibit positive charge while C6, C7, C8 atoms exhibit negative charges. The larger negative charge values are found in N1 atom for HL, N4 atom for NiL_2 of about -0.426 and -0.434 respectively. The maximum positive atomic is obtained in C2 atom for HL, Ni atom for NiL_2. The charge on H1A in the NH_2 group has the maximum magnitude among the hydrogen atoms present in the title compounds. However all the hydrogen atoms exhibit a net positive charge. The presence of large negative charge on S and N atom and net positive charge on H atoms may suggest the formation of intermolecular interaction in solid forms[60]. This is confirmed by Hydrogen-bonds found in crystalline structure (see Table 5).

Conclusion

Asymmetric unit includes two semi-molecules and one DMSO solvent molecule. In both molecules, the equivalent anionic CMTSC ligands coordinate to the central metal atom in a bidentate manner via the azomethine nitrogen N4 and the thiolato sulfur atoms S in a *trans* configuration forming a square-planar geometry. Although having similar geometries, the main molecules have significant differences in geometrical parameters around Ni(II) ions. Whereas Ni1 ion is located in a perfect square plane environment, Ni2 environment presents some distortions. This can be explained by the presence of solvent molecule (DMSO) implicated in hydrogen bonds with amine group.

The predicted structures of the ligand HL and complex NiL_2 at B3LYP/6-311G (d,p) level can well reproduce the structures of the two compounds. Global descriptors, such as ionization energy (I), electron affinity (A), HOMO-LUMO gap, dipole moment (μ), hardness (η) and softness (σ) were derived from the DFT calculations and used to identify the differences in the stability and reactivity properties of the studied compounds. Even though, calculations pertain to the gas phase, and the experimental data are for the solid state, in which crystal field effect may affect the relative energies and geometries, Molecular Modeling calculations results hold a good comparison between the theoretically predicted geometries and the experimental ones, which is clearly validating our methodology.

Acknowledgments: The authors gratefully acknowledge the support of the Algerian Ministry of Higher Education and Scientific Research.

References

1- G. Pelosi, F. Bisceglie, F. Bignami, P. Ronzi, P. Schiavone, M.C. Re, C. Casoli, E. Pilotti, J. Med. Chem., **2010**, 53, 8765.

2- U. Kulandaivelu, V.G. Padmini, K. Suneetha, B. Shireesha, J.V. Vidyasagar, T.R.Rao, A. Basu, V. Jayaprakash, Arch. Pharm., **2011**, 344, 84.

3- M.D. Hall, K.R. Brimacombe, M.S. Varonka, K.M. Pluchino, J.K. Monda, J. Li, M.J.Walsh, M.B. Boxer, T.H. Warren, H.M. Fales, M.M. Gottesman, J. Med. Chem., **2011**, 54, 5878.

4- S. Chandra, S. Bargujar, R. Nirwal, N.Yadav, SpectrochimicaActa. Part A: Molecular and Biomolecular Spectroscopy, **2013**, 106, 91-98.

5- M. Jagadeesha, M. Lavanya, S.K. Kalangi, Y. Sarala, C. Ramachandraiah, A. Varada Reddy,SpectrochimicaActa.Part A: Molecular and Biomolecular Spectroscopy, **2015**, 135, 180-184.

6- R.L. Arrowsmith, S.I. Pascu, H. Smugowski, Organomet. Chem., **2012**, 38, 1-35.

7- Giorgio Pelosi, The Open Crystallography Journal, **2010**, 3, 16-28.

8- S. Lhuachan, S. Siripaisarnpipat, N. Chaichat, Eur. J. Inorg. Chem., **2003**, 263.

9- D. Kovala-Demertzi, M.A. Demertzis, J.R. Miller, C.S. Prampton, J.P. Jasinski, D.X.West, J. Inorg. Biochem., **2002**, 92, 137.

10- P.N. Yadav, M.A. Demertzis, D. Kovala-Demertzi, S. Skoulika, D.X. West, Inorg. Chim. Acta., **2003**, 349, 30.

11- L.A. Ashfield, A.R. Cowley, J.R. Dilworth, P.S. Donnely, Inorg. Chem., **2004**, 43, 4121.

12- I. Paul, F. Basuli, T.C.W. Mak, S. Bhattacharya, Angew. Chem. Int. Ed., **2001**, 40, 2923.

13- S.I. Orysyk, V.V. Bon, O.O. Obolentseva, Yu.L. Zborovskii, V.V. Orysyk, V.I.Pekhnyo, V.I. Staninets, V.M. Vovk, Inorg. Chim. Acta., **2012**, 382, 127.

14- T.S. Lobana, G. Bhargava, V. Sharma, M. Kumar, Indian J. Chem., Sect. A: Inorg. Bio-inorg., Phys., Theor. Anal. Chem., **2003**, 42A, 309.

15- F.H. Allen, Acta Crystallogr. Sect. B, **2002**, 58, 380.

16- D. Pandiarajan, R. Ramesh, Inorg. Chem. Commun., **2011**, 14, 686.

17- N. SahaChowdhury, D. Kumar Seth, M.G.B. Drew, S. Bhattacharya, Inorg. Chim.Acta., **2011**, 372, 183.

18- T.S. Lobana, R. Sharma, G. Bawa, S. Khanna, Coord. Chem. Rev., **2009**, 253, 977.

19- D. Maity, A.K. Manna, D. Karthigeyan, T.K. Kundu, S.K. Pati, T. Govindaraju, Chem. Eur. J., **2011**, 17, 11152-11161.

20- R. Santhakumari, K. Ramamurthi, G. Vasuki, B.M. Yamin, G. Bhagavannarayana, Spectrochim. Acta , **2010**, 76, 369.

21- V. Ruangpornvisuti, B. Wanno, J. Mol. Mod.,**2004**,10, 418.

22- S.B. Jiménez-Pulido, F.M. Linares-Ordóñez, M.N. Moreno-Carretero, M. QuirósOlozábal, Inorg. Chem., **2008**, 47, 1096.

23- Y. Harek, L. Larabi, L. Boukli, F. Kadri, N. Benali-Cherif, M. M. Mostafa, Transition Metal Chemistry, **2005**, 30, 121-127.

24- Bruker, Smart and Saint Bruker AXS, Madison, **2004.**

25- A. Altomare, M. C. Burla, M. Camalli, G. L. Cascarano, C. Giacovazzo, A. Guagliardi, A. G. Moliterni, G. Polidori& R. Spagna, J.Appl. Cryst., **1999**, 32, 115–119.

26- G. M. Sheldrick,ActaCryst. A, **2008**, 64, 112–122.

27- L. J.Farrugia,J. Appl. Cryst., **1999**, 32, 837–838.

28- C. F.Macrae,I. J.Bruno, J. A.Chisholm, P. R.Edgington, P.McCabe, E.Pidcock, L.Rodriguez-Monge, R.Taylor, J.van de Streek and P. A.Wood, J. Appl. Cryst., **2008**, 41, 466-470.

29- L. J.Farrugia, J. Appl. Cryst., **2012**, 45, 849–854.

30- A.P. Scott, L. Radom, J. Phys. Chem., **1996**, 100, 16502-16513.

31- A. Ricca, C.W. Bauschlicher Jr., J. Phys. Chem., **1995**, 99, 9003-9007.

32- J. Song, F. Zhu, H. Wang, P. Zhao, Spectrochimica Acta Part A Molecular and Biomolecular Spectroscopy, **2014**, 129, 227-234.

33- J.J.P. Stewart, J. Comput. Chem., **1989**, 2, 209.

34- HyperChem version 8.0.10 Hypercube Inc, **2011.**

35- R.G. Parr, W. Yang, Density Functional Theory of Atoms and Molecules, Oxford University Press, New York **1989**.

36- A.D. Becke, Density-functional exchange-energy approximation with correct asymptotic behavior. Phys. Rev. A., **1988**, 38, 3098-3100.

37- A.D. Becke, J. Chem. Phys., **1993**, 98, 1372-1377.

38- C. Lee, W. Yang, R.G. Parr, Phys. Rev. B.,**1988**,37, 785-789.

39- E.V.R. de Castro, F.E. Jorge, Journal of Chemical Physics, **1998, 108**, 5225-5229.

40- M.J. Frisch, G.W. Trucks, H.B. Schlegel, G.E. Scuseria, M.A. Robb, J.R. Cheeseman, G. Scalmani, V. Barone, B. Mennucci, G.A. Petersson, H. Nakatsuji, M. Caricato, X. Li, H.P. Hratchian, A.F. Izmaylov, J. Bloino, G. Zheng, J.L. Sonnenberg, M. Hada, M. Ehara, K. Toyota, R. Fukuda, J. Hasegawa, M. Ishida, T. Nakajima, Y. Honda, O. Kitao, H. Nakai, T. Vreven, J.A. Montgomery, Jr, J.E. Peralta, F. Ogliaro, M. Bearpark, J.J. Heyd,

E. Brothers, K.N. Kudin, V.N. Staroverov, R. Kobayashi, J. Normand, K. Raghavachari, A. Rendell, J.C. Burant, S.S. Iyengar, J. Tomasi, M. Cossi, N. Rega, J.M. Millam, M. Klene, J.E. Knox, J.B. Cross, V. Bakken, C. Adamo, J. Jaramillo, R. Gomperts, R.E. Stratmann, O. Yazyev, A.J. Austin, R. Cammi, C. Pomelli, J.W. Ochterski, R.L. Martin, K. Morokuma, V.G. Zakrzewski, G.A. Voth, P. Salvador, J.J. Dannenberg, S. Dapprich, A.D. Daniels, O. Farkas, J.B. Foresman, J.V. Ortiz, J. Cioslowski, D.J. Fox, Gaussian 09, Revision A.1, Gaussian Inc., Wallingford, CT, **2009**.

41- R.D. Dennington II, T.A. Keith, J.M. Millam, GaussView 5.0, Wallingford CT, **2009**.

42- R.K Singh, S.K. Verma, P.D. Sharma, Int. J. Chem. Technol. Res., **2011**, 3, 1571-1579.

43- C.G. Zhan, J.A. Nichols, D.A. Dixon, J. Phys. Chem. A, **2003**, 107, 4184-4195.

44- R.G. Pearson, J. Am. Chem. Soc., **1985**, 107, 6801-6806.

45- E. M.Jouad, M.Allain, M. A.Khanand G. M.Bouet,J. Mol. Struct., **2002**, 604, 205.

46- Y. M.Chumanov, N.M.Samus, G.Bocelli, Y. K.Suponitski, V. I. Tsapkov,A. P. Gulya,Russian Journal of Coordination Chemistry, **2006**, 32, 14-20.

47- S. Sagdinc, B. Koksoy, F. Kandemirli, S.H. Bayari, J. Mol. Struct., **2009**, 91, 63-70.

48- A.A.R. Despaigne, J.G. Da Silva, A.C.M. Do Carmo, F. Sives, O.E. Piro, E.E. Castellano, H. Beraldo, Polyhedron, **2009**, 28, 3797–3803.

49- Z. Zhou, R.G. Parr, J. Am. Chem. Soc., **1990**, 112, 5720-5724.

50- R.G. Pearson, J. Am. Chem. Soc., **1988**, 110, 2092-2097.

51- J. Song, F. Zhu, H. Wang, P. Zhao, Spectrochimica Acta Part A Molecular and Biomolecular Spectroscopy, **2014**, 129, 227-234.

52- R.K. Singh, S.K. Verma, P.D. Sharma, Int. J. Chem. Technol. Res., **2011**, 3, 1571-1579.

53- C.G. Zhan, J.A. Nichols, D.A. Dixon, J. Phys. Chem. A, **2003**, 107, 4184-4195.

54- R.G. Pearson, J. Am. Chem. Soc., **1985**, 107, 6801-6806.

55- A.O Yüce, B.D Mert, G. Kardaş, B. Yazıcı, Corrosion Science, **2014**, 83, 310-316.

56- G.B. Saha, Fundamentals of nuclear pharmacy, Springer, **2010**, 6.

57- I. Sidir, Y.G. Sidir, M. Kumalar, E. Tasal, J. Mol. Struct., **2010**, 964, 134.

58- K. Jug, Z.B. Maksic, in: Theoretical Model of Chemical Bonding, Ed. Z.B. Maksic, Part 3, Springer, Berlin **1991**, p. 29, p. 233.

59- S. Fliszar, Charge Distributions and Chemical Effects, Springer, New York, **1983**.

60- L.X. Hong, L.X. Ru, Z.X. Zhou, Comput. Theor. Chem., **2011**, 969, 27.

6

Mechanochemical Synthesis of Kaolin-Potassium Phosphates Complexes for Application as Slow-Release Fertilizer

Rym Dhouib Sahnoun*, Kamel Chaari and Jamel Bouaziz

Laboratory of Industrial Chemistry, National School of Engineering, University of Sfax, BP 1173, 3038 Sfax, Tunisia

Abstract: In the current study we investigated the effects of the addition of KH_2PO_4, K_2HPO_4 and K_3PO_4 to kaolin by mechanochemical treatment, for the synthesis of new complex compounds. These compounds are offered as slow-release fertilizers. After, the characterization of the ground kaolin, we concluded that two hours of milling is sufficient to achieve the distortion of the crystalline network of kaolin in order to insert additives. The quantity of KH_2PO_4 inserted is not as much as those of K_2HPO_4 and K_3PO_4. This is due to the acid character of those compounds. The amounts of K^+ and PO_4^{3-} released when using KH_2PO_4 are larger than those when using the others additives.

Keywords: Kaolin; Phosphates; Characterization; Fertilizers; Mechanochemical treatment.

Introduction

Fertilizers are the essential input material for the sustainable development of crop production; they play an important role in ensuring food security. However, various environmental and economic drawbacks related to the use of conventional fertilizers became a focus of worldwide concern [1,2]. One way of overcoming these shortcomings involves the use of slow-release fertilizers which have demonstrated many advantages over the conventional types, such as decreasing the rate of fertilizer loss, supplying nutrients sustainably, lowering the frequency of their application, and minimizing the negative effects associated with over-dosage. Coated fertilizers are the major category of slow-release fertilizers [3] which are physically elaborated by coating granules of conventional fertilizers with various materials to reduce their dissolution rate [4-6]. Other types of fertilizers are elaborated through the use of mechanochemical treatment to insert the plant nutrients (potassium, ammonium, nitrate, phosphate), into other structures such as clay. The release rate of nutrients is thus controlled.

Mechanochemistry, term first defined by Ostwald [7] in the late 19th century, is considered as a branch of solid state chemistry dealing with intentional defects in the structure of solid substances by the application of mechanical forces. These defects are often created using intensive milling processes [8,9]. The formation of defects in solids during intense milling is marked by non-equilibrium conditions. After mechanical treatment, solids became metastable or in an *active* state [10]. The mechanochemical grinding process of a solid has been used to insert other molecules between their layers [11-18].

Kaolin is an important industrial material, used in engineering and construction applications, ceramic processing, environmental remediation and many other diverse applications. There is an ongoing interest in utilizing kaolin clay as a raw material for the manufacture of slow-release fertilizers by intercalating the nutrients in its structure using mechanochemical treatment. Recent studies of the mechanochemical treatment of kaolinite have explained the structural changes which occur upon the grinding of kaolinite [19-22]. It has been shown that the dry grinding of kaolinite along with the inserted molecule increases the degree of intercalation[16-18].

In this paper we characterize the kaolin ground for various times by XRD, FTIR and thermal analysis. In addition we examine the insertion, of potassium dihydrogen phosphate, KH_2PO_4, potassium monohydrogen phosphate K_2HPO_4 and potassium phosphate K_3PO_4 to kaolin by mechanochemical treatment, to obtain a slow-release fertilizers.

Experimental Section

Materials

A high grade kaolin supplied from BWW Minerals is used in the experiments. Its chemical composition in mass% is SiO_2, 52.41, Fe_2O_3, 3.48, Al_2O_3, 29.83, MgO, 0.81; K_2O, 0.73; SO_3, 0.07; CaO, 0.36 with L.O.I at 1000°C of 12.31.

In this study, the additives used for the synthesis of the new complex compounds are potassium dihydrogen phosphate KH_2PO_4 (Sigma Aldrich,

*Corresponding author: Rym Dhouib Sahnoun
E-mail address: rymdhouib@yahoo.fr

density 2.000 g.cm^{-3}), potassium monohydrogen phosphate K_2HPO_4 (Reidel de Haen, density 2.440 g.cm^{-3}) and potassium phosphate K_3PO_4 (Sigma Aldrich, density 2.546 g.cm^{-3}).

Mechano-chemical synthesis

Kaolin and kaolin mixed with different amounts of additives (KH_2PO_4, K_2HPO_4 and K_3PO_4) were ground (mechanochemically activated) using a planetary ball mill (Retsh PM 100). Each milling was carried out with a 100g air-dried sample in a pot having a capacity of 80 cm^3, using stainless-steel balls (10 mm in diameter). The applied rotation speed was 400 rpm.

Characterization

X-ray diffraction (XRD) patterns of the solid samples were obtained using the Philips X'Pert X-ray diffractometer equipment, operating with Cu-Kα radiation (λ= 1.54056 Å). The crystalline phases were identified from the Powder Diffraction Files (PDF) of the International Center for Diffraction Data (ICDD).

The infrared spectra of the samples were recorded in the wavenumber range of 400-4000 cm^{-1}; the spectral resolution was 4 cm^{-1} using a Perkin Elmer Spectrum BX LX 185255 instrument in KBr.

Thermal analysis (DTA and TG) were carried out by using (about) 20 mg of powder (DTA-TG, Seteram Model). The heating rate was 10°C.min^{-1} in a helium atmosphere.

The release of nutrients into water was determined by dissolving 1g of the ground product in 20 ml of distilled water for 24 hours at room temperature. After filtration, the obtained filtrate was determined by a liquid ion chromatography (Shimadzu L10 Series). The results were given in weight percentages.

Results and discussion

Characterization of milled kaolin relative to milling time

X-ray powder diffraction

One means of following the structural alteration resulting from mechanochemical treatment is to use XRD. The XRD patterns of kaolin and kaolin ground for various grinding times are shown in figure 1. As can be seen in this figure, the diffraction patterns for the raw and mechanically treated kaolin are consistent with the quick deterioration of the kaolinite structure during the grinding process. The diffractograms indicate that kaolinite suffers significant structural degradation during the grinding process while the quartz is not altered. After two hours of milling, almost no kaolinite reflections are present in the XRD pattern of kaolin (Figure1). The loss of intensity and the relative area in the peak at 0.71 nm ($2\theta = 12.4°$), which increased with grinding time, suggest the breaking of the bonds between kaolinite layers (001) (Figure 1). This means the demolition of the kaolinite structure happened because of the distortion and breakage of the crystalline network. We note also the fast deterioration of the kaolinite structure during the milling process, especially within the first 15 minutes.

The XRD patterns indicate that the quartz phase has not been altered because the position of peaks at 21.2° and 27.4° remains almost unchanged. This non-altered quartz content caused the acceleration of the mechanochemical activation of the kaolinite. We believe that the harder quartz grains act as additional grinding bodies during the dry grinding of kaolin [23].

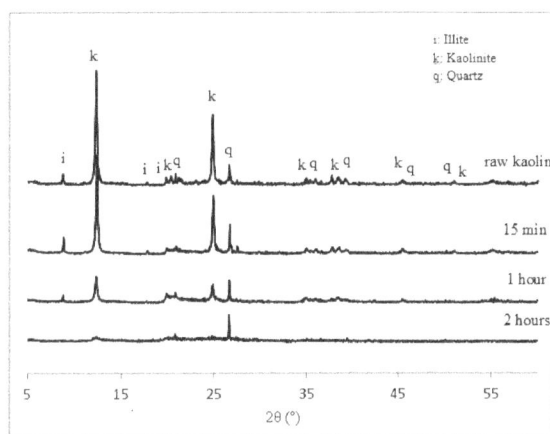

Figure 1. XRD spectra (Cu Kα) of kaolin milled for various grinding times.

FTIR spectrometry

Figure 2 shows the FTIR spectra of the kaolin clay in comparison to those of kaolin clay ground for 15 min, 1 h and 2 h, observing this figure, three regions are noted:

- *Hydroxyl-stretching mode*: This region is where most of the changes occur due to grinding. From the spectra we conclude:

- The intensity of all the hydroxyl stretching vibrations decreases with grinding time,

- The intensity of the bands attributed to the inner surface hydroxyl decrease faster than the band of the inner hydroxyl which means that the inner surface hydroxyls are lost before the inner hydroxyls,

- The out-of-phase behavior is removed by the mechanochemical treatment,

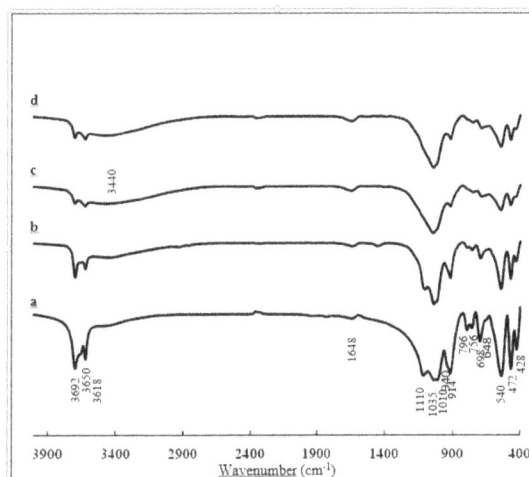

Figure 2. IR spectra of kaolin milled for various grinding times: a: raw kaolin; b: 15min; c: 1h; d: 2h.

- Hydroxyl deformation mode: Two bands are observed. First one at around 940 cm^{-1} is attributed to the hydroxyl deformation mode of the inner surface hydroxyl. Second one at around 914 cm^{-1} is ascribed to the hydroxyl deformation mode of the inner hydroxyl. In harmony with the loss of intensity of the hydroxyl-stretching vibrations, a decrease in intensity of the hydroxyl-deformation modes is observed. We note the decrease in intensity of those bands with the increase of grinding time. In fact after 15 min of grinding no intensity remains. Therefore, we conclude that the Inner surface hydroxyls are removed, by mechanochemical treatment, before the inner hydroxyls.

- Silicon-oxygen vibrations: In order to study the low wavenumber region of mechanochemically treated kaolinites, it is convenient to divide the spectra into two sections: (a) the SiO stretching vibrations and (b) the OSiO bending region. Band component analysis of the region 970 to 1170 cm^{-1} shows the presence of three bands in the IR spectra of the ground kaolin at 1110, 1035, and 1010 cm^{-1}. The spectral profile of this region shows a steady decrease in intensity with mechanochemical treatment and the bands appear to broaden at the same time. The band at 1010 cm^{-1} remains constant in position and intensity; however the bands at 1035 and 1110 cm^{-1} display changes in intensity with mechanochemical treatment. The intensities of those bands decrease with the increase of grinding time until they disappear. It is proposed that as the time of grinding increases, the hydroxyls are removed and the surface becomes similar to an alumina or siloxane surface.

The low-wavenumber region between 400 and 900 cm^{-1} may be divided into two parts: (a) the region from 650 to 850 cm^{-1} and (b) the region from 400 to 650 cm^{-1}. The first region shows the spectra of the hydroxyl translation modes and reflects changes in the SiO chains as the hydroxyl unit moves in and away from the siloxane chain. Bands are observed at 796, 756, 698 and 648 cm^{-1}. These bands reflect the degree of defect in the kaolinite stucture. As with all the other bands in the spectra of the kaolin, the bands decrease in intensity with mechanochemical treatment. After 1 h of grinding, little intensity remains. The second part of the spectra is attributed to OSiO and SiOAl vibrations. Bands are observed at 540, 472 and 428 cm^{-1}. The variation in intensity of these bands appears to decrease with mechanochemical treatment.

Thermal analysis

The DTA pattern for the natural sample (Figure 3) exhibits two peaks. The first is an endothermic one at 515 °C associated with the loss of hydroxyl groups from the kaolinite structure and the formation of a metakaolin phase. The second peak is an exothermic one at 998 °C attributed to the formation of a spinel and/or crystal mullite phases. The destruction of the kaolinite structure by the distortion and breakage of the crystalline network caused by milling, as mentioned above, was confirmed by the DTA analyses (Figure 3). Subsequently, the temperature

and the area of the endothermic peak (515°C) decrease with the increase of grinding time. Furthermore, after 2 h of grinding, this transformation almost disappears. These changes in the endothermic peak can be attributed to a reduction in the amount of heat required to break the structure of kaolinite. The latest may be destroyed before by mechanical treatment. This fact suggests that the hydroxyl groups are reduced in number and are more weakly bonded [28]. The dehydroxylation process is preceded by a dehydration step spanning the temperature range of 80-120 °C, where a small amount of water is detected in the case of the raw kaolin. However, we note an increase of the area of the broad peak assigned to the dehydration process for the ground kaolin. Then, we conclude that almost the whole quantity of hydroxyl groups is eliminated in the mechanical dehydroxylation step, especially in the case of the kaolin ground for 2 h.

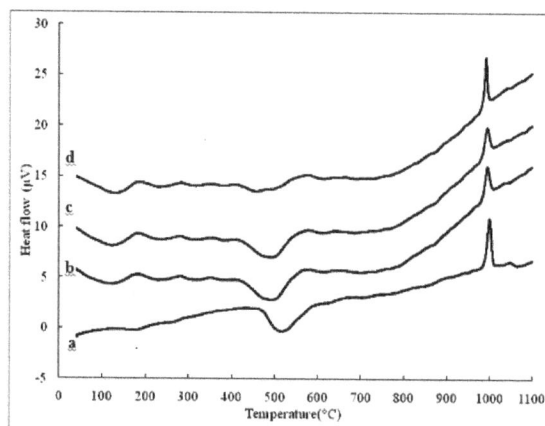

Figure 3. DTA curves of kaolin milled for various grinding times: a: raw kaolin; b: 15min; c: 1h; d: 2h.

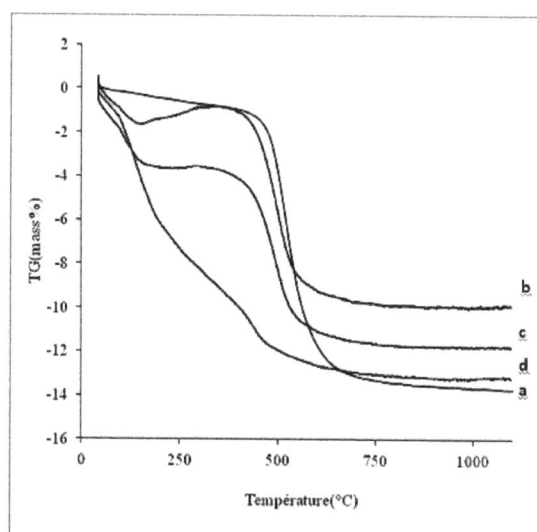

Figure 4. TG curves of kaolin milled for various grinding times: a: raw kaolin; b: 15min; c: 1h; d: 2h.

The thermogravimetric analysis of kaolin shows the biggest weight loss at 400-700°C due to the destruction of the kaolinite structure caused by its dehydroxylation (Figure 4). However, the TG curves of the mechanically treated kaolin exhibit two steps. The first step, between 40 and 250°C, can be assigned to the release of water formed through mechanical dehydroxylation of kaolinite. The second, above 400°C, is caused by the loss of water in the thermal dehydroxylation process [28]. As can be seen in figure 4, the first and second weight loss vary with the milling time. Therefore, a progressive increase of the weight loss in the low temperature range occurs relative to the milling time. The fact that the second weight loss decreased with the milling time suggests that the structure was partially destroyed and that the hydroxyl groups were more weakly bonded. This indicates that the decrease of the weight loss in the high temperature ranges is attributed to an anticipation of dehydroxylation towards low temperatures. This fact is a consequence of an increase of defects in the octahedral structure which is due to the progressive milling time [29, 30].

It seems that after the characterization of the ground kaolin, two hours of milling is enough to achieve the distortion and breakage of the crystalline network of kaolin in order to insert additives such as KH_2PO_4, K_2HPO_4 and K_3PO_4 for the synthesis of slow-release fertilizers.

Characterization of the mixtures kaolin-potassium phosphates (XRD and IR)

To study the insertion of phosphate and potassium in the network of kaolin to be used as slow-release fertilizers, the mixtures of kaolin-KH_2PO_4, K_2HPO_4 or K_3PO_4 in different proportions were ground for 2 h.

Figure 5 shows the XRD patterns of the kaolin-KH_2PO_4 mixtures prepared with different amounts of KH_2PO_4. As it can be noticed, the XRD patterns of the mixtures prepared with 25 mass% and 35 mass% of KH_2PO_4 contain almost only the characteristic peaks corresponding to the quartz. However, the peaks attributed to KH_2PO_4 start to appear only when its quantity in the mixtures exceeds 55 mass%.

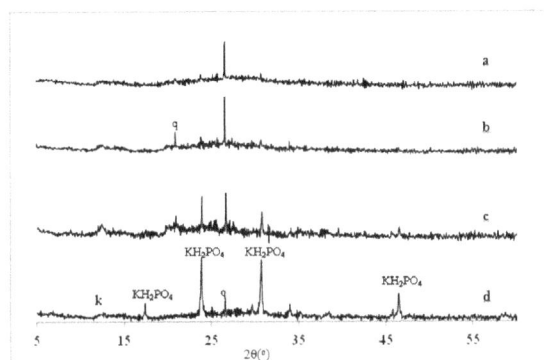

Figure 5. XRD patterns of kaolin-KH_2PO_4 sample system with varying amounts of KH_2PO_4 (mass %) milled for 2h at 400 rpm a: 25 mass%; b: 35 mass%; c: 55 mass% and d:75 mass%.

The figure 6 reports the IR spectra of raw kaolin and kaolin-KH_2PO_4 sample system with varying

amounts of KH_2PO_4 (mass %) milled for 2h at 400 rpm. While observing figure 6, we note the following conclusions:

a-A further decrease in intensity until the total disappearance of the 3692 cm^{-1} band is observed. This is expected as the intercalation results from the hydrogen bonding with the phosphate of the inner surface hydroxyls of kaolinite;

b-The increase of the amount of KH_2PO_4 causes the decrease of the intensity of the 3618 cm^{-1} band;

c-A broad band attributed to the adsorption of water appears at 3270 cm^{-1};

d-The broadening of the band at 1040 cm^{-1} is assigned to the SiO stretching mode. This change is also due to the existence of a new vibration band of PO from KH_2PO_4.

e-The characteristic bands of KH_2PO_4 (2440, 1730, 1300 and 890 cm^{-1}) begin to appear when the addition of KH_2PO_4 reaches 55 mass%.

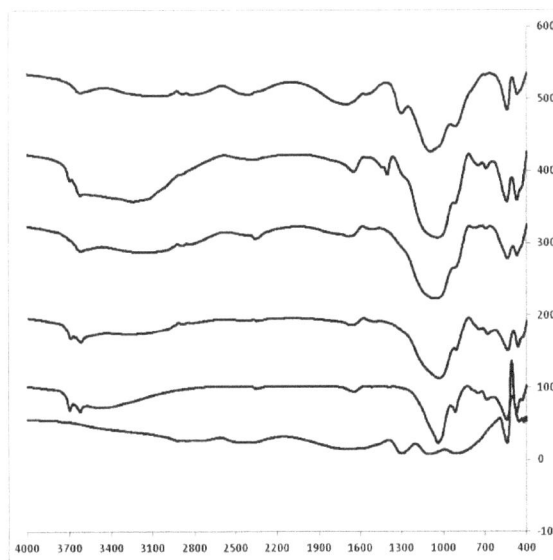

Figure 6. IR spectra of raw kaolin and kaolin-KH_2PO_4 sample system with varying amounts of KH_2PO_4 (mass %) milled for 2h at 400 rpm.

As a result from the IR and XRD analyses, we conclude that KH_2PO_4 is intercalated in the structure of kaolin to the point where its quantity is under 55 mass%. This means that at a low content of kaolin, the distorted crystal of kaolinite was not large enough to receive all the K^+ and PO_4^{3-} atoms from the starting material. However when the content of kaolin increased, the capacity of kaolin to receive KH_2PO_4 improved. Thus all K^+ and PO_4^{3-} atoms were incorporated into the structure of kaolin.

Characterization by XRD and IR analyses of the kaolin-K_2HPO_4 sample mixtures with varying K_2HPO_4 contents (mass %) ground for 120 minutes is shown in figures 7 and 8. With the addition of 25 mass% of kaolin, clear peaks corresponding to K_2HPO_4 remain in the milled product. When the quantity of kaolin increases to 35 mass% both starting samples were reduced to a state of poor diffraction.

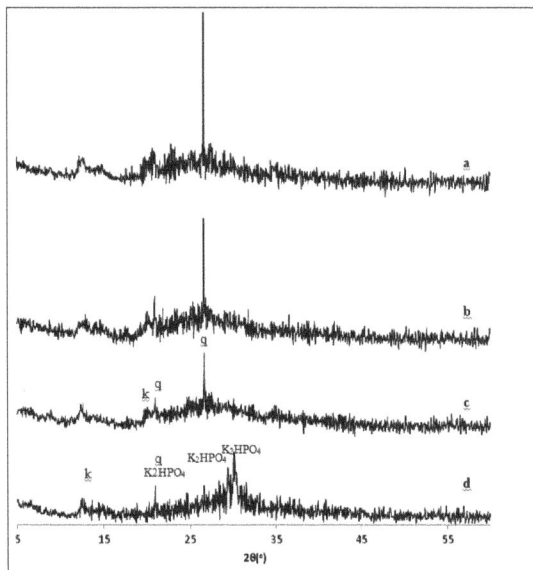

Figure 7. XRD patterns of kaolin-K_2HPO_4 sample system with varying amounts of K_2HPO_4 (mass%) milled for 2h at 400 rpm. a: 25 mass%; b:35 mass%; c:55 mass% and d:75 mass%.

By observing the IR spectra illustrated in figure 8, we note the presence of the broad band centered at 3400 cm^{-1}, indicating the hydrogen bonding, while the characteristic bands of K_2HPO_4 appear in the IR spectrum of the complex kaolin-75 mass% K_2HPO_4. This means that at low kaolin content, and as discussed in the case of the kaolin-KH_2PO_4 sample, the amorphous structure was not large enough to incorporate K^+ and PO_4^{3-}. Thus, we note that a quantity between 55 mass% and 75 mass% of K_2HPO_4 can be intercalated in the network of the kaolin.

The same results were obtained in the case of the kaolin-K_3PO_4 mixtures. This is well represented in figures 9 and 10 supporting this conclusion. In fact, modification of the XRD and IR patterns of the mixtures is noted only when the quantities of added K_3PO_4 reach 75 mass%.

The quantity of KH_2PO_4 included in the structure of kaolin is not equal to those of K_2HPO_4 and K_3PO_4. This is probably due to the acid character of those compounds. Also, we believe that the alumina octahedral layer in the structure of kaolinite protonated to form the surface of the complex $AlOH^{2+}$ with a positive charge; meanwhile, the silica tetrahedral layer in the structure of kaolinite deprotonated to form the complex SiO^- with a negative charge. Consequently, two adjacent layers in the kaolinite are held together by electrostatic forces, which is not favorable to intercalation. As a result of this hypothesis, the amounts of the intercalated K_3PO_4 and K_2HPO_4 are higher than that of KH_2PO_4 which has a greater acidic character than that of the other phosphates.

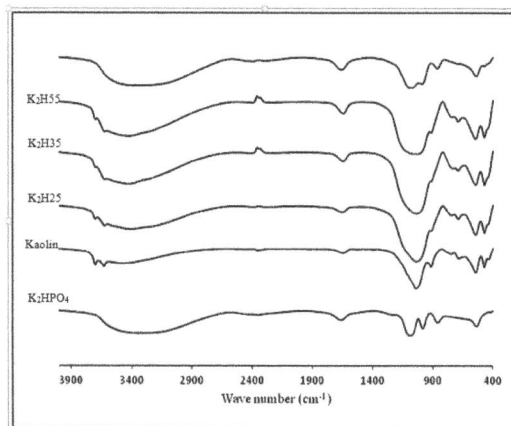

Figure 8. IR spectra of raw kaolin and kaolin-K_2HPO_4 sample system with varying amounts of K_2HPO_4 (mass%) milled for 2h at 400 rpm.

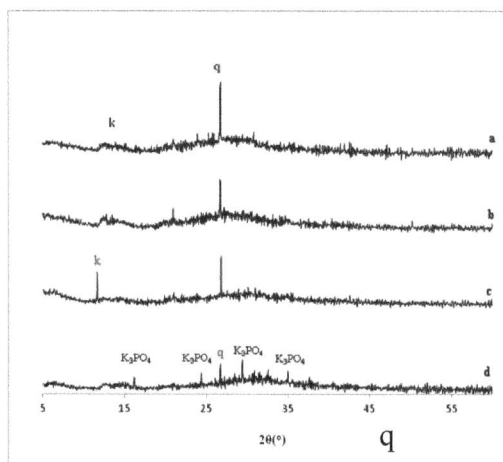

Figure 9. XRD patterns of kaolin-K_3PO_4 sample system with varying amounts of K_3PO_4 (mass%) milled for 2h at 400 rpm (a: 25 mass%; b:35 mass%; c:55 mass% and d:75 mass%).

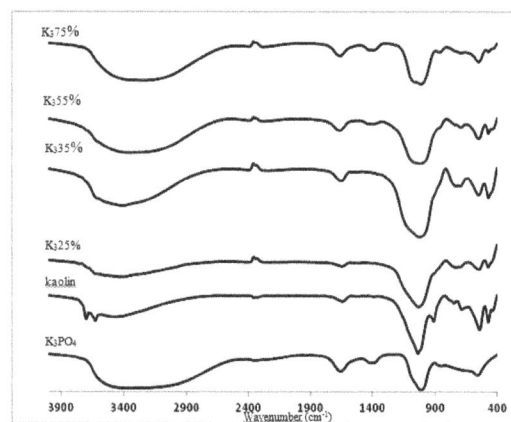

Figure 10. IR spectra of raw kaolin and kaolin-K_3PO_4 sample system with varying amounts of K_3PO_4 (mass%) milled for 2h at 400 rpm.

The behavior of the release of nutrients

The main purpose of this work is to develop the kaolin-KH_2PO_4, K_2HPO_4 or K_3PO_4 complexes used as slow-release fertilizers. This implies that the slower the release of K and P, the better the performance of the fertilizers. Figures 11 and 12 illustrate the following points:

1- The release rate of the nutrients decreases when the content of kaolin increases.

2- For all complexes, the released quantities of K and P are higher than those in the case of the mixtures with 25 mass% of kaolin. This can be explained by the fact that, and as mentioned in the previous section, when the content of kaolin is 25 mass%, the capacity of the distorted structure of amorphous kaolin to incorporate the phosphate additives is insufficient and the excess of KH_2PO_4, K_2HPO_4 or K_3PO_4 in the milled mixtures readily dissolves when dispersed in water.

3- Sufficient kaolin content in the mixtures increases its capacity to allow the incorporation of KH_2PO_4, K_2HPO_4 or K_3PO_4 into the amorphous structure. When dispersed in water, the PO_4^{3-} and K^+ ions could not readily pass through the kaolinite structure or network before dissolving into water, resulting in the sharp decrease of the rate of the release of nutrients.

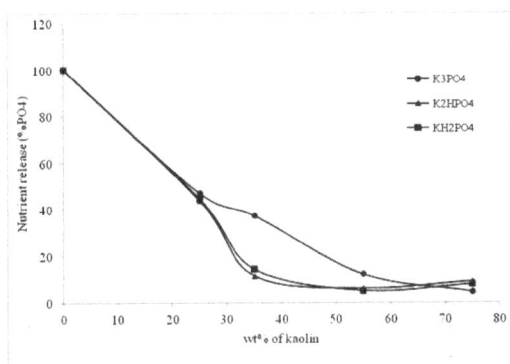

Figure 11. Nutrients release profile of PO_4^{3-} kaolin-potassium phosphates samples system with varying amounts of kaolin (mass%) milled for 2h at 400 rpm.

4- The amounts of K^+ and PO_4^{3-} released when using KH_2PO_4 are higher than those when using the other additives. We believe that this occurs in connection with the rate of the incorporated amount of each additive.

5- The liberation of potassium is easier than that of phosphate because potassium is smaller.

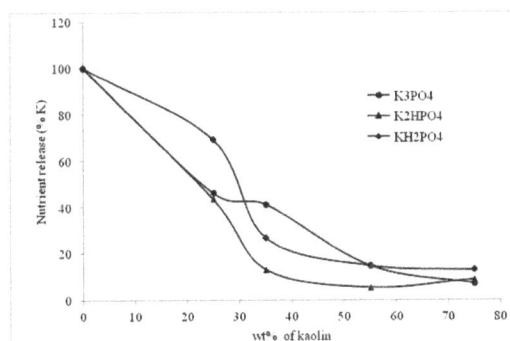

Figure 12. Nutrients release profile of K^+ kaolin-potassium phosphates samples system with varying amounts of kaolin (mass%) milled for 2h at 400 rpm.

Conclusion

The insertion of K^+ and PO_4^{3-} ions as plant nutrients from KH_2PO_4, K_2HPO_4 or K_3PO_4 compounds into the structure of kaolin, through mechanochemical treatment, was successfully reached. The results in the current work can be summarized as follows:

a- It seems that two hours of milling is sufficient to achieve the distortion and breakage of the crystalline network of kaolin.

b- The quantity of KH_2PO_4 introduced in the structure of kaolin is smaller than those of K_2HPO_4 and K_3PO_4. This is probably due to the acid character of those compounds.

c- The release rate of the nutrients decreases when kaolin content increases.

d- The amounts of K^+ and PO_4^{3-} released when using KH_2PO_4 are higher than those when using the other additives. We believe that this occurs in connection with the rate of the incorporated amount of each additive.

References

1- A.E. James, R.E. Sojka, J. Environ. Manage. 87, **2008**, 364-372.

2- B.L. Ni, M.Z. Liu, S.Y. Lü, Chem. Eng. J., 155, **2009**, 892-898.

3- A.A. Ibrahi, B.T. Jibril, Ind. Eng. Chem. Res., 44, **2005**, 2288-2291

4- O.A. Salman, Ind. Eng. Chem. Res. 28, **1989**, 630-632.

5- S.M. Al-Zahrani, Ind. Eng. Chem. Res., 39, **2000**, 367-371.

6- C. Ma, J.A. Garcia, Ind. Eng. Chem. Res., 35, **1998**, 245-249.

7- W. Ostwald, Lehrbuch der Allgemeinen Chemie. Bd. 2. Stochiometrie, Engelmann, Leipzig, **1891**, 1163 S.

8- A. Alacova, J. Ficeriova, M. Golja, Metalurgija, 43(4), **2004**, 305-309

9- H. Masuda, K. Higashitani, H. Yoshida, Powder technology handbook, (3rd Ed.)CRC Press, Taylor & Francis Group, New York, **2006**.

10- V.V. Boldyrev, Thermochim Acta, 110 (**1987**), 303-317.

11- A. Baccour, R. Dhouib Sahnoun, J. Bouaziz, Powder Technology 264, **2014**, 477-483.

12- A. Mitrović, M. Zdujić, International Journal of Mineral Processing 132, **2014**, 59-66.

13- E. Mako, J. Kristof, E. Horvath, V. Vagvolgyi, Journal of Colloid and Interface Science 330, **2009**, 367-373.

14- K. Shahverdi-Shahraki,T. Ghosh, K. Mahajan, A. Ajji, P. J. Carreau, Applied Clay Science 105-106, **2015**, 100-106.

15- H. Cheng, Q. Liua, J. Yang, S. Mac, R. L. Frost, Thermochimica Acta, 545, **2012**, 1- 13.

16- J. E. Gardolinski, F. Wypych and M. P. Cantão, Exfoliation and hydration of kaolinite after intercalation with urea, Quím. Nova vol.24 no.6 São Paulo Nov. /Dec. **2001**.

17- P. Hu, H. Yang, Applied Clay Science 48, **2010**, 368-374.

18- Solihin, Q. Zhang, W. Tongamp, F. Saito, Powder Technology 212, **2011**, 354-358.

19- W. Ching-Wei, S. Chang-Jung, G. Sue-Huai, H. Cheng-Lin, C. Cheng-Gang, Journal of Hazardous Materials 244-245, **2013**, 412-420.

20- F. Dellisanti, G. Valdrè, International Journal of Mineral Processing 102-103, **2012**, 69-77.

21- C. Vizcayno, R.M. de Gutiérrez, R. Castello, E. Rodriguez and C.E. Guerrero, 49 (4), **2010**, 405-413.

22- D. Sun, B. Li, Y. Li, C. Yu , B. Zhang , H. Fei, Materials Research Bulletin 46, **2011**, 101-104

23- E. Mak´o, R. L. Frost, J. Kristof, and E.Horvath, Journal of Colloid and Interface Science 244, **2001**, 359-364.

24- E. Kristóf, A. Z Juhász, I. Vassány, Clay Clay Miner. 41, **1993**, 608

25- C. Vizcayno, R. M. de Gutiérrez, R. Castello, E. Rodriguez, C. E. Guerrero, Appl. Clay Sci. 49, **2010**,   405-413.

26- M. Valášková, K. Barabaszová, M. Hundáková, M. Ritz, E. Plevová, Appl. Clay Sci. 54, **2011**, 70.

27- Pauling L., Nature de la liaison chimique. Paris : PUF, **1949**, p 112.

28- C. Vizcaynoa, R. Castello, I. Ranza, B. Calvo, Thermochimica Acta 428, **2005**, 173-183.

29- F. Dellisanti, G. Valdrè, International Journal of Mineral Processing 102-103, **2012**, 69-77.

30- F. Dellisanti, V. Minguzzi, G. Valdrè, Applied Clay Science 31, **2006**, 282-289.

7

Tin electrodeposition from sulfate solution containing a benzimidazolone derivative

Said Bakkali [*], Abdelillah Benabida and Mohammed Cherkaoui

Laboratoire de Matériaux d'Electrochimie et d'Environnement, Faculté des Sciences, Université Ibn Tofaïl, 14000 Kenitra, Maroc

Abstract: Tin electrodeposition in an acidic medium in the presence of N,N'-1,3-bis-[N-3-(6-deoxy-3-O-methyl-D-glucopyranose-6-yl)-2-oxobenzimidazol-1-yl)]-2-tetradecyloxypropane as an additive was investigated in this work. The adequate current density and the appropriate additive concentration were determined by gravimetric measurements. Chronopotentiometric curves showed that the presence of the additive caused an increase in the overpotential of tin reduction. The investigations by cyclic voltammetry technique revealed that, in the presence and in absence of the additive, there were two peaks, one in the cathodic side attributed to the reduction of Sn^{2+} and the other one in the anodic side assigned to the oxidation of tin previously formed during the cathodic scan. The surface morphology of the tin deposits was studied by scanning electron microscopy (SEM) and XRD.

Keywords: Tin, electrodeposition, N,N'-1,3-bis-[N-3-(6-deoxy-3-O-methyl-D-glucopyranose-6-yl)-2-oxobenzimidazol-1-yl)]-2-tetradecyloxypropane, chronopotentiometry, voltammetry, gravimetry.

Introduction

Tin is one of the metals most often used in electroplating. This is due to its corrosion resistance, non-toxicity, ductility, excellent wetting properties, high electrical conductivity and its electrical reliability [1].

Thus, in the food industry, tinplate used for packaging, is often obtained by plating tin on iron. In the field of electronics, tin plating improves corrosion resistance and weldability of copper.

Another interesting application of tin is in the field of lithium ion battery (LIB). Indeed, several studies show that Sn-based materials anodes in LIBs improve their electrochemical performance [2-4].

Tin deposit is generally obtained from acidic tin (II) or alkaline tin (IV) solutions. Alkaline baths present some difficulty since they work at high temperature and must be protected from Sn^{2+} ions. By contrast, acidic tin baths are easy to control; however, tin from these electrolytes is strongly deposited even at low polarization. The deposit obtained is then porous and dendritic. The addition of organic additives in the tinning bath improves the quality of the deposit. Thus, to slow the deposition rate, various organic compounds have been used such as aromatic carbonyl [5-7], reaction products of amine - aldehyde [8-9] or others compounds [10].

In this study, we were interested in the effect of a new additive; N,N'-1,3-bis-[N-3-(6-deoxy-3-O-methyl-D-glucopyranos-6-yl)-2-oxobenzimidazol-1-yl)]-2-tetradecyloxypropane [11] (Fig. 1) on the electro-deposition of tin from an acidic solution of tin sulfate. The tin electrodeposition was studied by voltammetric and gravimetric techniques. The deposit was characterized by SEM and XRD.

Experimental Section

A conventional three-electrode cell was used for the electrochemical investigation. The working electrodes were iron and copper disks with an area of 1 cm². The potential was measured versus a saturated calomel electrode. The counter electrode is a platinum wire.

Prior each experiment, the solution was deoxygenated by bubbling nitrogen gas for 20 minutes. The working electrode was firstly polished on abrasive paper of grain size ranging from 100 to 1200, then with alumina and cleaned in an ultrasonic bath for 3 minutes. The Potentiostat used was type VoltaLab PGZ 100 interfaced with a microcomputer.

Corresponding author: Said. Bakkali
Email address: sbak6@yahoo.fr

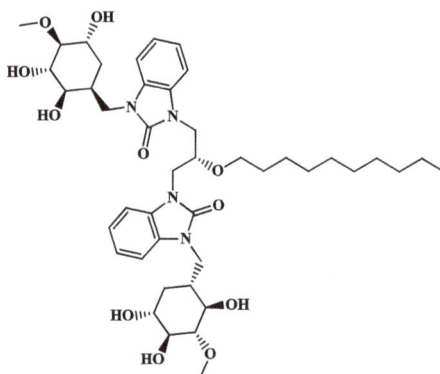

Figure 1. Molecular structure of *N,N'*-1,3-bis-[*N*-3-(6-deoxy-3-*O*-methyl-D-glucopyranos-6-yl)-2-oxobenzimidazol-1-yl)]-2-tetradecyloxypropane [11]

The SEM observations were made with a scanning electron microscope LEO 1530 FEG kind. XRD measurements were made on a PANalytical diffractometer type X'PERT3 POWDER with Cu Kα1 radiation.

The bath temperature was maintained at $(20 \pm 1)\,°C$. The pH of the bath solution was measured using a digital pH meter 510 from Eutech and adjusted with sulphuric acid or sodium bicarbonate solution. All the solutions were prepared from analytical grade chemicals and double-distilled water.

Results and discussion

Gravimetric measurements

The Effects of the additive concentration and current density on the deposition rate was determined by gravimetric measurements. For this, deposits were performed on iron rotating electrode (2000 rpm) at variable additive concentrations and different current densities. The electrode was weighed before and after the realization of the deposit. The electrodeposition rate was calculated from the Weight-gain by the formula below [12].

$$v = \frac{m_f - m_i}{\Delta t . S . \rho_{Sn}} \quad (eq\ 1)$$

Where v is deposition rate ($\mu m.h^{-1}$), m_f is the final mass of the electrode covered with deposit (g), m_i is the initial mass of the electrode (g), S is the electrode surface (cm^2), Δt is the time (h) and ρ_{Sn} is the tin density ($g.cm^{-3}$).

The results obtained are summarized in Figure 2, which shows that with a concentration of 1.8×10^{-5} M of additive and a current density of 15 $mA.cm^{-2}$ the electrodeposition rate reaches its maximum, which is $37.5 \mu m.h^{-1}$. In fact, at higher current densities the rate of evolution of hydrogen gas becomes prominent and exceeds metal deposition, which results in lowering of rate of deposition. The presence of additive inhibits the reduction of hydrogen, which improves the deposition rate of tin. As the concentration of the additive becomes important, its blocking effect increases and affects the rate of deposition of tin.

Based on these results the concentration of the additive in the bath solution was fixed at 1.8×10^{-5} M.

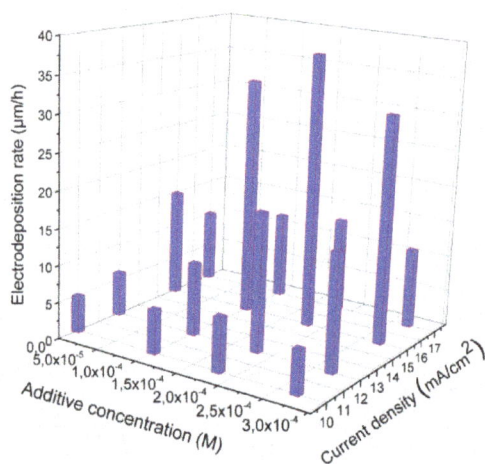

Figure 2. Electrodeposition rate curves obtained at variable current densities in solutions containing 0.014M of $SnSO_4$, 0.056M of H_2SO_4 and different additive concentrations. The working electrode is Iron rotating disk (2000 rpm).

Chronopotentiometric measurements

In Figure 3 were summarized the chronopotentiometric curves obtained in both electrolytes at 15mA.cm^{-2} on iron rotating electrode. As it can be seen, the presence of the additive in the

solution increased the overpotential of the metal deposits. This result was obtained with others additives [13,16]. This is caused by the adsorption of the additive on the working electrode surface [12-16], which impeded the charge transfer [17,18].

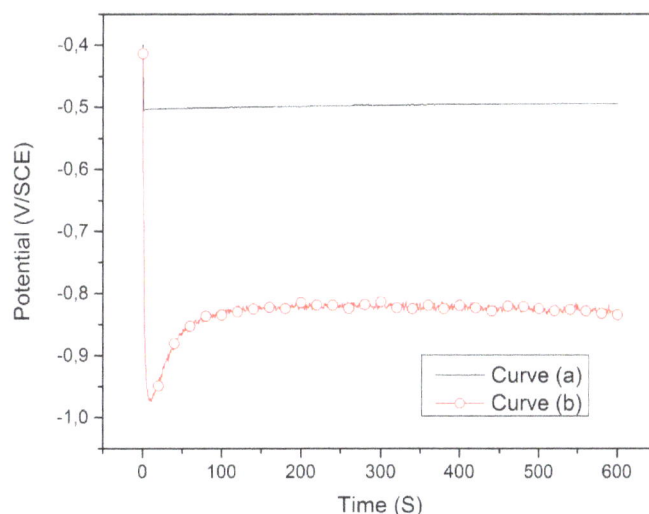

Figure 3. Chronopotentiometric curves recorded using iron rotating electrode (2000 rpm) at 15mA.cm^{-2} obtained during the electroreduction of Sn^{2+} in solutions: (a) 0.014 M SnSO$_4$ + 0.056M H$_2$SO$_4$; (b) 0.014 M   SnSO$_4$ + 0.056M H$_2$SO$_4$+ 1.8x10^{-5} M additive.

Voltammetric measurement

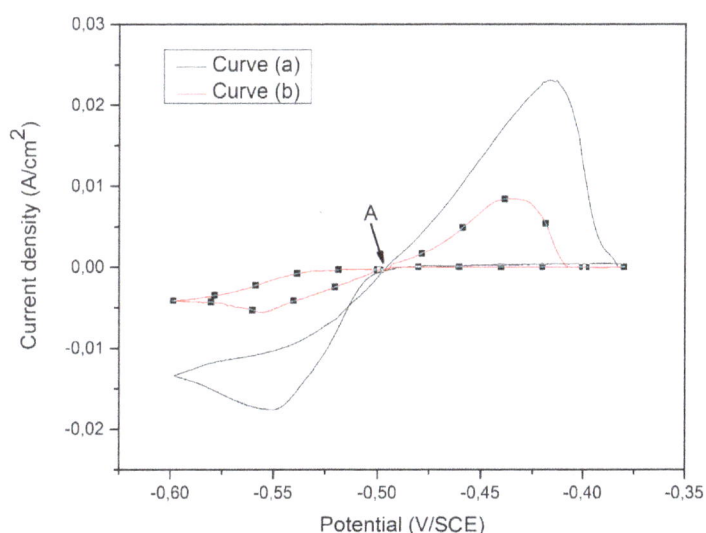

Figure 4. Cyclic voltammetric curves on a fixed copper disc electrode in electrolytes: (a) 0,014 M SnSO$_4$ + 0,056M H$_2$SO$_4$; (b) 0.014 M SnSO$_4$ + 0.056M + 1.8x10^{-5} M additive. The scan rate was 20mV.S^{-1}.

To prevent the oxidation of the working electrode during cyclic voltammetry measurements, iron was replaced by copper working electrode. The voltammograms, recorded at a scan rate of 20 mV s^{-1} in both unstirred electrolytes, are shown in Figure 4. They were initiated at −0.380 V vs SCE, scanned in the negative direction and reversed at −0.6 V in the positive direction. In both electrolytes, during the forward scan the increase in current observed at the point 'A', is assigned to the commencement of the electroreduction of Sn^{2+} at the cathode. The point 'A' refers to the nucleation potential where tin electrodeposition occurs on the copper electrode. At a more negative potential, a cathodic peak was formed which indicated that the reduction of tin was a diffusion controlled process [19]. When the scan was reversed, the current decreased and a crossover occurred between the cathodic and anodic current traces. The presence of the crossover is diagnostic for the nuclei formation on the electrode [19,20]. At a more positive potential, an anodic peak appeared which corresponds to the dissolution of the deposited Tin previously formed. The current density past the peak was null, indicating the completion of oxidative dissolution of metallic tin at the electrode surface.

The charge density due to cathodic (Q c) and anodic (Q a) processes can be obtained by integrating the cathodic and anodic branches of the cyclic voltammogram curves, respectively. Table 1 shows the charge density associated with the reduction and oxidation processes in the absence and presence of additives.

The ratio of the anodic and cathodic charge density (Qa/Qc) (Table 1) was significantly less than 1 in the additive free bath. This can be explained by the contribution of the reduction of H^+ to the cathodic current. This ratio increases in the bath with additive. This can be related to the inhibitory effect of the additive on the hydrogen evolution [12].

Table 1. Cathodic Qc and anodic Qa charges density calculated from the cyclic voltammograms of tin electrodeposition and dissolution.in absence and presence of additive.

	Qc (mC/cm^2)	Qa (mC/cm^2)	Qa/Qc
Solution without additive	109.45	66.79	61%
Solution with additive	22.07	20.06	91.9%

Effect of the temperature

As shown in Figures 5 and 6 in both solutions the current density at a given potential increases as the temperature increases.

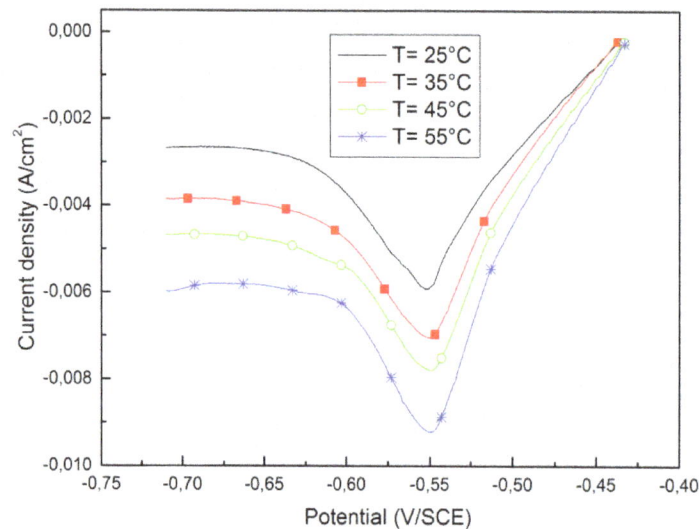

Figure 5. Linear sweep voltammograms in solution containing 0.014 M SnSO$_4$ + 0.056M H$_2$SO$_4$ on fixed iron disk at different temperatures. The scan rate was 20mV.S^{-1}.

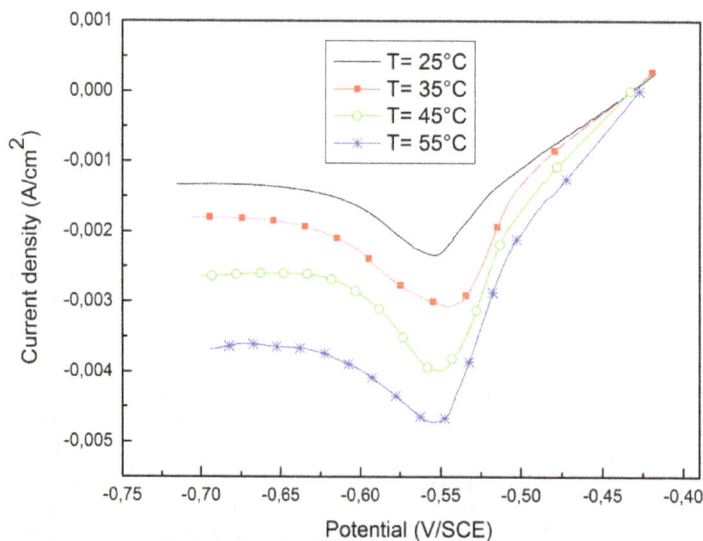

Figure 6. Linear sweep voltammograms in solution containing 0.014 M SnSO$_4$ + 0.056M H2SO4 + 1.8x10^{-5} M additive on fixed iron disk at different temperatures. The scan rate was 20mV.S^{-1}.

The relationship between current density at peak potential and T^{-1} in each electrolyte is shown in Figure 7. The curves are straight lines; their slopes allowed us to calculate the activation energy of the process using the Arrhenius equation (eq 2).

$$\frac{d(\log(i_p))}{d(\frac{1}{T})} = -\frac{Ea}{2.3*R} \quad (eq\ 2)$$

Where E_a is the activation energy of the process and i_p is the peak current density.

The results obtained are summarized in Table (2).
The activation energy Ea is an important parameter for determining the rate-controlling step. In fact, the diffusion of the electroactive specie is the rate-controlling step if E_a is less than 28 kJ.mol^{-1}. Values higher than 40 kJ.mol^{-1} characterize the adsorption of the species on the reaction surface and subsequent chemical reaction [21]. In all solutions, E_a is less than 28 kJ.mol^{-1}, which indicates a diffusional control of the electrodeposition process in both electrolytes. The results showed also an increase in the activation energy in the solution with the additive. This indicates that the additive increases the energy barrier of the tin reduction. The rise in activation energy value in presence of the additive could be interpreted by physical adsorption occurring in the first stage [22,23] which explains the nature of organic molecules -metal interactions.

Table 2. activation energy values for each electrolyte.

	Without Additive	With Additive
Correlation coefficient (R^2)	0.9984	0.9928
E_a(KJ/mol)	10.98	19.41

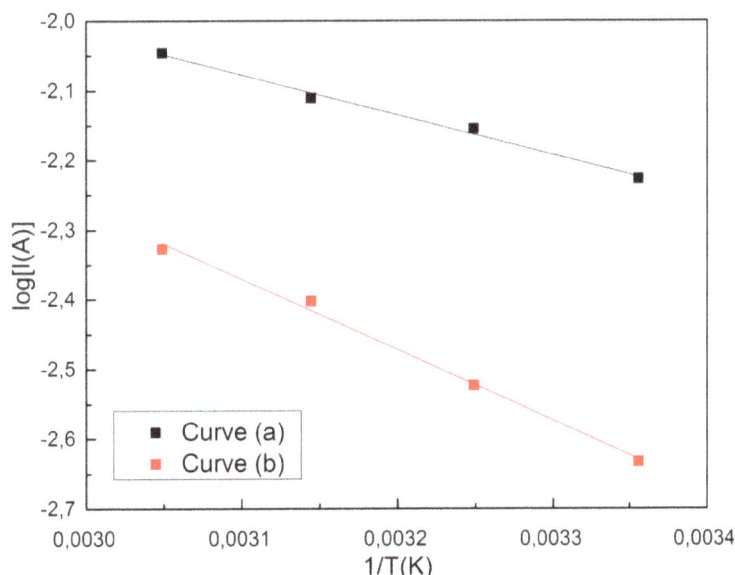

Figure 7. The relation between ln i vs. 1/T on fixed iron disk at peak potential (-550mV/SCE) in solutions: (a) 0.014 M $SnSO_4$ + 0.056M H_2SO_4; (b) 0.014 M $SnSO_4$ + 0.056M H_2SO_4 + 1.8x10^{-5} M additive.

Electroplating tin characterisation

It is well known that the crystal structure of an electrodeposited metal is usually affected by the application of an organic additive [24-26]. To characterise the deposit and to understand the effect of N,N'-1,3-bis-[N-3-(6-deoxy-3-O-methyl-D-gluco-pyranose-6-yl)-2-oxobenzimidazol-1-yl)]-2-tetra-decyloxypropane on the crystal structure of the layer electrodeposited, SEM and XRD characterisation were carried out on samples with thickness of about 10μm of deposit performed on iron substrate, from solutions containing or not the additive. The deposit thickness was estimated by the formula:

$$e = \frac{m_f - m_i}{S.\rho_{Sn}} \quad (eq\ 3)$$

Where e is the thickness (cm), m_f is the final mass of the electrode covered with deposit (g), m_i is the initial mass of the electrode (g), S is the electrode surface (cm^2), and ρ_{Sn} is the tin density (g.cm^{-3}).

Crystallographic structure

The observed crystallographic distances d(hkl) obtained from the XRD analysis were compared with the expected values given in JCPDS. In both solutions, the peaks in the XRD patterns (Figure 8) can be clearly indexed as the tetragonal Sn (JCPDS card No. 00-004-673) indicating that the deposits Deposits contained only pure phase of Sn.

For evaluating the preferred orientation, which measures the relative degree of preferred orientation among crystal planes of Tin samples, texture coefficient Tc was used. It was calculated by the following equation [27,28].

$$Tc = \frac{\frac{I_{(hkl)}}{I_{0(hkl)}}}{\frac{1}{n}\sum\frac{I_{(hkl)}}{I_{0(hkl)}}} \quad (eq\ 4)$$

Where I (hkl) is the relative peak intensity of tin electrodeposits and n is the number of reflections. The index 0 refers to the relative intensities for the standard tin powder sample (JCPDS card No. 00-004-673). If the texture coefficient (Tc) is greater than 1.0, it indicates the existence of a preferred orientation.

The Tc values of tin deposits from both solutions are given in table 3. This Figure shows a strong (200) preferred orientation of the film deposited from additive free solution. Whereas the layer deposited from solution with additive showed a strong (220) preferred orientation. Note that it has been reported that the formation of tin grains oriented along (220) lead to an improved corrosion resistance of the tin film [29].

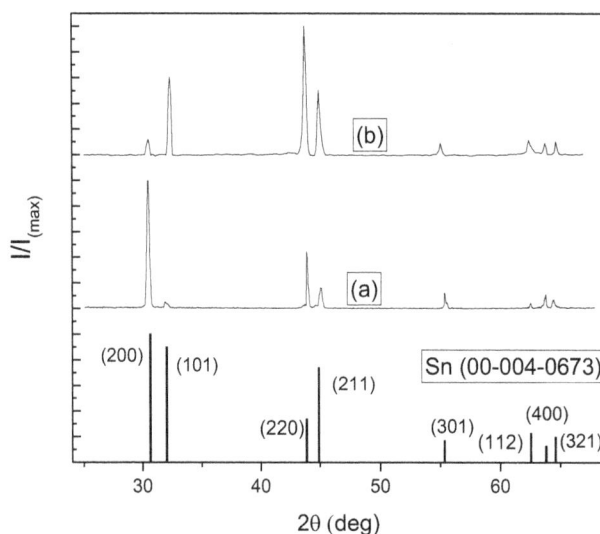

Figure 8: XRD patterns of tin films electrodeposited on iron rotating electrode (2000 rpm) at 15mA/cm^2 (a) 0,014 M SnSO$_4$ + 0,056M H$_2$SO$_4$; (b) 0,014 M SnSO$_4$ + 0,056M + 1.8x10^{-5} M additive.

Pattern of tin (JCPDS card No. 00-004-673) has been given for reference.

Table 3. Texture coefficients of different planes of tin coatings electrodeposited from both solutions.

.Planes	Solution without additive	Solution with additive
(200)	2,19882	0,14873
(101)	0,12216	0,82626
(220)	1,61678	3,64528
(211)	0,46948	0,83743
(301)	1,36558	0,65231
(112)	0,38407	0,54122
(400)	1,29342	0,72906
(321)	0,5497	0,6197

Morphology of the deposits

The changes in Tin deposit occurring in the presence of the additive were also confirmed with the help of morphological studies (Figure. 8). In fact, in additive free solution, the tin deposit was porous, dendritic and had an irregular form, Cu substrate can also be seen in some places confirming the poor cathode coverage; the deposit had also different and slightly larger crystal sizes. (Figure. 10a). In contrast, the SEM photomicrograph of the deposit from the bath with additive showed a uniform arrangement of crystals, refinement in the crystal size which was around 3-4μm and a bright deposit. (Figure. 10b). This is due to the adsorption of additive molecules on the electrode surface, which decreased the surface diffusion of anions toward the active growth sites and thus retarded grain growth.

Figure 10. SE Micrographs of the deposits obtained on iron rotating electrode (2000 rpm) at 15mA/cm^2 (a) 0,014 M SnSO$_4$ + 0,056M H$_2$SO$_4$; (b) 0,014 M SnSO$_4$ + 0,056M + 1.8x10^{-5} M additive.

Conclusion

Electrodeposition of tin from acidic bath in the absence and presence of *N,N'*-1,3-bis-[*N*-3-(6-deoxy-3-*O*-methyl-D-glucopyranose-6-yl)-2-oxobenz-imidazol-1-yl)]-2-tetradecyloxypropane as additive was investigated, in order to study the influence of this additive on the mechanism and kinetics of Sn(II) ions reduction, as well as the morphology of the resulting Sn coatings. A concentration of 1,8.10^{-5}M of the additive and current density of 15mA/cm^2 were optimum to improve the deposition speed. The additive has increased the overpotential of tin reduction and affected the hydrogen evolution was but has not modified the shape of the voltammogram In fact, two peaks were observed, one in the cathodic range and another in the anodic one. The reduction of Sn^{2+} ions was controlled by diffusion in both solutions.

The study of X-ray diffraction patterns showed that the films deposited from both solution were formed from pure tin and that the presence of additive affected the preferred growth orientation of tin film and improves the quality of deposited film which became regular and smooth. All these results could be explained by the adsorption of the additive on the working electrode surface.

References

1- J. M. Bureka, A. S. Budiman, Z. Jahed, N. Tamura, M. Kunz, S. Jin, S. Min J. Han, G. Lee, C. Zamecnik, T. Y. Tsui, Materials Science and Engineering, 2011, A528, 5822-5832.

2- F. Wanga, L. Chen, C. Deng, H. Ye, X. Jiang, Ga. Yang Electrochimica Acta **2014**, 149, 330-336.

3- K. Zhuo, Myung-Gi Jeong, C-Hwa Chung Journal of Power Sources, **2013**, 244, 601-605.

4- D. T. Mackay, M. T. Janish, U. Sahaym, P. G. Kotula, K. L. Jungjohann, C. B. Carter, M. G. Norton Journal of Materials *Science*, **2014**, 49, 1476-1483.

5- N. Kaneko, N.Shinohara, H. Nezu, Electrochem. Acta, **1992**, 37, 2403-2409.

6- Y. Nakamura, N.Kaneko, H.Nezu, Journal of Applied Electrochemistry, **1994**, 24, 569-574.

7- N. Dohi, Journal of the Metal Finishing Society of Japan, 1970, 21, 240.

8- M. Clarke, S. C. Britton, Transactions of the Institute of Metal Finishing, **1962**, 39, 5.

9- M. Clarke, J. A. Bernie, Electrochimca. Acta, **1967**, 12, 205.

10- S. Bakkali, T. Jazouli M. Cherkaoui, M. Ebntouhami, N. El Hajjaji., E. Chassaing Plating and Surface finishing, **2003**, 90, 46- 49.

11- B.Lakhrissi, L.Lakhrissi, M.Massoui, E.M.Essassi, F.Comelles, J.Esquena, C.Solans, C.Rodrÿiguez-Abreu, Journal of Surfactants and Detergents, **2010**, 13, 329-338.

12- S. Bakkali, R. Touir, M. Cherkaoui, M. Ebn Touhami Surface & Coatings Technology **2015**, 261, 337-343.

13-M. Charrouf, S.Bakkali, M.Cherkaoui, M. EL Amrani, Journal of the Serbian Chemical Society. **2006**, 71(6), 661-668.

14- S. Bakkali, M. Charrouf, M. Cherkaoui, Physical and Chemical News, **2005**, 25, 110-117.

15- S.Bakkali, M. Cherkaoui, M. Belfaquir, ScienceLib Editions Mersenne, juillet **2013**, 5, 130705, 1-13.

16 -S. Tzeng, S. H. Lin, Y. Y. Wang, C.C.Wan, Journal of applied electrochemistry 1996, 26, 419-423.

17- J. O'M Bockris and K. N. Reddy, 'Modern
Electrochemistry', Vol. 2, Plenum, New York,
1975, 969.

18- J Long, Z. Zhang, ZC Guo, X.Y. Zhu, H. Huang
Advanced Materials Research, **2012**, 460, 7-10.

19- I. Petersson, E. Ahlberg Journal of
Electroanalytical Chemistry, **2000**, 485, 166-177.

20- A.M. I. Magdy, M. A. O. Enam, Surface &
Coatings Technology, **2013**, 226, 7-16.

21- H. M.A. Soliman Applied Surface Science, **2002**,
195, 155-165.

22- N.P. Clark, E. Jakson, M. Robinson, British
Corrosion Journal. **1979**,14, 33.

23- Lj. M. Vracar, D.M. Drazic Corrosion Science,
2002, 44, 1669-1680.

24- J.Y. Lee, JW Kim, MK Lee J Electrochem Soc,
2004, 151, 25-25.

25- G Trejo, H Ruiz, R OrtegeBorges et al J Appl
Electrochem, **2001,** 31, 685.

26- DS Baik, DJ Fray J Appl Electrochem., **2001**, 31,
1141.

27- K.S. Kim, C.H. Yu, J.M. Yang, Microelectronics
Reliability, **2006**, 46, (7), 1080-1086.

28- A. Dilawar, M.Z. Butt, I. Muneer, F. Bashir, M.
Saleem, International Journal for Light and
Electron Optics, **2017**, 128, January, 235-246.

29- P. Eckold, R. Niewa, and W. Hügel,
Microelectronics Reliability, **2014**, 54, (11),
2578-2585.

8

Insight into the interaction between α-lapachone and bovine serum albumin employing a spectroscopic and computational approach

Otávio Augusto Chaves [1], Edgar Schaeffer [1], Carlos Maurício R. Sant'Anna[1], José Carlos Netto-Ferreira[1,2,*], Dari Cesarin-Sobrinho[1] and Aurelio Baird Buarque Ferreira [1,*]

[1] Departamento de Química, I.C.E. Universidade Federal Rural do Rio de Janeiro, Rodovia BR-465, Km 7, Seropédica/RJ, Brazil.
[2] Instituto Nacional de Metrologia, Qualidade e Tecnologia (INMETRO), Divisão de Metrologia Química, Duque de Caxias/RJ, Brazil.

Abstract: Serum albumin is the most abundant protein in blood plasma; among its functions is the transport of a high variety of drugs in the body. Quinones show several biological and pharmacological activities, such as anti-malarial, antitumor, anti-microbial, anti-inflammatory and anti-parasitic. We report fluorescence and circular dichroism (CD) spectroscopic studies to try to understand the interaction process between α-lapachone (α-LAP) and bovine serum albumin (BSA). Studies using computational methods, such as molecular docking, were performed to identify the main cavity in which this interaction occurs as well as the type of intermolecular interactions between the amino acid residues from albumin and the ligand. The BSA fluorescence quenching by added α-LAP is a static process, indicating an initial association BSA: α-LAP. The K_a and K_b values for the interaction BSA: α-LAP are in the range 10^5-10^4 L·mol^{-1}, indicating a strong binding between these two species. CD data show that there is no significant perturbation on the secondary structure of the protein with binding. The negative ΔG^o values are consistent with spontaneous binding occurring endothermically ($\Delta H^o = 127$ kJ·mol^{-1}), and possibly driven by hydrophobic factors ($\Delta S^o = 0.526$ kJ·mol^{-1}·s^{-1}). The number of binding sites (*n*) indicates the existence of just one main binding site in BSA for α-LAP, with molecular docking results showing that it binds preferentially to the albumin in the domain IIA, where the Trp-212 residue is located. The ligand interacts *via* hydrogen bond with Arg-259 and Tyr-149 residues and *via* T-stacking with the fluorophore Trp-212 residue.

Keywords: Bovine Serum Albumin, α-Lapachone, spectroscopy, molecular docking.

Introduction

Quinones form a large and diverse group of secondary metabolites, being present in several families of higher plants, fungi, lichens, bacteria, arthropods and echinoderms[1,2]. α-Lapachone (α-LAP, 2,2-dimethyl-3,4-dihydro-2H-benzo[g]chromene-5,10-dione, Fig. 1A) is a natural naphthoquinone present in small amounts in trees of the *Tabebuia* species (family *Bignoniaceae* family), which occurs in most of Central and South America. They are commonly called "*ipê*" or "*pau d'arco*" in Brazil and "*lapacho*" in Argentina and other Spanish-speaking countries[3]. In addition, it can be obtained from the isomerization of lapachol (2-hydroxy-3-(3-methyl-2-butenyl)-1,4-naphthoquinone), a quinone that is more abundant and readily extracted from the same sources[4]. The biological activities of these compounds and simple derivatives have been investigated from 1940 (anti-malarial) up to the present (anti-tumor, anti-microbial, anti-inflammatory and anti-parasitic)[5-8]. It is worth noting that "*ipê-roxo*" has been traditionally used in folk medicine[6].

Serum albumin is a protein present in the circulatory system with a variety of physiological functions: maintenance of osmotic pressure; transport, distribution and participation in the metabolism of several endogenous and exogenous ligands (such as, drugs, metabolites, fatty acids, amino acids and hormones). The interaction between drugs and serum albumin is a major and important factor for understanding the interaction of the organism with drugs, since it influences their distribution and excretion[9,10].

The Bovine Serum Albumin (BSA) structure consists of three structurally similar domains (I, II and III), each containing two subdomains, A and B[11,12]. Each of these subdomains has a major binding pocket.

Fluorescence spectroscopy, especially quenching of fluorescence from tryptophan residues – Trp, is

*Corresponding authors: José Carlos Netto-Ferreira, Aurelio Baird Buarque Ferreira
E-mail address: *jcnetto.ufrrj@gmail.com, aureliobf@uol.com.br*

often applied to the binding study in biomolecules. The BSA structure has two tryptophan residues, one in the region IB (Trp-134) and another in the region IIA (Trp-212)[13] (Fig. 1B). Human Serum Albumin (HSA) is commonly replaced by BSA in laboratory experiments due to the higher availability and lower cost of the later. On a first approach, much of binding capacity of BSA may be inferred by its similarity with the human counterpart, since the BSA structure shares 76% identity and 88% similarity in protein sequences with HSA[14].

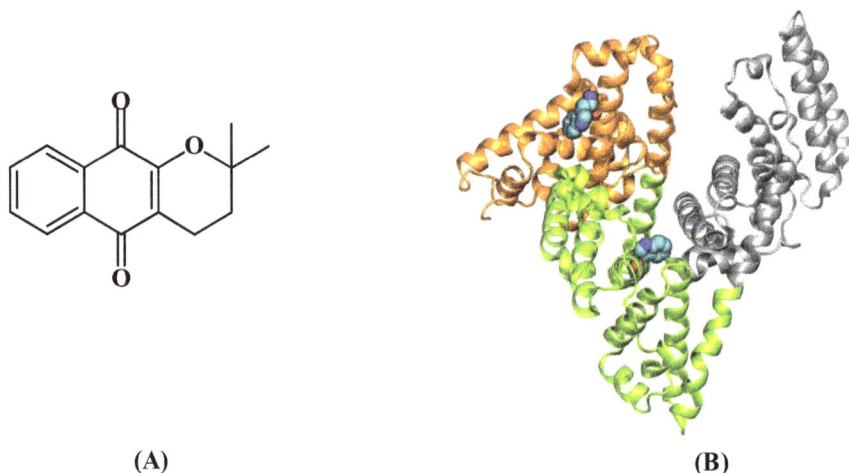

(A)   (B)

Figure 1: **(A)** Chemical structure of α-LAP. **(B)** Crystallographic structure of BSA (pdb: 4F5S) showing the domain I (**brown**), II (**green**) and III (**gray**). Trp-134 is located in domain I (**brown**) and Trp-212 in domain II (**green**).

Results and Discussion

Fluorescence Spectroscopy and Fluorescence Quenching

Fluorescence quenching can be employed to evaluate the binding affinities between macromolecules and ligands (quenchers)[15]. Fig. 2 shows quenching of the tryptophan residue of BSA (Conc.$_{BSA}$ = 1.0 x 10^{-5} mol·L^{-1}) by addition of aliquots of known concentration of α-LAP solution at 298K, one of the three temperatures employed in this experiment (288K; 293K and 298K). This quenching process indicates that the naphthoquinone is located inside the protein and next to a tryptophan residue[16]. The absence of significant changes in the maximum of the fluorescence emission for BSA is clear evidence that the presence of α-LAP does not exert any influence on the polarity of the microenvironment inside the cavity containing the tryptophan residue[17].

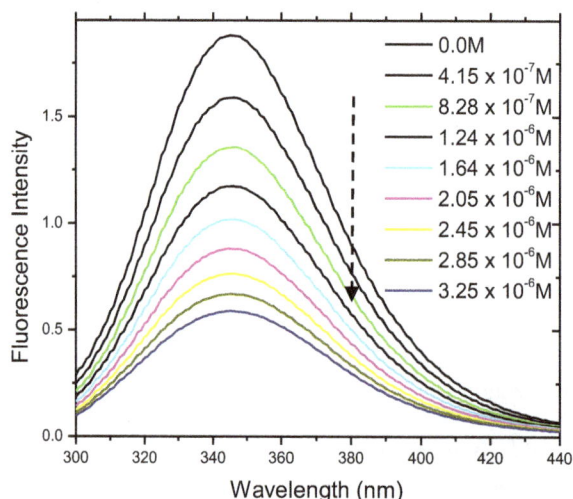

Figure 2. Fluorescence spectra of BSA (conc.: 1.00 x 10^{-5} mol·L^{-1} in PBS buffer solution – pH 7.4) and fluorescence quenching by incremental addition of α-LAP (conc.: 4.15 x 10^{-7} to 3.25 x 10^{-6} mol·L^{-1}) T = 298K, λ_{exc} = 280 nm.

Applying equation 1 to each of the quenching experiments performed at different temperatures (288K; 293K and 298K), the Stern-Volmer quenching constant (K_{sv}) and the quenching rate constant (k_q) for the interaction BSA: α-LAP at each temperature can be obtained (Figure 3 and Table 1)[18].

$$(A) \quad \frac{F_0}{F} = 1 + k_q \tau_0 [Q] = 1 + K_{SV}[Q] \qquad (B) \quad k_q = \frac{K_{SV}}{\tau_0} \qquad (1)$$

(Where, F_0 and F are the fluorescence intensities of BSA without and with the quencher (α-LAP), respectively; K_{sv} is the Stern-Volmer quenching constant, k_q is the quenching rate constant of BSA fluorescence, [Q] the quencher concentration and τ_0 is the lifetime of BSA without the quencher (10^{-8} s)[17].)

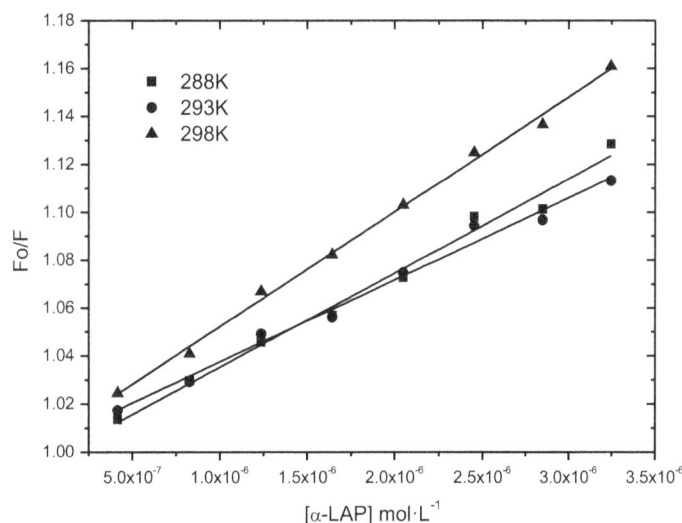

Figure 3. Stern-Volmer plots for the fluorescence quenching of BSA by α-LAP at different temperatures.

Table 1. Stern-Volmer quenching constant (K_{sv}) and quenching rate constant (k_q) values for the association BSA: α-LAP at 288K, 293K and 298K.

T (K)	K_{SV} (L·mol^{-1})	k_q (L·mol^{-1}·s^{-1})	r^2
288	(3.93±0.18) x 10^4	3.93 x 10^{12}	0.9855
293	(3.42±0.16) x 10^4	3.42 x 10^{12}	0.9854
298	(4.79±0.11) x 10^4	4.79 x 10^{12}	0.9962

The fluorescence quenching can be induced at a certain distance between the fluorophore and the quencher and does not require contact between them (Förster's theory). Quenching of a fluorophore can also occur as a result of the formation of a non fluorescent complex between the fluorophore and a non-fluorescent molecule in the ground state. This mechanism is known as "ground-state complex formation"[15]. The quenching rate constants ($k_q \approx 10^{12}$ L·mol^{-1}·s^{-1}) have higher values than the diffusion rate constant ($k_d \approx 5$ x 10^9 L·mol^{-1}·s^{-1} in water at 25°C), indicating that the probable mechanism of fluorescence quenching is static[17-19]. In this case, the initial formation of a non-fluorescent association in the ground-state between the fluorophore (tryptophan residue in BSA) and the quencher can be expected. The small and irregular variation in K_{SV} values with temperature does not show a significant trend.

Binding Constant (K_a) from a Stern-Volmer Modified Plot

To obtain further information about the interaction between BSA and α-LAP, if it can be considered strong, moderate or weak, we calculated the modified Stern-Volmer binding constant (K_a)[18,20] (Figure 4 and Table 2). This constant can be obtained according to equation 2:

$$\frac{F_0}{F_0 - F} = \frac{1}{fK_a} \frac{1}{[Q]} + \frac{1}{f} \qquad (2)$$

(Where, F_o and F are the fluorescence intensities of BSA without and with the quencher (α-lapachone) at 350 nm, respectively; K_a is the modified Stern-Volmer binding constant; f the fraction of the initial fluorescence that is accessible to the quencher and [Q] the quencher concentration.)

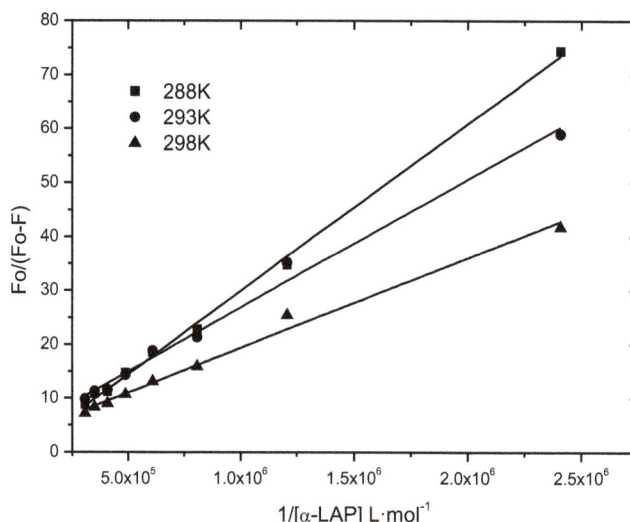

Figure 4. Modified Stern-Volmer plots for the quenching of BSA fluorescence by α-LAP at different temperatures.

Table 2. Modified Stern-Volmer binding constant (K_a) values for the association BSA: α-LAP at 288K, 293K and 298K.

T (K)	K_a (L·mol^{-1})	r^2
288	$(3.1 \pm 0.1) \times 10^4$	0.9975
293	$(1.3 \pm 0.1) \times 10^5$	0.9896
298	$(1.6 \pm 0.1) \times 10^5$	0.9896

The K_a values are in the range 10^4-10^5 L.mol^{-1}, indicating a strong interaction between albumin and α-LAP [21,22]. The binding constant values depend on the charge and structure of α-LAP, as well as on the conformation and charge of BSA [23].

$$\textbf{(A)} \quad \ln K_a = -\frac{\Delta H^0}{RT} + \frac{\Delta S^0}{R}$$

(Where, ΔH^0, ΔS^0, ΔG^0 are the enthalpy, entropy and Gibbs free energy, respectively; R is the gas constant ($R = 8.314 \times 10^{-3}$ kJ·mol^{-1}·K^{-1}),

Thermodynamic Parameters (ΔG^0, ΔH^0, ΔS^0)

To get some insight on the thermodynamic parameters ΔG^0, ΔH^0, ΔS^0 which control the interaction BSA: α-LAP, data from Table 2 were plotted according to the van't Hoff equation (3A) (Figure 5) and ΔG^0 was calculated employing the Gibbs free energy equation[24] (3B) (Table 3):

$$\textbf{(B)} \quad \Delta G^0 = \Delta H^0 - T\Delta S^0 \qquad \textbf{(3)}$$

T is the temperature (288K, 293K and 298K) and K_a the binding constant.)

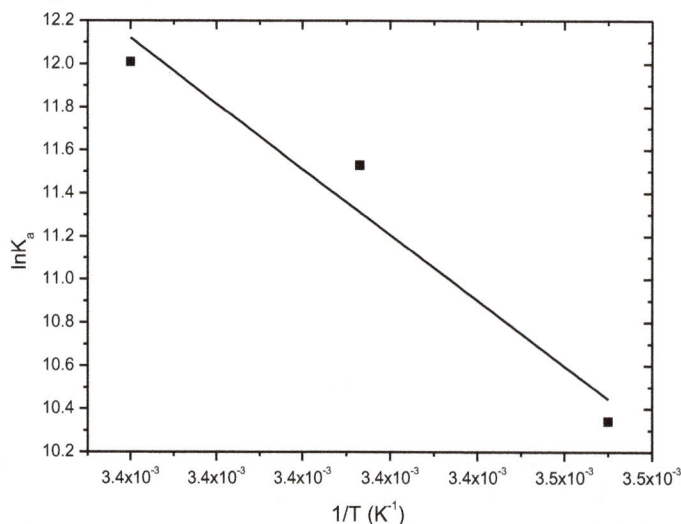

Figure 5. Van't Hoff plot employing K_a values obtained from modified Stern-Volmer plots at 288K, 293K and 298K.

Table 3. Thermodynamic parameters (ΔH°, ΔS°, ΔG°) for the interaction between BSA and α-LAP at 288K, 293K and 298K.

T (K)	ΔH° (kJ·mol^{-1})	ΔS° (kJ·mol^{-1}·K^{-1})	ΔG° (kJ·mol^{-1})	r^2
288			-24.9	
293	127	0.526	-27.6	0.9056
298			-30.2	

The negative values of ΔG° shown in Table 3 are consistent with a spontaneous binding between BSA and α-LAP, whereas the positive value of ΔH° indicates that the binding process is endothermic. Finally, the positive value of ΔS° shows that the interaction is mainly due to hydrophobic factors[25], which can be related to the influence of hydration molecules. There are two possible contributions that may explain the increase in entropy: hydration molecules can be expelled from the protein cavity as a consequence α-LAP entry into it, or the desolvation of the α-LAP, as it enters the cavity, can cause an increase in the number of micro-states of the system[23-25].

Number of Binding Sites (*n*)

The binding site number (*n*) and the binding constant (K_b) can be obtained from the binding site model which assumes that there are similar and independent binding sites for a quencher in the biomolecule[26], as expressed by equation 4[27]. These results are shown in Figure 6 and Table 4.

$$\log \frac{F_0 - F}{F} = \log K_b + n \log[Q] \qquad (4)$$

(Where F_0 and F are the fluorescence intensities in the absence and presence of α-LAP, K_b is the binding constant with BSA, and *n* is the number of binding sites).

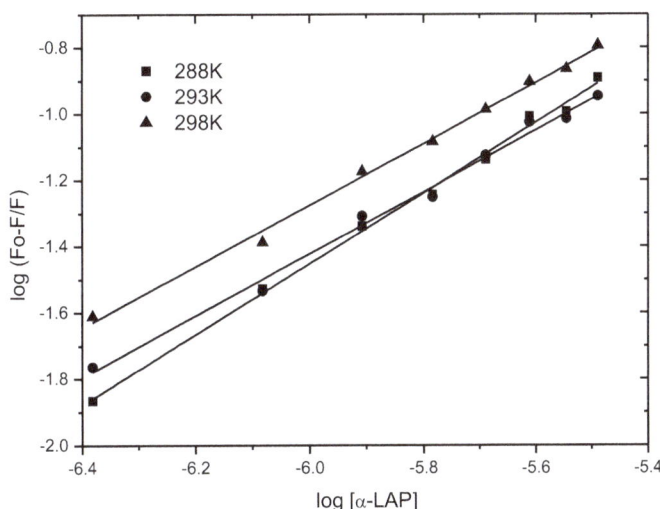

Figure 6. Linear fit for the values of log [(F_0–F)/F] *vs.* log [Q] for BSA: α-LAP at 288K, 293K and 298K.

Table 4. Number of binding sites (*n*) and binding constant (K_b) values for the association BSA: α-LAP at 288K, 293K and 298K.

T (K)	*n*	K_b (L·mol^{-1})	r^2
288	1.07±0.03	(8.8±0.2) x 10^4	0.9952
293	0.93±0.03	(1.5±0.2) x 10^4	0.9905
298	0.93±0.02	(1.9±0.1) x 10^4	0.9951

As can be seen in Table 4, K_b values are in the range 10^4 L·mol^{-1}, comparable to the range of the modified Stern-Volmer binding constant (K_a) values (Table 2). In both cases there is no consistent trend in the variation with temperature. The *n* values were approximately 1 at different temperatures (288K, 293K and 298K), which indicates the existence of just one main binding site in BSA for α-LAP [17,26].

Circular Dichroism (CD)

Circular dichroism spectra were obtained to evaluate the changes in the secondary structure of

BSA induced by the addition of α-LAP. The CD spectra of BSA exhibit two minus signs (negative Cotton effects) at 208 nm and 222 nm, which are characteristic of the α-helix protein structure. These bands originate from n-π* protein transitions[28-30].

Initially the molar residual ellipticity (MRE) was calculated applying equation 5[27], and used to obtain the α-helix % (equation 6).

$$MRE = \frac{\theta}{(10.n.l.C_P)} \quad (5)$$

(Where, θ is the observed ellipticity (mdeg); n is the number of amino acid residues (582 for BSA)[31]; l is the optical length of the optical cuvette (1 cm) and C_p is the molar concentration of BSA (1,00 x 10^{-6} mol·L^{-1})).

$$(A) \% \, \alpha - helix = \left[\frac{(-MRE_{208} - 4000)}{(33000 - 4000)}\right] x \, 100 \quad (B) \% \, \alpha - helix = \left[\frac{(-MRE_{222} - 2340)}{30300}\right] x \, 100 \quad (6)$$

(Where, MRE_{208} and MRE_{222} are the significant molar residual ellipticity at 208 nm and 222 nm (deg.cm^2.dmol^{-1}), respectively.)

Fig. 7 shows the circular dichroism spectra at 298K for free BSA (1.00 x 10^{-6} mol·L^{-1}) and in the presence of α-LAP.

The circular dichroism signals are similar, indicating that the naphthoquinone does not make a significant perturbation on the secondary structure of BSA[32,33].

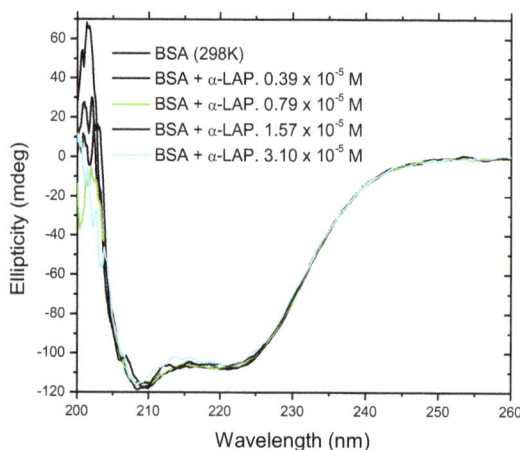

Figure 7. Circular dichroism spectra of BSA and BSA: α-LAP at 298K Conc.$_{BSA}$ = 1.00 x 10^{-6} mol·L^{-1}.

Table 5. Values for α-helix % at 208 nm and 222 nm resulting from the association BSA: α-LAP at 298K.

	C$_{BSA}$ (mol·L^{-1})	T (K) C$_{\alpha\text{-lap}}$ (mol·L^{-1})	298 208	222
	1.00 x 10^{-6}	-	55.24	53.36
BSA	1.00 x 10^{-6}	0.39 x 10^{-5}	54.56	53.04
α-	1.00 x 10^{-6}	0.79 x 10^{-5}	54.46	52.73
LAP	1.00 x 10^{-6}	1.57 x 10^{-5}	52.32	52.14
	1.00 x 10^{-6}	3.10 x 10^{-5}	48.37	51.77

Results (at Table 5) show that the addition of α-LAP to BSA results in a change of 1.6% to 6.9% in α-helix % for BSA, demonstrating again that the naphthoquinone does not make a significant perturbation on the secondary structure of the protein.

Molecular Docking Studies

HSA is commonly substituted for BSA in laboratory experiments due to its higher availability and lower cost, and on a first approach much of its binding capacity may be inferred by similarity with its human counterpart. Like HSA, with which it

shares a 76% identity and 88% similarity in sequence, the globular protein BSA consists of three structurally similar domains (I, II, and III), each containing two sub-domains (A and B)[34], with the domains I (Sudlow I) and II (Sudlow II) being the most important. Site I, named *warfarin binding site*, is located in the IIA subdomain, while site II, named *indole/benzodiazepine binding site* is located in subdomain IIIA[35]. From the studies of BSA fluorescence quenching described above, it is known that α-LAP can interact with a tryptophan residue, either the Trp-134 or the Trp-212. A molecular docking study was performed to analyze the interaction sites as well as the nature of the intermolecular interactions between α-LAP and the amino acid residues in each site.

The docking score results suggest that α-LAP has more favorable interactions with the Trp-212 site (docking score 57.4) than in the Trp-134 site (docking score 44.9). The docking results clearly suggested that α-LAP is more probably bound to the Sudlow's site I (subdomain IIA) where the Trp-212 residue is located. Analysis of the interaction between BSA and α-LAP in this site shows that the ligand can also interact via hydrogen bonds with an arginine and a tyrosine amino acid residue (Fig. 8). The Tyr-149 residue is able to form a hydrogen bond

with the carbonyl group of α-LAP nearest to the pyran oxgen with a distance between the donor and acceptor atoms of 1.74 Å. The oxygen of the pyranic group of α-LAP can also interact through hydrogen bond with the Arg-259 residue, with a distance of 2.77 Å. Besides these interactions, the ligand interacts with BSA through hydrophobic interactions. The molecular docking results show that the ligand aromatic ring makes a T-stacking interaction with the Trp-212 residue, with a distance of 3.49 Å. This proximity between the quencher and the fluorophore can explain the efficiency of the fluorescence quenching[15].

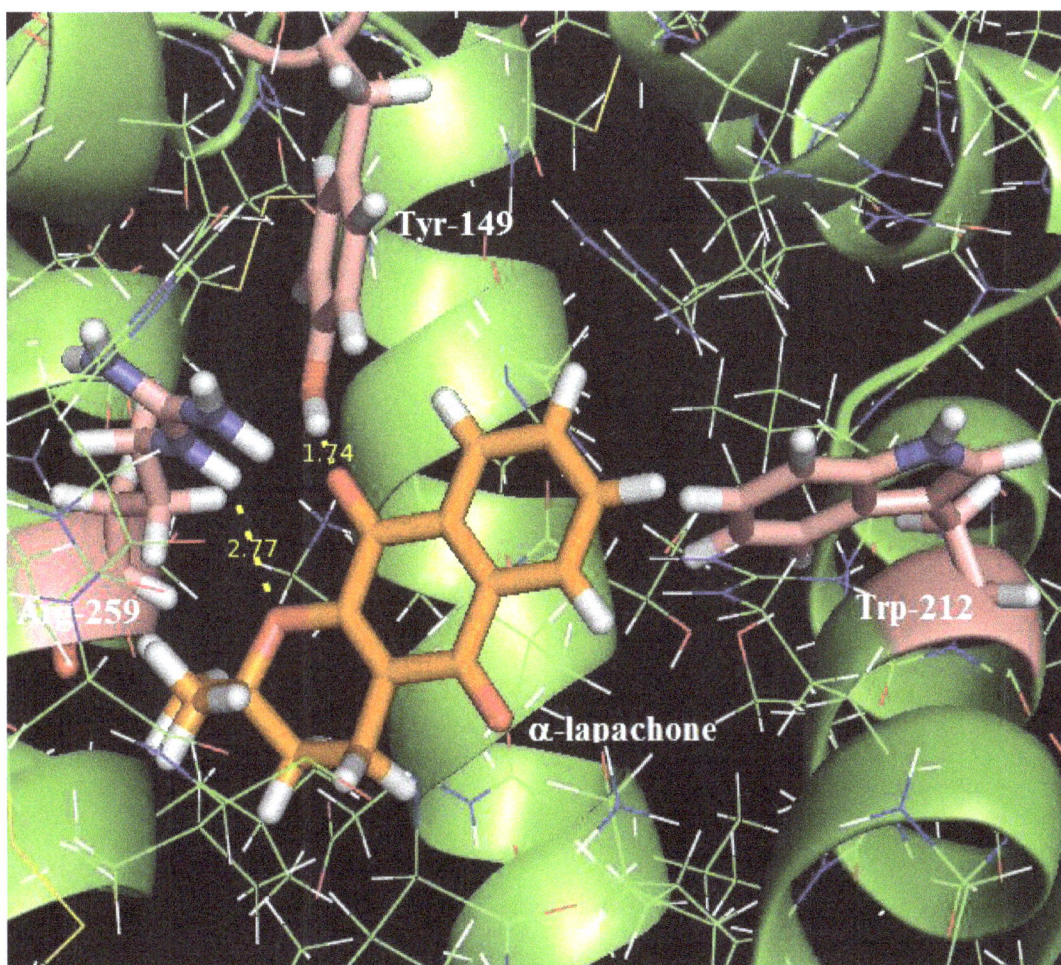

Figure 8. Best score pose for α-LAP in BSA in the cavity containing the Trp-212 residue, obtained by molecular docking (ChemPLP function). Carbon: orange (α-LAP), green (BSA), light beige (selected residues); hydrogen: white; oxygen: red; and nitrogen: blue. (figure generated with the PyMOL software).

Conclusion

α-Lapachone binds strongly with bovine serum albumin, but this binding does not make a significant perturbation on the secondary structure of BSA. This interaction is spontaneous (ΔG° < 0) and the other thermodynamic parameters suggest that the binding is endothermic (ΔH° = 127 kJ·mol^{-1}) and occur mainly by hydrophobic factors (ΔS° = 0.526 kJ·mol^{-1}·s^{-1}) due to the hydration molecules effect. The quenching rate constant (k$_q$≈10^{12} L·mol^{-1}·s^{-1})

obtained employing fluorescence spectroscopy indicates that the tryptophan fluorescence quenching is a static process, as a consequence of a ground-state interaction BSA: α-LAP. The number of binding sites (*n*) indicates that there is just one main interaction site and the docking results are indicative that α-LAP preferentially binds in the domain IIA, where the Trp-212 residue is located. The naphthoquinone interacts via hydrogen bond with Arg-259 and Tyr-149 residues and via a T-stacking interaction with the fluorophore Trp-212 residue.

Acknowledgements

This research was supported by the Brazilian agencies: CAPES (Coordenação de Aperfeiçoamento de Pessoal de Nível Superior), CNPq (Conselho Nacional de Desenvolvimento Científico e Tecnológico) and FAPERJ (Fundação de Amparo à Pesquisa do Estado do Rio de Janeiro).

Experimental Section

Spectroscopic Experiments:
Chemicals

Commercially available Bovine Serum Albumin (BSA) and PBS buffer (pH = 7.4) were obtained from Sigma-Aldrich Chemical Company. Water used in all experiments was millipore water. Ethanol (spectroscopic grade) was obtained from Vetec. α-LAP was obtained from the Photochemistry Group of the Chemistry Department/Universidade Federal Rural do Rio de Janeiro (Seropédica, RJ, Brazil). Its spectroscopic and spectrometric properties are in full accord with the structure proposed[36].

Instruments

The fluorescence spectra were measured on a Jasco J-815 fluorimeter, in a 1 cm quartz cell and employing a thermostatic cuvette holder Jasco PFD-425S15F. The circular dichroism spectra were measured in a Jasco J-815 spectropolarimeter. All spectra were recorded with appropriate background corrections.

Methodology

In a 3.0 mL solution, containing appropriate concentration of BSA (1.00×10^{-5} mol·L^{-1}, in PBS buffer, pH = 7.4), successive aliquots from a stock solution of α-LAP (1.00×10^{-3} mol·L^{-1} in ethanol) were added, to obtain concentrations ranging from 0 to 3.25×10^{-6} mol·L^{-1}. The addition was done manually by using a micro syringe. Fluorescence spectra were measured in the range 300-440 nm, at 288K, 293K and 298K with excitation wavelength at 280 nm. The circular dichroism spectra were recorded for free BSA (1.00×10^{-6} mol·L^{-1}) and α-LAP (0,39; 0,79; 1.57; 3.10×10^{-5} mol·L^{-1}), in the range of 200-260 nm, at −298K.

In order to compensate for the inner filter effect, the fluorescence intensity values of the samples were corrected for their absorption at excitation and emission wavelengths using the equation 7[15,37].

$$F_{cor} = F_{obs} 10^{[(A_{ex}+A_{em})/2]}$$

(7)

(Where F_{cor} and F_{obs} are the corrected and the observed fluorescence intensity values, while A_{ex} and A_{em} represent changes in the absorbance values of the samples at the excitation ($\lambda = 280$ nm) and emission wavelengths ($\lambda = 345$ nm), respectively.)

Computational Experiments

The crystallographic structure of bovine serum albumin was obtained from the Protein Data Bank (PDB) whose access code is 4F5S[14]. This structure has a resolution of 2.47 Å. The α-LAP structure was built and energy-minimized with the semiempirical method AM1[38], available in the Spartan'14 program (Wavefunction, Inc., Irvine, CA, USA).

The molecular docking was performed with Gold 5.2 program (CCDC). Hydrogen atoms were added to the protein according to the data inferred by the program on the ionization and tautomeric states[39]. A docking interaction cavity in the protein was established with a radius of 10 Å and 15 Å from Trp-134 and Trp-212. The best result was obtained with radius of 10 Å. The number of genetic operations (crossover, migration, mutation) in each docking run used in the searching procedure was set to 100,000. The program optimizes hydrogen-bond geometries by rotating hydroxyl and amino groups of amino acid side chains. The scoring function used was 'ChemPLP'[40], which is the default function of the GOLD program. The score of each pose identified is calculated as the negative of the sum of a series of energy terms involved in the protein-ligand interaction process, so that the more positive the score, the better is the interaction. The figure of the best score was generated by PyMOL 1.leval program (Delano Scientific LLC).

References

1- T.L.G. Lemos, F.J.Q. Monte, A.K.L. Santos, A.M. Fonseca, H.S. Santos, M.F. Oliveira, S.M.O. Costa, O.D.L. Pessoa, R. Braz-Filho, Nat. Prod. Res. 2007, 21, 529-550.

2- L. F. Fieser, E. Berliner, F. Bondhus, F. C. Chang, W. G. Dauben, M. G. Ettlinger, G. Fawaz, M. Fields, M. Fieser, C. Heidelberger, H. Heymann, A. M. Seligman, W. R. Vaughan, A. G. Wilson, E. Wilson, M.I. Wu, M. T. Leffler, K. E. Hamlin, R. J. Hathaway, E. J. Matson, E. E. Moore, M. B. Moore, R. T. Rapala, H. E. Zaugg, J. Am. Chem. Soc. 1948, 70, 3151-3162.

3- A.R. Burnett, R.H. Thomson, J. Chem. Soc. C, 1967, 2100-2104.

4- H. Hussain, K. Krohn, V.U. Ahmad, G.A. Miana, I.R. Green, ARKIVOC, 2007, II, 145-171.

5- E. Pérez-Sacau, R.G. Diaz-Peñate, A. Estévez-Braun, A.G. Ravelo, J.M. García-Castellano, L. Pardo, M. Campillo, J. Med. Chem. 2007, 50, 696-706.

6- A. S. Cunha, E. L. S. Lima, A. C. Pinto, A. Esteves-Souza,. A. Echevarria, C. A. Camara, M. D. Vargas, J. C. Torres, J. Braz. Chem. Soc. 2006, 17, 439-442.

7- E. N. Silva Júnior, M. C. B. V. de Souza, A. V. Pinto, M. C. F. R. Pinto, M. O. F. Goulart, F. W. A. Barros, C. Pessoa, L. V. Costa-Lotufo, R. C. Montenegro, M. O. de Moraes, V. F. Ferreira, Bioorg. Med. Chem. **2007**, 15, 7035-7041.

8- Kim, S. O.; Kwon, J. I.; Jeong, Y. K.; Kim, G. Y.; Kim, N. D.; Choi, Y. H.; Biosci. Biotechnol. Biochem. **2007**, 71, 2169-2174.

9- K. A. Majorek, P. J. Porebski, A. Dayal, M. D. Zimmerman, K. Jablonska, A. J. Stewart, M. Chruszcz, W. Minor, Mol. Immunol. **2012**, 52, 174–182.

10- K. Taguchi, V.T.G. Chuang, T. Maruyama, M. Otagiri, J. Pharm. Sci. **2012**, 101, 3033-3046.

11- B. K. Paul, A. Samanta, N. Guchhait, J. Phys. Chem. B, **2010**, 114, 6183–6196.

12- D. C. Carter, X. M. He, S. H. Munson, P. D. Twigg, K. M. Gernert, M. B. Broom, T. Y. Miller, Science, **1989**, 244, 1195-1198.

13- S. Sugio, S. Kashima, S. Mochizuki, M. Noda, K. Kobayashi, Protein Eng. Des. Sel. **1999**, 12, 439-446.

14- A. Bujacz, Acta Cryst. **2012**, D68, 1278-1289.

15- J.R. Lakowicz. Principles of Fluorescence Spectroscopy, 1st ed.; Springer New York, U.S.A., **2006**; pp. 923–928.

16- J. Liu, J.N. Tian, J. Zhang, Z. Hu, X. Chen, Anal Bioanal Chem. **2003**, 376, 864-867.

17- J. Tian, X. Liu, Y. Zhao, S. Zhao, J. Luminesc. **2007**, 22, 446-454.

18- M.R. Eftink, C.A. Ghiron, Anal Bioanal Chem. **1981**, 114, 199-227.

19- D. Brune, S. Kim, Biophysics, **1993**, 90, 3835-3839.

20- M.R. Eftink. Fluorescence Quenching Reactions: Probing Biological Macromolecular Structures. In: Biophysical Biochemical Aspects of Fluorescence Spectroscopy, 1st ed.; T.G. Dgurvy,; Plenum Press, New York, U.S.A., **1991**, Vol. 1, pp. 1-41.

21- A. Satheshkumar, K.P. Elango, Spectrochim. Acta Mol. Biomol. **2014**, 130, 337-343.

22- O. A. Chaves, A. P. O. Amorim, L. H. E. Castro, C. M. R. Sant'Anna, M. C. C. de Oliveira, D. Cesarin-Sobrinho, J. C. Netto-Ferreira, A. B. B. Ferreira, Molecules. **2015**, 20, 19526-19539.

23- I.E. Borissevitch, T.T. Tominaga, H. Imasato, M. Tabak, J. Luminesc. **1996**, 69, 65-76.

24- W. He, Y. Li, J. Tian, H. Liu, Z. Hu, X. Chen, J. Photochem. Photobiol. A: Chem. **2005**, 174, 53-61.

25- P. D. Ross, S. Subramanian, Biochemistry **1981**, 20, 3096-3102.

26- Z. Cheng, R. Liu, X. Jiang, Spectrochim. Acta Mol. Biomol. **2013**, 115, 92-105.

27- J. Li, J. Li, Y. Jiao, C. Dong, Spectrochim. Acta Mol. Biomol. **2014**, 118, 48-54.

28- S. Y. Venyaminov, J. T. Yang. In Determination of protein secondary structure. Circular dichroism and the conformational analysis of biomolecules, ed. by G. D. Fasman, Plenum Press, New York, EUA, **1996**, pp. 69-80.

29- P. Yang, F. Gao. The principle of bioinorganic chemistry, Science Press, Beijing, **2002**, pp. 349-360.

30- W.Y. He, Y. Li, H. Z. Si, Y. M. Dong, F. L. Sheng, X. J. Yao, Z. D. Hu, J. Photochem. Photobiol. A: Chem. **2006**, 182, 158-165.

31- A. Varlan, N. Hillebrand, Mol. **2010**, 15, 3905-3919.

32- Y. Yue, Y. Zhang, J. Qin, X. Chen, J. Mol. Struc. **2008**, 888, 25-30.

33- Y. Yue, Y. Zhang, Y. Li, J. Zhu, J. Qin, X. Chen, J. Luminesc. **2008**, 128, 513-516.

34- B.K. Paul, A. Samanta, N. Guchhait, J. Phys. Chem. B, **2010**, 114, 6183–6196.

35- M. Fasano, S. Curry, E. Terreno, M. Galliano, G. Fanali, P. Narciso, S. Notari, P. Ascenzi, IUBMB Life **2005**, 57, 787-796.

36- M.J.S. Dewar, E.G. Zoebisch, E.F. Healy, J.J.P. Stewart, J. Am. Chem. Soc. **1985**, 107, 3902-3909.

37- S.R. Feroz, S.B. Mohamad, G.S. Lee, S.N.A. Malek, S. Tayyab, Phytomedicine **2015**, 22, 621-630.

38- C.A.C. Ferreira, V.F. Ferreira, A.V. Pinto, R.S.C. Lopes, M.C.R. Pinto, A.J.R. Silva, An. Acad. Bras. Ci. **1987**, 59, 5-8.

39- G. Jones, P. Willett, R.C. Glen, A.R. Leach, R. Taylor, Journal of Molecular Biology **1997**, 267, 727-748.

40- O. Korb, T. Stützle, T.E. Exner, J. Chem. Inf. Model. **2009**, 49, 84-96.

9

Chemistry, synthesis and progress report on biological activities of thiadiazole compounds

Mohammad Asif

Department of Pharmacy, GRD (PG) Institute of Management and Technology, Dehradun, (Uttarakhand), 248009, India

Abstract: Thiadiazoles are an important class of heterocyclic compounds that exhibit diverse applications in organic synthesis, pharmaceutical and biological applications. They are also useful as oxidation inhibitors, cyanine dyes, metal chelating agents, anti-corrosion agents. Researchers across the globe are working on this moiety due to their broad spectrum of applications of thiadiazole chemistry. This article provides information about developments, exploration, synthetic strategies, techniques for the synthesis of thiadiazoles and their diverse biological activities, structure-activity relationship of the compounds and physical properties. This article is an important tool for organic and medicinal chemists to develop newer thiadiazole compounds that could be better agents in terms of efficacy and safety.

Key words: Thiadiazoles, synthesis, biological activities.

Introduction

The five-member heterocyclic compounds; particularly nitrogen and sulphur heterocycles; thiadiazoles have been successfully tested against several diseases and therefore received special attention in pharmaceutical chemistry due to their diverse potential applications [1-2]. Among the different thiadiazoles; more information about the synthesis and applications of 1,3,4-thiadiazoles is available in the literature, relatively less about 1,2,5-thiadiazoles. But there is a scanty of information is there about 1,2,3-thiadiazoles and 1,2,4-thiadiazoles. The resistance towards available drugs is rapidly becoming a major worldwide problem. The need to design new compounds to deal with this resistance has become one of the thrust areas of research today. Thiadiazoles continuously draws interest for development of newer drug moiety. Researchers have demonstrated a broad spectrum of biological properties of thiadiazoles in both pharmaceutical and agrochemical fields. Compounds having thiadiazole nucleus have wide spectrum of pharmacological activities such as antimcirobial, antitubercular, antileishmanial, anti-inflammatory, analgesic, CNS depressant, anticonvulsant, anticancer, antioxidant, antidiabetic, molluscicidal, antihypertensive, diuretic, analgesic properties. For instances, 1,3,4-thiadiazole derivatives have demonstrated a broad spectrum of biological properties in both pharmaceutical and agrochemical fields. They have known to exhibit diverse biological activities such as in vitro inhibition of cyclooxygenase and 5-lipoxygenase activities [3]. New acylated 5-thio-beta-D-glucopyranosylimino-disusbstituted 1,3,4-thiadiazoles prepared by cycloaddition of the glycosyl isothiocyanate with the reactive intermediates 1-aza-2-azoniaallene hexachloro antimonates, and have been tested in vitro antiviral activity against HIV-1, HIV-2, human cytomegallovirus (HMCV) [1-20].

Thiadiazole compounds show various types of biological activity among them 2,5-disubstituted 1,3,4-thiadiazoles are associated with diverse biological activity probably virtue of –N=C-S-grouping. Therapeutic importance of these rings prompted us to develop selective molecules in which substituent could be arranged in a pharmacophoric pattern to display higher pharmacological activities. Thiadiazoles have occupied an important place in drug industry, 1,3,4-thiadiazoles have wide applications in many fields [5]. 1,3,4-thiadiazole derivatives possess interesting biological activity probably conferred to them due to strong aromaticity of the ring system which leads to great *in vivo* stability and generally, a lack of toxicity for higher vertebrates, including humans when diverse functional group that interact with biological receptor are attached to aromatic ring [6]. Approach to practice of medicinal chemistry has developed from an empirical one involving synthesis of new organic compounds based on modification of chemical compounds of known biological activities could be better explored. It is well established that slight alteration in the structure of certain compounds are

Corresponding author: Mohammad Asif
Email address: aasif321@gmail.com

able to bring drastic changes to yield better drug with less toxicity to the host it observed that chemical modification not only alters physiochemical properties but also pharmacological properties [7].

The development of 1,3,4-Thiadiazole chemistry is linked to the discovery of phenylhydrazines and hydrazine in the late nineteenth century. The first 1,3,4-Thiadiazole was described by Fischer in 1882 but the true nature of the ring system was demonstrated first in 1890 by Freund and Kuh. There are several isomers of thiadiazole, that is 1,2,3-Thiadiazole (1), 1,2,5-Thiadiazole (2), 1,2,4-Thiadiazole (3) and 1,3,4-Thiadiazole (4). 1,3,4-Thiadiazole is the isomer of thiadiazole series. A glance at the standard reference works shows that more studies have been carried out on the 1,3,4 Thiadiazole than all the other isomers combined. Members of this ring system have found their way in to such diverse applications as pharmaceuticals, oxidation inhibitors, cyanide dyes, metal complexing agents [8-10]. The ending -azole designates a five membered ring system with two or more heteroatoms, one of which is Nitrogen. The ending –ole is used for other five membered heterocyclic ring

without Nitrogen. The numbering of monocyclic azole system begins with the heteroatom that is in the highest group in the periodic table and with the element of lowest atomic weight in that group. Hence the numbering of 1,3,4-Thiadiazole (4) is done in following manner. This designates that one sulphur group is present in the ring [11,12]. Apart from the pharmacological applications, thiadiazoles and their derivatives have been known to exhibit varied physical properties such as exhibit anticorrosion, liquid crystal, optical brightening and fluorescent properties which were discussed in this review article.

Chemistry of Thiadiazole

Thiadiazole moiety act as a "hydrogen binding domain" and "two-electron donar system". Thiadiazole act as a bioisosteric replacement of thiazole moiety. So, it acts as third and fourth generation cephalosporin. Thiadiazole is a five membered ring system containing sulphur and nitrogen atom. They occur in four isomeric forms (1-4). Its dihydro derivative provides bulk of literature on thiadiazole [13].

1,2,3-Thiadiazole (1)   1, 2,4-Thiadzole (2)   1,2,5-Thiadiazole (3) 1,3,4-Thiadiazole (4)

Physical properties of -1, 3, 4-thiadiazoles

Structure and Aromatic Properties

Microwave spectra of 1,3,4-thiadiazole and three isotopically substituted species. They could determine the structure of the molecule with an uncertainty of 0.03 A^o in the coordinates of the hydrogen atom and of less than 0.003 A^o in the coordinates of the other atoms. By an analysis of difference between the measured bond lengths and covalent radii, the author came to the conclusion that the aromatic character, as measured by the π-electron delocalization decreases in the order –1,2,5-thiadiazole > thiophene > 1,3,4-thiadiazole > 1,2,5-oxadiazole [14].

Dipole Moment

The dipole moment of 1,3,4-thiadiazole in the gas phase by microwave technique and found a value

of 3.28+-0.03 D. By use of geometry, the π-electron distribution and the bond moment, dipole moment of 3.0 D can be calculated, directed from the sulphur atom towards the center of the nitrogen-nitrogen bond [15-17].

Recent Strategies in the Synthesis of 1, 3, 4-thiadiazoles

Recent strategies on the synthesis of 1,3,4-Thiadiazole derivatives can be summarized in to following points:
Many synthesis of the 1,3,4-Thiadiazole proceed from thiosemicarbazide or substituted thiosemicarbazide.
Thiosemicarbazide cyclizes directly to 2-amino-5-methyl-1,3,4-thiadiazole with acetyl chloride. This simple route to 2-amino 5-substituted-1,3,4-thiadiazole seems to be quite general [18]. In the example shown R may be methyl [18], norhydnocarpyl [19], benzyl [20], cyclopropyl [21] and many others.

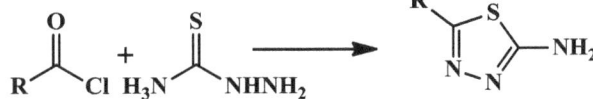

Scheme 1

The acetyl chloride could bring about the cyclization of alkyl- or arylsubstituted thiosemicarbazide. The action of acetyl chloride on

4-methylthiosemicarbazide produces 5-methyl-2-methylamino-1,3,4-thiadiazole [22].

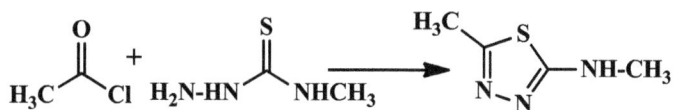

Scheme 2

Formic acid could cyclize the alkanoyl halides by acylation. He found that by heating 4-phenylthiosemicarbazide with formic acid, 2-anilino-1,3,4-thiadiazole was formed [22].

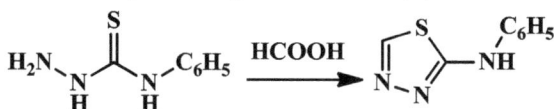

Scheme 3

A number of 2-amino-5-aryl-1,3,4-thiadiazole using phosphoric acid as the dehydrating agents. An example of smooth cyclization in high yield by phosphoric acid is the formation of 2-benzamido-5-phenyl-1,3,4-thiadiazole from 1,4-dibenzoylthio-semicarbazide [23].

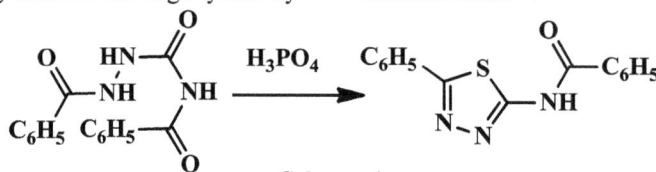

Scheme 4

2-amino-5-mercapto-1,3,4-thiadiazole was developed. When thiosemicarbazide is treated with carbon disulphide and potassium hydroxide, the potassium salt of thiosemicarbazide-4-dithiocarboxylic acid is formed. Heating this potassium salt of thiosemicarbazide-4-dithiocarboxylic acid to 140°C causes cyclization to the salt of 2-amino-5-mercapto-1,3,4-thiadiazole [24].

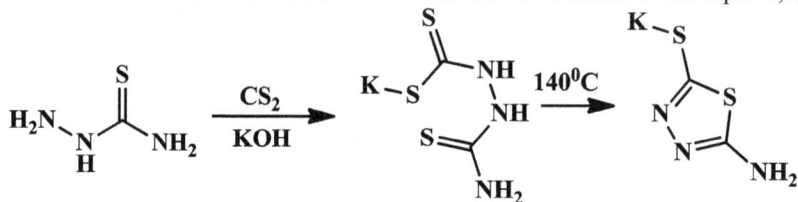

Scheme 5

In certain instances neutral carbon disulphide react directly with thiosemicarbazide to form aminomercaptothiadiazoles. A modification of the carbon disulphide-thiosemicarbazide procedure which results in higher yield of 2-amino-5-marcapto-1,3,4-thiadiazole is carried out in dimethylformamide at 80°, the yield is over 90% [25].

The benzalthiosemicarbazones could be oxidatively cyclize to form 2-amino-5-phenyl-1,3,4-thiadiazole by ferric chloride [25]. A large number of 5-substituted 2-amino-1,3,4-thiadiazole have been prepared by this procedure [26].

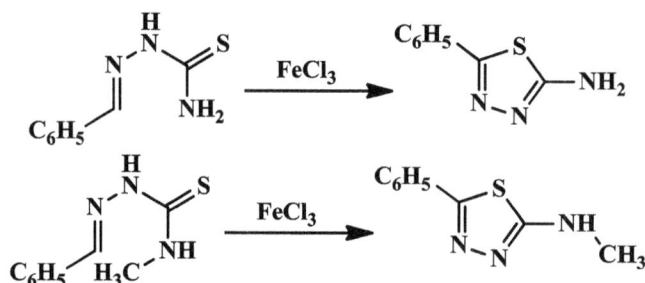

Scheme 6

A number of aldose thiosemicarbazones could be converted to thiadiazole derivatives by Young and Eyre method [27].

There are two method by which 1, 3, 4-thiadiazole can be prepared from thiocarbazides.
If 1-phenylthiocarbazide is heated with formic acid, it is converted to 2-phenylhydrazino-1,3,4-thia-diazole [28].

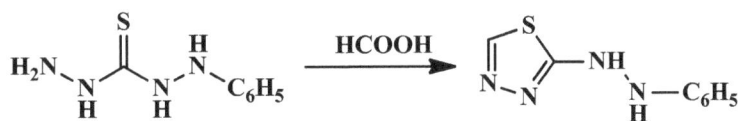

Scheme 7

This method is related to the oxidation of 1-phenylbenzalthiocarbazone to 2-phenyl-5-phenyl hydrazino-1,3,4-thiadiazole [29].

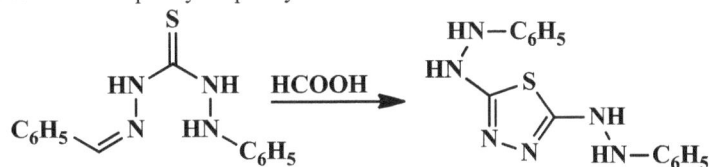

Scheme 8

Following methods have been reported for the preparation of 1,3,4-thiadiazole from dithiocarbazates.

Another route to 1,3,4-thiadiazole is via substituted dithiocarbazic acid and their esters. A reaction which belongs in this group is the formation of 2,5-dimercapto-1,3,4-thiadiazole by action of carbon disulphide on hydrazine in basic medium [30,31].

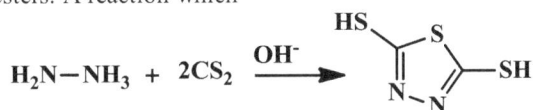

Scheme 9

When 3-acyldithiocarbazic esters are treated with acids, they cyclize to form substituted thiadiazoles. This is a quite general reaction. Both benzyl and methyl 3-acyldithiocarbazates have been employed [32,33].

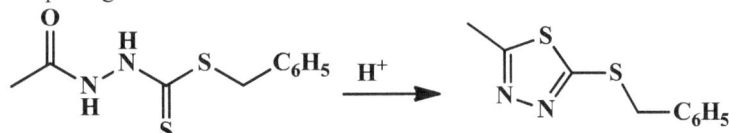

Scheme 10 A

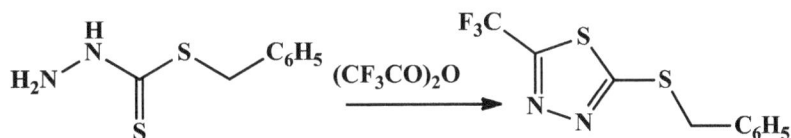

Scheme 10 B

Thioacylhydrazines may often serves as starting materials for the preparation of 1,3,4-thiadiazole. If thiobenzoylhydrazine is heated with ethyl orthoformate, 2-phenyl-1,3,4-thiadiazole is formed. If ethyl orthoacetate is substituted for the orthoformate, 2-methyl-5-phenyl-1, 3, 4-thiadiazole is obtained [34,35].

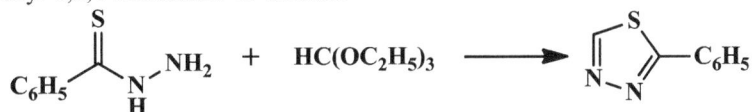

Scheme 11

Thiobezhydrazide is smoothly converted to 2-phenyl-1, 3, 4-thiadiazole by the action of formic acid [27]. Thiobenzhydrazide is form 2, 5-diphenyl-1,3,4-thiadiazole (33) in small amount when warmed in benzene [36].

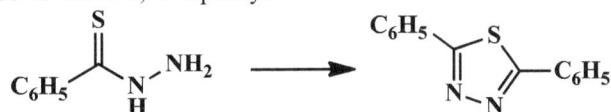

Scheme 12

Stolle obtained 2,5-diphenylthiadiazole (36) by a variety of methods. He found that benzoyl-hydrazine [37] or N,N'-dibenzoylhydrazines [38] react with phosphorus pentasulfide to form 2,5-diphenyl-1,3,4-thiadiazole.

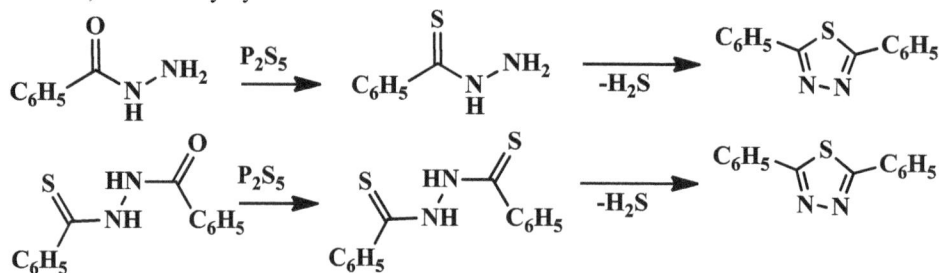

Scheme 13

The reaction of N, N'-diacylhydrazine with phosphorus pentasulfide was used by Stolle and his students for the preparation of a large number of 2,5-disubstituted 1,3,4-thiadiazole [39,40].

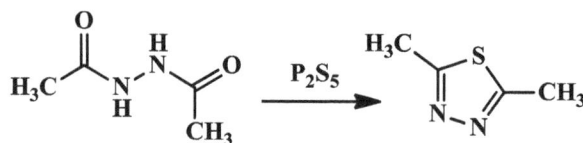

Scheme 14

Bithiourea and substituted bithiourease have been converted to 1,3,4-thiadiazole by several methods.

Bithiourea, when treated with 3% hydrogen peroxide is cyclized to 2,5-diamino-1,3,4-Thiadiazole [41].

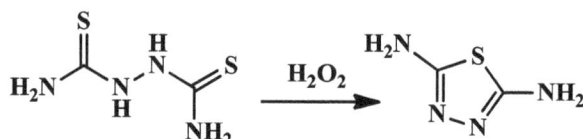

Scheme 15

Acetic anhydride acts on bithiourea to form a diacetyl derivative of 2, 5-diamino-1, 3, 4-Thiadiazole. The acetyl group is easily removed by hydrolysis to give the parent thiadiazole [42].

Synthesis of 1,3,4-thiadiazoles

The usual or classical method of synthesis of thiadiazoles involves the condensation of thiosemicarbazides with carboxylic acids or carboxylic acid chlorides or carboxylic acid esters with cyclising or condensing agents such as phosphorus oxychloride, phosphorus pentachloride, acetic anhydride, sulphuric acid etc. For instance; The reaction of 6-chloro-1,3-benzothiazol-2-yl semicarbazide, aromatic acid in $POCl_3$ produces 2-aryl-5-(6-chloro-1,3-benzothiazol-2-yl-amino-1,3,4-thiadiazoles in good yield. The precursor 6-chloro-1,3-benzothiazol-2-yl semicarbazide was obtained by the reaction of 6-Chloro-2-amino benzothiazole, CS_2 and hydrazine hydrate in ethanol and ammonia solution. The synthesized thiadiazoles have showed significant antimicrobial activities [43].

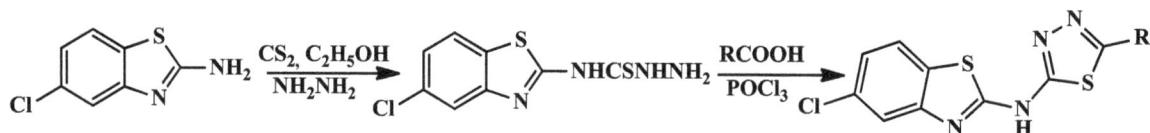

Scheme 16

A series of N-(5-phenyl)-1,3,4-thiadiazole-2-ylbenzamide derivatives synthesized from thiosemicarbazide and benzoyl chloride in phosphorous penta chloride. The synthesized compounds have been evaluated for their analgesic activity, study revealed that all the animals receive 0.6%v of 10ml/kg body weight of acetic acid intraperitonially and number of writhing was recorded after 10 min upto next 15 min. the same groups animals were used next day for evaluating analgesic activity [44]. 2-Amino-5-aryl-1,3,4-oxadiazoles were prepared by heating a mixture of aromatic carboxylic acids, thiosemicarbazine and conc. sulphuric acid, then these were converted to schiffs bases by irradiating a mixture of 2-amino-5-aryl-1,3,4-oxadiazoles and aldehydes for 3 min at 40% power. The products showed promising antidiabetic activity [45].

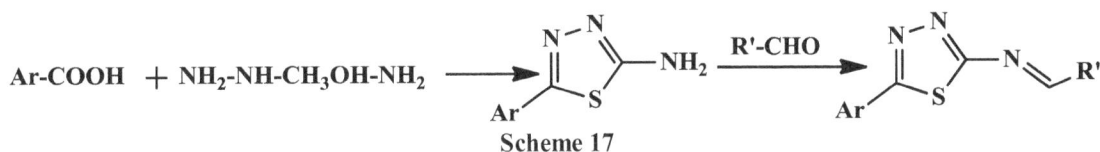

Scheme 17

A series of S-[5-(phenylamino)-1,3,4-thiadiazole-2- yl] benzenecarbothioate and S-[5-(phenyl amino)-1,3,4-thiadiazole-2-yl] ethanethioate were prepared by refluxing benzoyl chloride and acetyl chloride in presence of potassium carbonate with 5-(phenyl amino)-1,3,4-thiadiazole-2-thiol.

5-(Phenylamino)-1,3,4-thiadiazole-2-thiol were prepared by cyclization of arylthio-semicarbazide with carbondisulphide. Some of these thiadiazole derivatives exhibited significant antibacterial and antifungal activities [46].

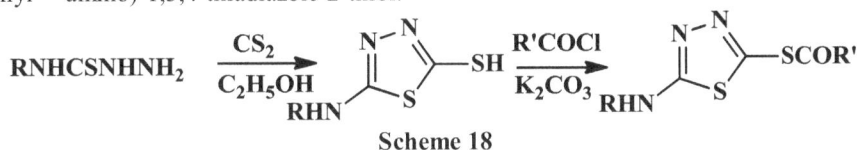

Scheme 18

Cyclization of the thiosemicarbazones with acetic anhydride produced 4,5-dihydro-1,3,4-thiadiazolyl derivatives. These compounds were evaluated for inhibitory effect on tyronase enzyme and results indicated some of these thiadiazole derivatives possess moderate inhibitory effect on

tyronase enzyme [47]. Thionation of N,N'-acylhydrazines with the use of a fluorous Lawesson's reagent leads to 1,3,4-thiadiazoles in high yields. The isolation of the final products is achieved in most cases by a simple filtration [48].

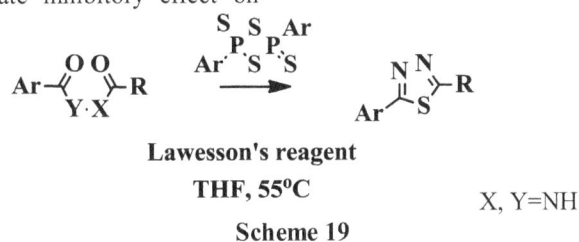

Lawesson's reagent

THF, 55°C

X, Y=NH

Scheme 19

In order to to improve the yield and purity of the products, easy isolation or work up; researchers developed the new synthetic strategies, innovative methods, new reagents for the synthesis of thiadiazoles. For instance, Rai and co-workers introduced thiourea as a new reagent for the direct conversion of 2.5-diaryl- 1,3,4-oxadiazole to 2.5-diaryl-1,3,4-thiadiazole. They observed that, when

the reaction of 1,3,4-oxadiazoles with thiourea was carried out at retlux temperature for 3 to 4 days, only 2 to 5% of oxadiazoles gets converted to thiadiazoles. In order to reduce the reaction time and to increase the yield, they carried out in a sealed tube at water bath temperature for 10-15 hr and obtained the yield in 65-72% [49].

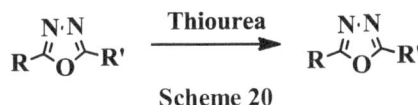

Scheme 20

A method of using thiourea as thionating agent for the transformation of oxadiazoles to thiadiazoles has been widely accepted and implemented. For instance, the unsymmetrical 1,3,4-oxadiazole when

treated with two fold excess thiourea in tetrahydrofuran produced 2-(benzylsulfonylmethyl)-5-(arylsulfonylmethyl)-1,3,4-thiadiazole [50].

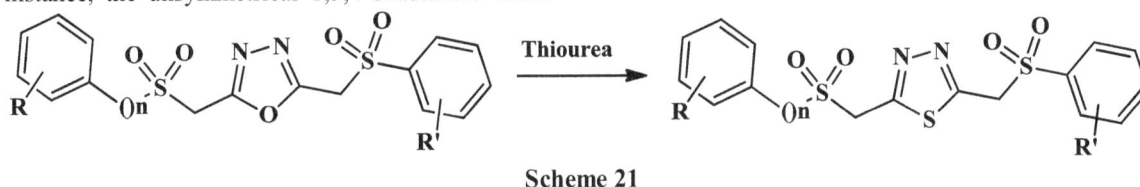

Scheme 21

A series of fluorine-containing thiadiazoles were synthesized from thiosemicarbazides by conventional method by heating mixture of thiosemicarbazide and 2N sodium hydroxide, by green synthesis such as ultrasonification and

microwave irradiation. The ultrasonication method, the reaction mixture was subjected to ultrasonic irradiated for 30-35 min at room temperature. The products obtainned in all the three methods were compared, and the study reports that the green synthesis yielded more percentage of yield. Other than this these methods are environment frienadly and economically cheaper. The thiadiazoles synthesized have exhibited antimicrobial activity [51]. A series of thiadiazole derivatives synthesized found

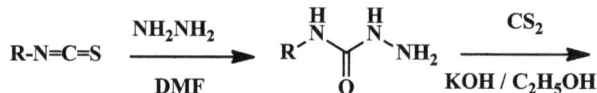

potential allosteric, substrate competitive inhibitors of the protein kinase JNK. The study showed that these compounds are potent and selective JNK inhibitors targeting its JIP-1 docking site [52].

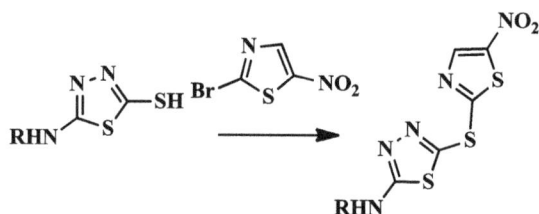

The microwave (MW) irradiation provide enhanced reaction rate and improved product field in chemical synthesis and has been extending to modern drug discovery in complex multi-step synthesis and it is proving quite successful in the formation of a variety of carbon-heteroatom bonds. For instance, using this MW irradiation technique; 4-(Substituted benzylidene)-1-(5-mercapto-1,3,4-thiadiazol-2-yl)-2-phenyl-1H-imidazol-5(4H)-one was prepared by the condensation reaction of 4-arylidene-2-phenyloxazol-5(4H)-one and 5-amino-1,3,4-thiadiazole-2-thiol [53].

Thiosemicarbazides reacted with tetracyano-ethene in ethyl acetate with admission of air to form the 7-amino-2-organylimino-2,3-dihydro-1,3,4-thiadi azepine-5,6-dicarbonitriles), 7-amino-1-organylimino-3-oxo-pyrazolo[1,2-c]-1,3,4-thiadiazole-5,5,6-tricarbonitriles, 7-amino-1-organyl-imino pyra zolo[1,2-c]-1,3,4-thiadiazole-3,3,5,5,6-pentacarbo nitriles in moderate yields. Rationales for the observed conversations are presented [54].

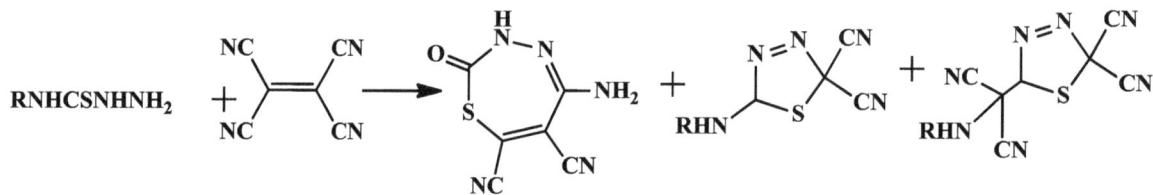

Scheme 22

5-(4-Fluoro-3-nitrophenyl)-1,3,4-thiadiazol-2-ylamine, on reflux with 4-methoxyphenacyl bromide in ethanol as solvent yielded 2-(4-fluoro-3-

nitrophenyl)-6-(4-methoxyphenyl)-imidazo[2,1-b]-1,3,4-thiadiazole [55].

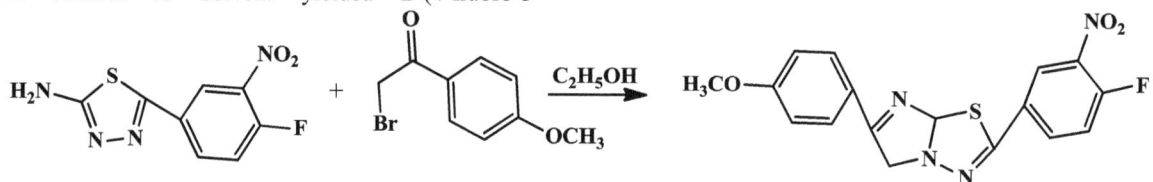

Scheme 23

A large number of 1,3,4-thiadiazoles have been reported to exhibit antidiabetic properties. For instance; 2-Amino-5-aryl-1,3,4-thiadiazole synthesized by the reaction of thiosemicarbazide, aromatic carboxylic acid in conc. sulphuric acid. Then the compound 2-Amino-5-aryl-1,3,4-

thiadiazole was converted to chloroacetyl derivative by its reaction with chloroacetyl chloride in the presence of sodium acetate in acetic acid. Finally it was transformed in to N-(5-(4-aminophenyl)-1,3,4-thiadiazole-2-yl)-2-chloroacetamide [56].

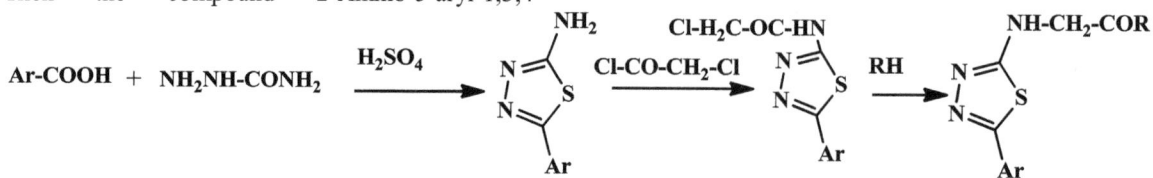

Scheme 24

The compounds synthesized were evaluated for their antidiabetic activity using wistor albino rats by Alloxan induced tail tipping method. The results of

the study revealed that the synthesized compounds exhibited significant antidiabetic activities.

Synthesis of 1.3,4-thiadiazoles

Thiadiazoles can be synthesized from mainly thiosemicarbazide or hydrazide that is thiadiazole can be cyclized from thiosemicarbazide or hydrazide by methods like conventional method, ultrasound or microwave using catalyst like H_2SO_4, $POCl_3$, CS_2, polyphosphoric acid and HCl. Important new general routes of 1,3,4-thiadiazole have been reported, The

major routes are: 2-(2-(4-chlorophenyl)acetyl)-N-aryl hydrazine carbothioamides were prepared by reacting 4-chlorophenyl acetyl hydrazide and aryl isothiocyanate in the presence of ethanol. Various 5-(4-chloro-benzyl)-4-aryl-4H-1,2,4-triazole-3-thiols 2,5-(4-chloro benzyl)-N-aryl-1,3,4-thiadiazole-2-amine have been prepared by the cyclization with sodium hydroxide, sulphuric acid and iodine in potassium iodide in presence of sodium hydroxide [57].

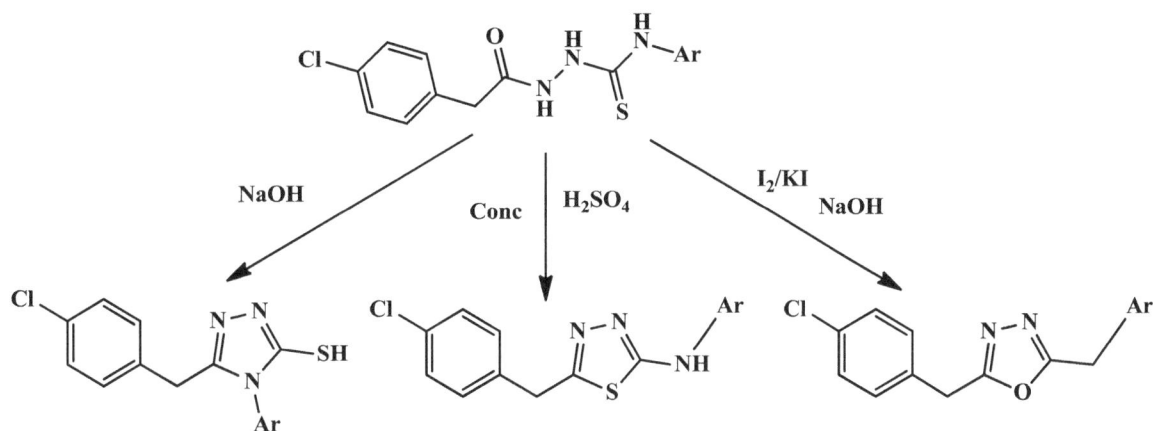

Scheme 25

The 5-(pyridine-4yl)-1,3,4-thiadiazole-2-amine has been synthesized by reacting isonicotino-hydrazide with potassium thiocyanate on further cyclo condensation with concentrated sulphuric acid.

The compound reacted with various aromatic aldehydes in the presence ethanol which on further cycloaddition with chloroacetyl chloride and triethylamine in DMF [58].

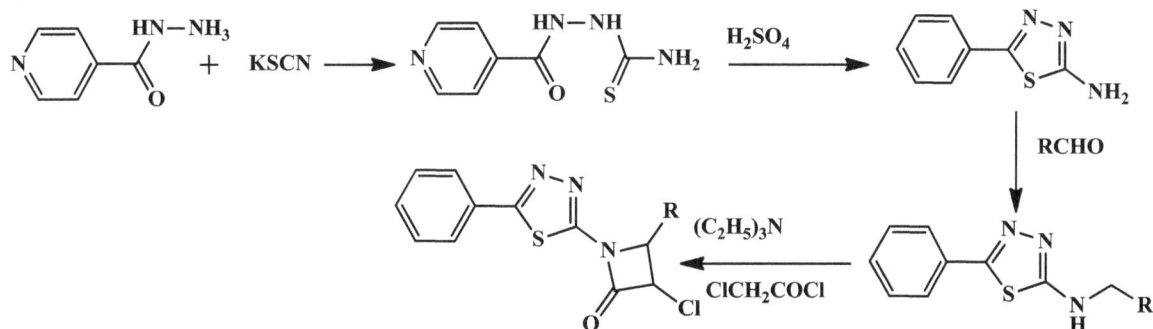

Scheme 26

The 3,6,-disubstituted 1,2,4-triazolo(3,4-b)-1,3,4-thiadiazole from 3-substitued-4-amino-5-mercapto-1,2,4-triazoles and 3-substituted 4-caboxy pyrozoles, naphthyl oxymethyl and flurophenyl group as substituent. Presence of fluorosubstituent

and aromatic naphthalene ring was found to enhance activity. The difference in electro negativity between fluorine and carbon created a large dipole moment which contributed to the molecule ability to be engaged in intermolecular interactions [59].

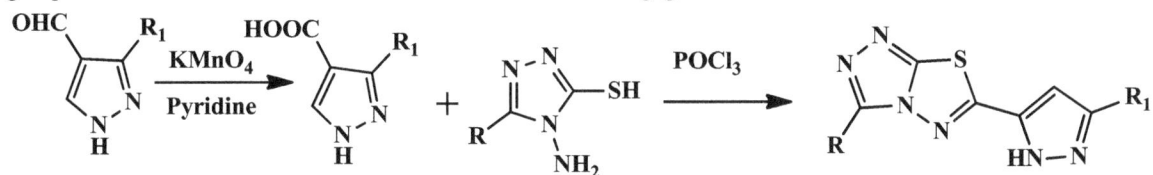

Scheme 27

The 4-amino-5-benzyl-4H-1,2,4-triazole-3-thiol with 5- methyl-1-aryl-1H-1,2,3-trazole-4-carboxylic acids in phosphorus oxychloride. It was established reaction performed with closuring thiadiazole ring. Thus by the reaction of 4-amino-5-benzyl-4H-1,2,4-

triazole-3-thiol with 5-methyl-1-aryl-1H-1,2,3-triazole-4-carboxylic acid new 3-benzyl-6-(5-methyl-1-phenyl-1H-1,2,3-triazole-4-yl)(1,2,4)-triazolo(3,4-b)(1,3,4)thiadiazole [60].

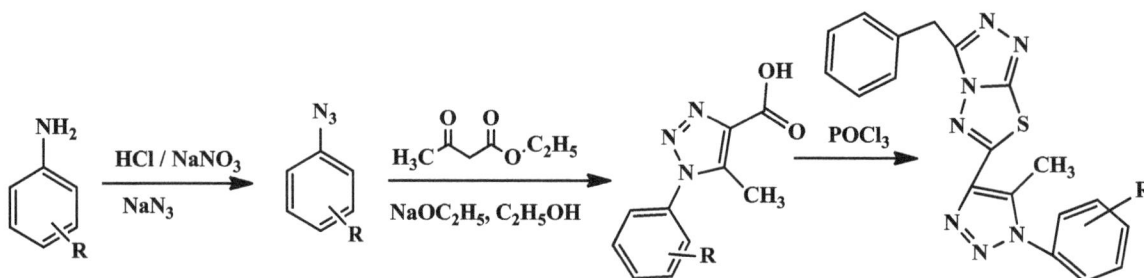

Scheme 28

The thiosemicarbazide with carbon disulphide and DMF under result formation of 5-(3-aryl-1*H*-pyrazole-5-yl)-2-mercapto-1,3,4-thiadiazoles [61].

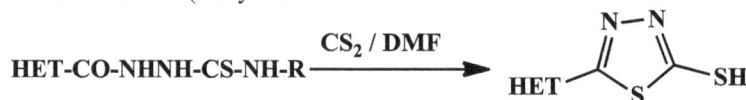

Scheme 29

The thioamides were treated with hydrazine hydrate followed by carbon disulphide solution. The reaction mixture was irradiated in a microwave oven to yield 5-substituted-2-mercapto-1,3,4-thiadiazoles [62].

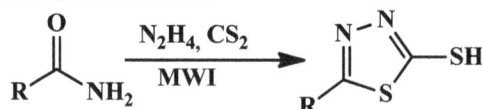

Scheme 30

Reactivity of the 1, 3, 4-thiadiazoles

Rearrangements and Ring Opening Reaction

The 1,3,4-thiadiazole ring is rather susceptible to attack by strong neucleophile. Thus the parent compound is stable to acids but is readily cleaved by bases [63]. 2-Amino- and 2-hydrazino-1, 3, 4-thiadiazole can be rearranged to 1,2,4-triazoline-3(2)-thiones. Goerdeler and Galinke43 showed that 2-amino- and 2-methylamino-1, 3, 4-thiadiazole (R=H and CH$_3$) are rearranged by methylamine in methanol at 150 °C to the isomeric triazolinethiones.

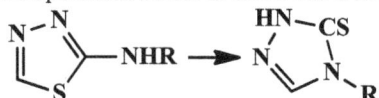

2-Amino-1,3,4-thiadiazole (R=H), when refluxed with benzyl amine in xylene, gave a mixture of about equal amount of 2-benzylamino-1,3,4-thiadiazole (R=CH$_2$Ph) and 4-benzyl-1,2,4-triazolin-3(2)-thione (R=CH$_2$Ph). The same two compounds were formed in the reaction between 2-chloro-1, 3, 4-thiadiazole and benzylamine [64,65].

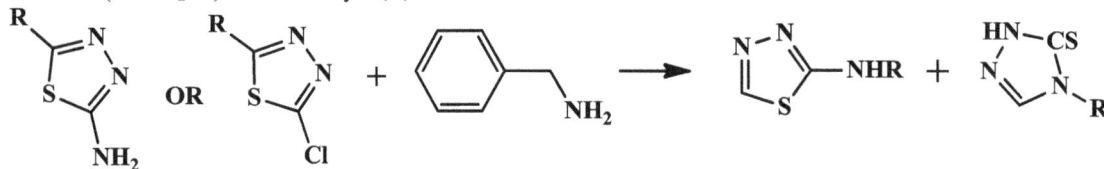

Similarly, 2-alkyl-5-chloro-1,3,4-thiadiazole reacted with a large excess of hydrazine hydrate on heating to give 4-amino-1,2,4-triazolin 4-amino-1,2,4-triazolin-3(2)-thiones.

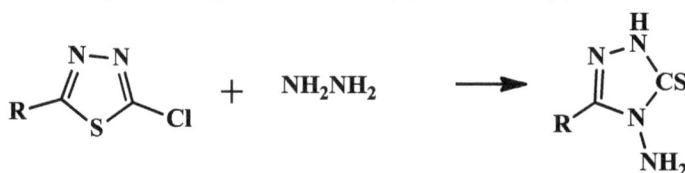

Under the same conditions, 2-amino-5-chloro-1,3,4-thiadiazole and 2-amino-1,3,4-thiadiazolin-5(4)-thione gave a mixture of 3,4-diamino-1,2,4-triazoline-5(1)-thione and 3-hydrazino-4-amino-1,2,4-triazoline-5(1)-thione. 2,5-Dichloro- and 2,5-dimercapto-1,3,4-thiadiazole gave only [66].

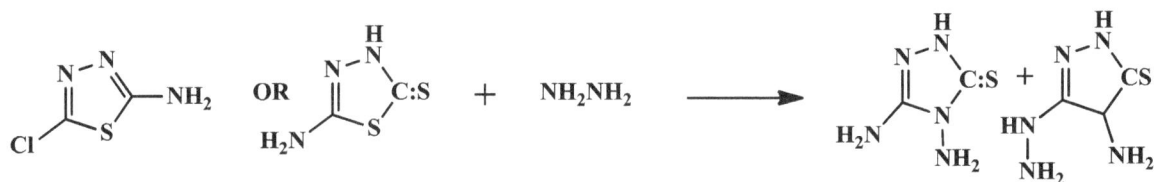

Similar rearrangements can be affected by acids. When 1-benzyl-1-(1,3,4-thiadiazole-2-yl) hydrazine was refluxed with dilute hydrochloric acid, the triazolinethion (R=H) was formed in quantitative yield. When the reaction was performed in the presence of some acetic acid, a mixture of (R=H) and (R=CH₃) was formed [67].

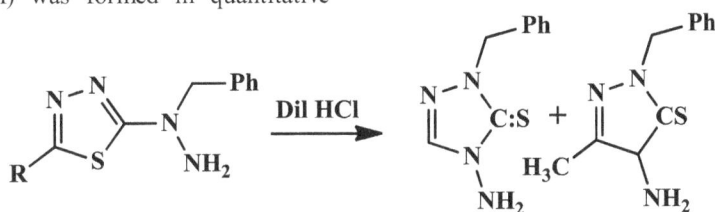

In this acid catalyzed rearrangement

2-benzylthiocarbohydrazide is likely an intermediate.

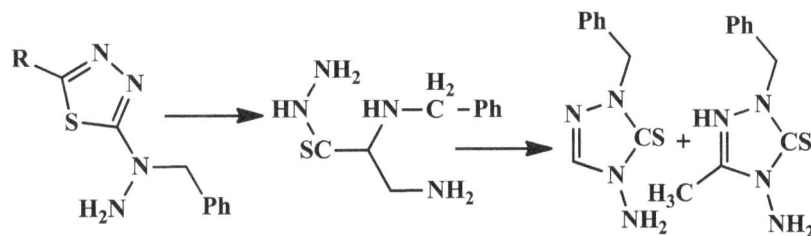

The rearrangement of by benzyl amine probably proceeds with ring opening to an amidrazone

followed by recyclization to (R=CH₂Ph).

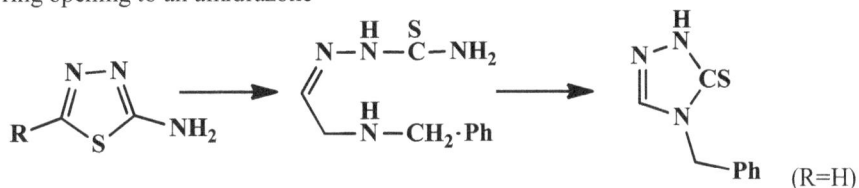

Substitution Reaction

Although the 1,3,4-thiadiazole ring is classed as π-excessive according to Albert [68], the presence of two nitrogen atoms of pyridine type in the ring leaves the carbon atoms with rather low electron density, and consequently no electrophilic substitution in the unsubstituted 1,3,4-thiadiazole ring have been recorded. A bromine adduct of the simple 1,3,4-thiadiazole, but it decomposed and lost bromine in the air. Nitration, even under drastic condition could not be achieved [63]. The 2-phenyl-1, 3, 4-thiadiazole to a mixture of concentrated nitric acid and sulphuric acid at 0 °C and obtained a mixture of the three isomeric 2-nitrophenyl-1,3,4-thiadiazole in the ratio p: m: o = 2:3:1, but no 2-phenyl-5-nitro-1,3,4-thiadiazole [69]. A 2-amino group does activate the ring towards electrophilic agents, prepared 2-amino-5-bromo-1,3,4-thiadiazole by bromination of 2-amino-1,3,4-thiadiazole in 40% hydro-bromic acid. The product was not isolated but was diazotized to give 2,5-dibromo-1,3,4-thia-diazole [70].

Recent advancement in the therapeutic potential of thiadiazole derivatives

During recent years there has been intense investigation of different classes of thiadiazole compounds, many of which known to possess interesting biological properties such as antimicrobial, antituberculosis, anti-inflammatory, anticonvulsants, antihypertensive [71-76], antioxidant, anticancer and antifungal [77-81] activity.

Analgesic and Anti-inflammatory Activity

Several 3,6-disubstituted-1,2,4-triazolo[3,4-b]-1,3,4-thiadiazole and their dihydro analogues showed anti-inflammatory and analgesic activity. Compounds **1** showed good anti-inflammatory and analgesic activities [72]. The 2-substituted-1,3,4-thiadiazoles, is 5-(2',4'-Difluoro-4-hydroxybiphenyl-5-yl)-4-(4-methoxyphenyl)-1,3,4-thiadiazole **(2)** presented good analgesic activity [82].

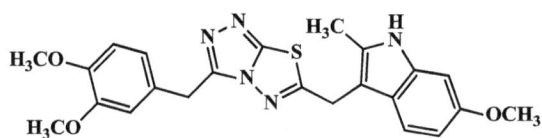

Compound 1

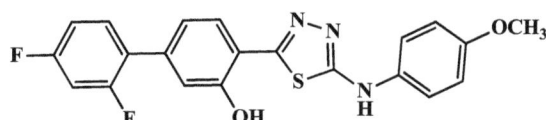

Compound 2

2A series of 2-trifluoromethyl/sulfonamido-5,6-diarylsubstituted imidazo [2,1-b]-1,3,4-thiadiazole derivatives, compounds **3** and **4** showed selective inhibitory activity toward COX-2 over COX-1. These compounds also exhibited significant anti-

inflammatory activity, which is comparable to that of celecoxib. Some 1,3,4-thiadiazole derivatives were found as analgesic effects of compound **5** were higher than those of both morphine and aspirin [83-85].

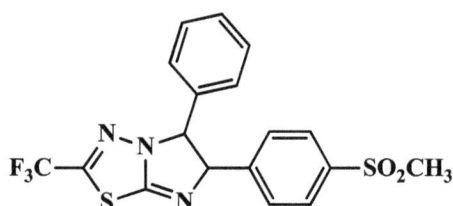

Compound 3

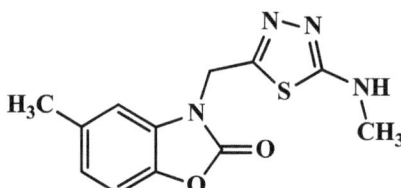

Compound 4

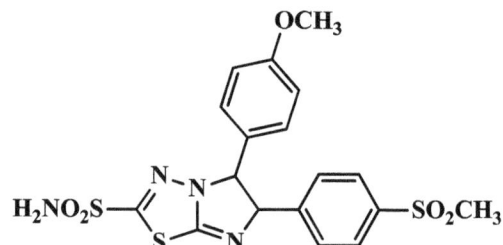

Compound 5

Currently available non-steroidal anti-inflammatory drugs (NSAIDs) like ibuprofen, flurbiprofen, fenbufen and naproxen exhibit gastric toxicity. Modification of the carboxyl function of representative NSAIDs resulted in increased anti-inflammatory activity with reduced ulcerogenic effect [86,87]. Certain compounds bearing 1,2,4-triazole and 1,3,4-thiadiazole nuclei possess significant anti-inflammatory activity with reduced

GI toxicity. Replace the carboxylic acid group of 2-(4-isobutylphenyl) propanoic acid and biphenyl-4-yloxy acetic acid by a composite system, which combines both the triazole and the thiadiazole nucleus. Seven cyclized compounds 6a, 6b, 7a, 7b, 8a, 8b and 8c were found to have anti-inflammatory properties comparable to their standard reference drugs ibuprofen and flurbiprofen [88].

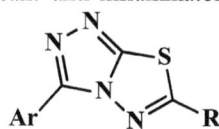

Compound 6

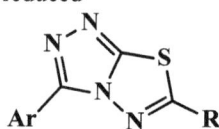

Compound 7

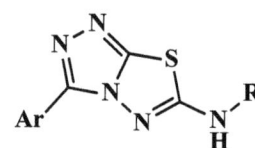

Compound 8

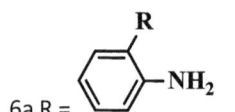

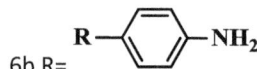

All compounds exhibited moderate to good analgesic activity. These compounds were also showed superior GI safety profile along with reduction in lipid peroxidation as compared with ibuprofen and flurbiprofen. Two series of N-[5-oxo-4-(arylsulfonyl)-4,5-dihydro-1,3,4-thiadiazol-2-yl]-amides (**9**) possess good analgesic activity and also fair anti-inflammatory activity. Ulcerogenic and

irritative action on the gastrointestinal mucosa, in comparison with indomethacin is low [89]. The 1,3,4-thiadiazole derivatives of diclofenac showed anti-inflammatory activity from 79.04% to 82.85%. The maximum activity (82.85%) was shown by thiadiazole derivative that is compound **10** having p-fluoro phenyl amino group at second position [90].

Compound 9

Compound 10

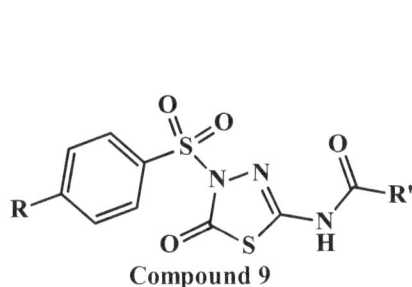

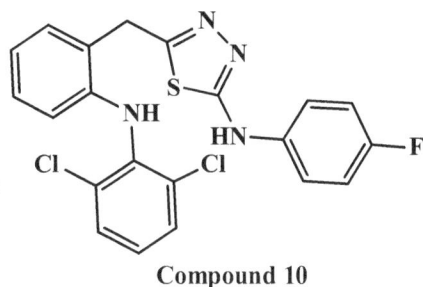

2-(2-oxobenzothiazolin-3-yl)methyl)-5-aminoalkyl/aryl-1,3,4-thiadiazoles (**11**) were showed analgesic and anti-allergic activity [91]. 2-trifluoromethyl/Sulfonamido-5,6-diaryl substituted imidazo(2,1-b)-1,3,4-thiadiazoles (**12**) showed

cyclooxygenase-2-inhibitiors activity and used as potential the anti-inflammatory activity using standard drug Celecoxib at concentration 10 mg/kg [92].

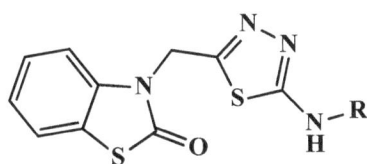

11

R=ethyl, methyl, allyl, phenyl,cyclohexyl

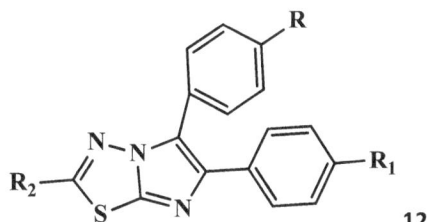

12

R=H,OCH$_3$; R$_1$= SCH$_3$, SO$_2$CH$_3$; R$_2$=CF$_3$, SO$_2$NH$_2$

Antimicrobial and Antifungal Activity

The methylene bridged benzisoxazolyl imidazo [2,1b] [1,3,4]-thiadiazoles were investigated as antibacterial and some compounds showed moderate to good bacterial inhibition. Particularly compounds **13a**, **13b**, **14a**, **14b** and **15a** have shown very good antibacterial activity. Compound **15a** has exhibited

highest antibacterial activity. The high activity is attributed to the presence of electron withdrawing chloro- and bromo- functional groups. Antifungal results indicated that compounds **13b**, **13c** and **15b** have shown good activity. Compound **13b** showed very good antifungal activity comparable to that of standard [93].

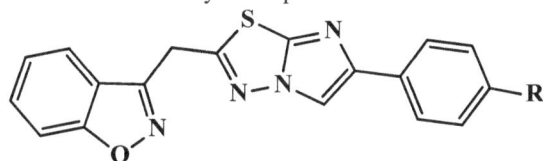

Compound 13a-c
13aR=Cl, 13bR=Br, 13cR=3-coumarinyl

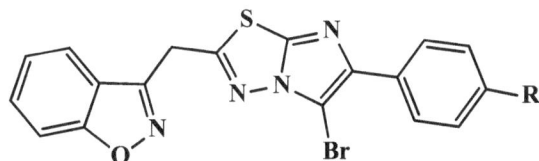

Compound 14a-b
14a R=Cl, 14b, R=NO$_2$

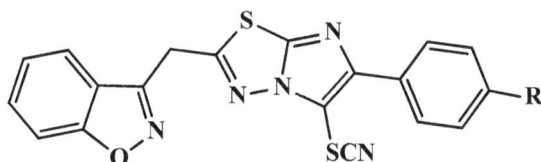

Compound 15a-b 15a, R=Cl, 15b, R=OCH$_3$

The 2-(arylmethanesulfonylmethyl)-5-aryl-1,3,4-thiadiazoles **16a–d**, 3-(arylmethane-sulfonyl methyl)-5-aryl-4H-1,2,4-triazol-4-amines 71a–d exhibited high activity (22–39 mm) on both Gram

(+ve) and Gram (-ve) bacteria. In fact, compounds **16d** and **17d** showed pronounced activity (31–39 mm) towards Gram (+ve) bacteria [94].

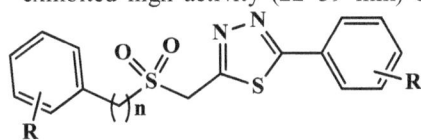

Compound 16a-d

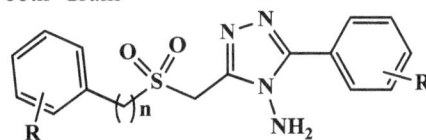

Compound 17a-d

a: R = H, R' = H
b: R = H, R' = 2-Cl
c:R=4-Cl,R'=H
d: R = 4-Cl, R' = 2-Cl
n = 0 (1,3,5,7,9,11)
n = 1 (2,4,6,8,10, 12)

Compounds 2-(4-chlorobenzylsulfonylmethyl)-5-(2-chlorophenyl)-1,3,4-thiadiazole (16d) displayed greater activity against spore germination of tested fungi A. niger, F. solani and C. lunata. Some sulfone-linked bis heterocycles, compounds 19a showed excellent activity against Gram-positive bacteria (inhibitory zone >25 mm), good activity against Gram-negative bacteria (inhibitory zone >20 mm). The compounds (19a-c) showed high inhibitory effect towards tested fungi [95]. Biological studies of *bis* thiadiazole/triazole (20) by sonication as potential antibacterial activity using standard drug Ampicillin Trihydrate at concentration 50μg/ml [96].

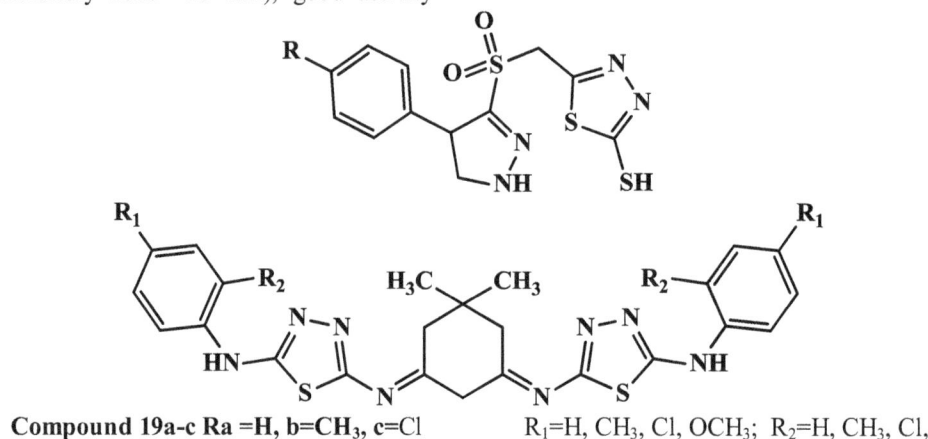

Compound 19a-c Ra =H, b=CH₃, c=Cl   R₁=H, CH₃, Cl, OCH₃; R₂=H, CH₃, Cl,

Antimicrobial activity of some pyridyl and napthyl substituted 1,2,4-triazole and 1,3,4-thiadiazole derivatives (21) against S. *aureus* and E. *coli* [97]. Antibacterial activity of some N,N-(5-(6-(4-subsitutedphenyl) imidazo(2,1-b) (1,3,4)-thiadiazole-2-yl)-pyrimidine-2,4-diyl) di acetamide derivatives (22) against E. *coli*, *Staphylococcus aureus* and *Bacillus subtillis* using cupplate-agar diffusion method using standard drug Methotrexate at concentration 50 μg/ml [98].

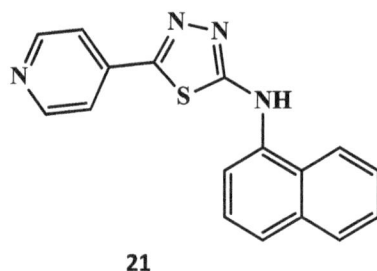

21

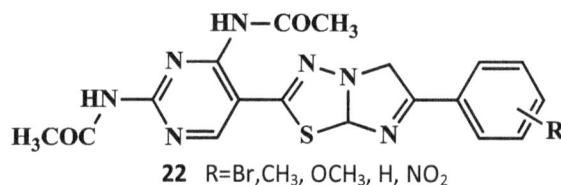

22   R=Br,CH₃, OCH₃, H, NO₂

Antifungal activity

Antifungal activity of 5-(3,5-diphenyl pyrazol-4-yloy methyl)-2-(4-oxo-2-substituted phenyl-3-thiazolidinyl)-1,3,4-oxidiazoles/Thiadiazoles and related compounds (23) against F. *oxysporum*, C. *capsicum* and R. *solani* using standard drug Dimethyl formamide at concentration 25 μg/ml [99]. Fungicidal activities of 3-aryloxymethyl-6-substituted- 1,2,4-triazolo(3,4-b)-1,3,4-thiadiazoles (24) against species A. *flavus* and A. *niger* using standard drug Dithane at concentration 100 ppm [100].

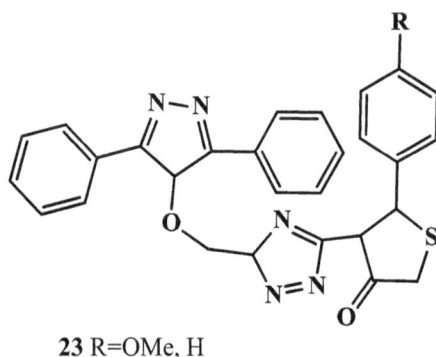

23 R=OMe, H

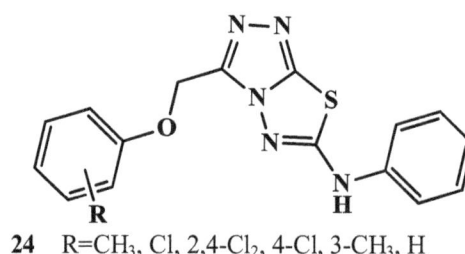

24   R=CH₃, Cl, 2,4-Cl₂, 4-Cl, 3-CH₃, H

Antitubercular activity

The 3-(2-sulphido-1,3,4-thiadiazolium-4-carbonylphenyl) syndones and 4-(4-(2-sulphido-1,3,4-thiadiazolium)benzoyl)-1,3,4-thiadiazolium-2-thiolates from 3-(4/3-(hydrazine carbonyl)phenyl) syndones (25), and their antimicrobial and antitubercular (anti-TB) activity against M. *tuberculli* using standard drug Cotrimoxazole and Fluconazoleat concentration 100 μg/ml [101]. Bioactivity of *s*-triazolo

(3,4-b)(1,3,4)thiadiazoles, *s*-triazolo (3,4-b)(1,3,4) thiadiazines and *s*-triazolo(3,4:2,3)-thiadiazino(5,6-b)quinoxaline (**26**) as potential anti-TB activity

against *M. tuberculli* using standard drug Rifampicin at concentration 0.03 μg/ml [102].

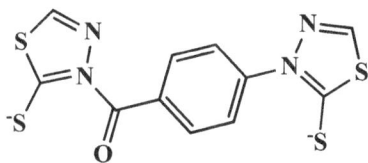

25

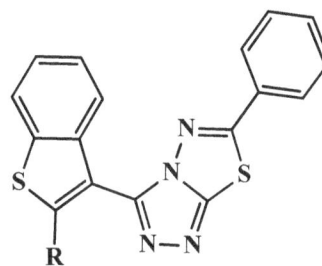

26 R=NHCOCH₃

Thiadiazole derivative 2-(4-chlorophenylamino)-5-(4-aminophenyl)-1,3,4-thiadiazole **27a** showed 57% inhibition against *M. tuberculosis*. Further they found that compound **27b** has exhibited the highest inhibitory activity (69%) against in vitro growing *M. tuberculosis* [103,104]. This compound while not active enough to be considered as therapeutics, are definitely lead compounds in the search for novel

agents to combat resistance. Two series of 2- and 3-[5-(nitro aryl)-1,3,4-thiadiazol-2-ylthio, sulfinyl and sulfonyl] propionic acid alkyl esters and screened for anti-tuberculosis activity against *M. tuberculosis* and found that the compound **28** that is Propyl 3-[5-(5-nitrothiophen-2-yl)-1,3,4-thiadiazol-2-ylthio]-propionate was the most active one [105].

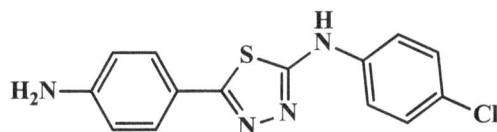

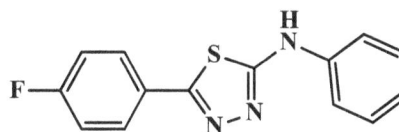

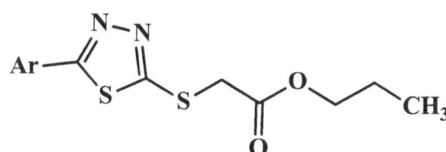

Compound 27a-b

Compound 28 Ar= 5-nitro-2-thienyl

Anticancer Activity

A series of 5-substituted 2-(2,4-dihydroxy-phenyl)-1,3,4-thiadiazoles were evaluated for their antiproliferative activity against the cells of human

cancer lines. Compounds **29** and **30** of different structures prove to be the most active. They exhibited higher inhibitory activity against T47D cells (human breast cancer cells) than cisplatin [106].

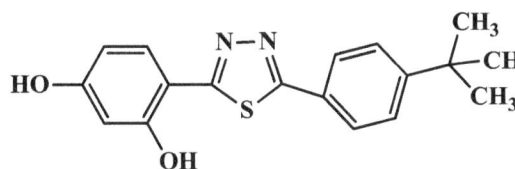

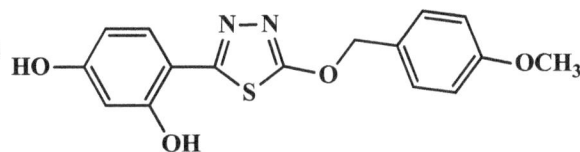

Compound 29

Compound 30

Antitumor Activity

2-acylamino 2-aroylamino and ethoxycarbonyl imino-1,3,4-thiadiazoles (**31**) as antitumor agents [107].

Cytotoxic activity of 3,6-disubstituted 1,2,4- triazole-(3,4-b)-1,3,4-thiadiazoles (**32**) as potential antileishmanial activity aganist standard drug Doxorubicin at concentration 10μM [108].

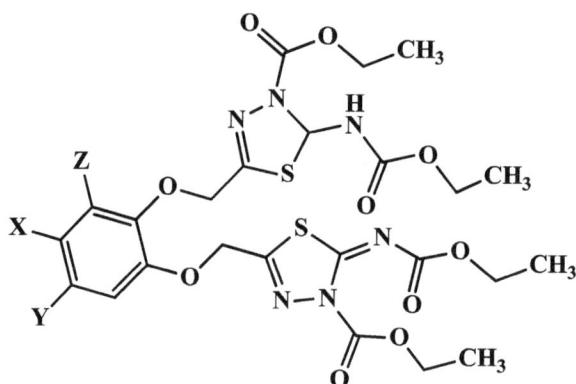

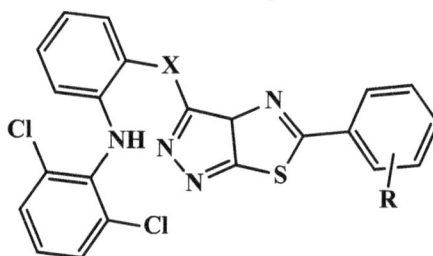

31 X=H, Y=H, Z=H,OCH$_3$ **32** R=-3-Chloro, -4-Chloro, -4-nitro, -2-methoxy,
                                                X= -CH$_2$, -CH$_2$COOCH$_2$

Antiviral Activity

The *N*-(5-Aryl/aryloxymethyl-1,3,4-thiadiazole-2-yl)glyoxylamide thio-semicarbazone (**33**) as

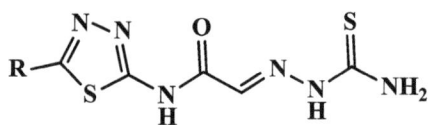

potential antiviral and antifungal agents against *Alternaria brassicae* and *Helminthosporium oryzae* using standard drug Bavistin and Dithane at concentration 45µg/ml [109].

R=aryloxy methyl

Anti-Helicobacter pylori activity

Helicobacter pylori, is a Gram-negative bacterium, causes gastric, duodenal ulcers, and gastric cancer. In vitro anti-*H. pylori* activity of N-[5-(5-nitro-2-heteroaryl)-1,3,4-thiadiazol-2-yl]thiomorpholines and some related compounds. They found that nitrofuran analog (**34**) containing thiomorpholine S, S-dioxide moiety was the most potent compound tested [110]. A series of 5-(nitroaryl)-

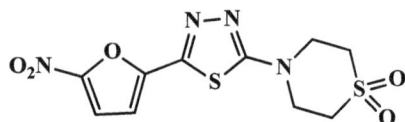

1,3,4-thiadiazoles bearing certain sulfur containing alkyl side chain similar to pendent residue in tinidazole molecule were evaluated against *H. pylori*. The compound **35** containing 2-[2-(ethylsulfonyl)ethylthio]-side chain from nitrothiophene series was the most potent compound tested against clinical isolates of *H. pylori*, however, nitroimidazoles **35b** and **36c** were found to be more promising compounds because of their respectable anti-*H.pylori* activity [111].

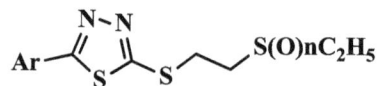

Compound 34

Compound 35a-c   36a, Ar=5-NO$_2$-thiophene, n=2;
36b, Ar=1-Me-5-NO$_2$-imidazole n=2;
36c, Ar=1-Me-5-NO$_2$-imidazole n=0

Anticonvulsants Activity

The anticonvulsants properties of a number of substituted 2-hydrazino-1,3,4-Thiadiazole, compound 2-(aminomethyl)-5-(2-biphenylyl)-1,3,4-Thiadiazole (**37**) possess potent anticonvulsants properties in rat and mice and compared with

phenytoin, phenobarbital and carbamazepine in a number of test situations [112]. A number of compounds such as compound **38a** and **38b** showed anticonvulsants activity. These two compounds may be considered promising for the development of new anticonvulsant agents [113].

Compound 37   **Compound 38a-b:**   **38aR=Ethyl; 38b: R=*m*-fluorophenyl**

Anticonvulsant Activity

Anticonvulsant activity of substituted oxidiazole and thiadiazole derivatives (**39**) using eletrocovulsometer using standard drug Phenytoin Sodium

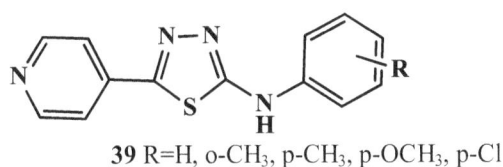

at concentration 25 mg/kg [114]. The 2,5-disubstituted 1,3,4-thiadiazoles (**40**) as potential anticonvulsant activity using standard drug Carbamazepine and Phenytoin at concentration 30 and 100 mg/kg [115].

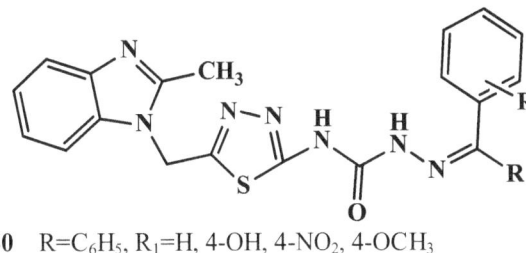

39 R=H, o-CH₃, p-CH₃, p-OCH₃, p-Cl

40   R=C₆H₅, R₁=H, 4-OH, 4-NO₂, 4-OCH₃

Antidepressant Activity

A imine derivatives of 5-amino-1, 3, 4-thiadiazole-2-thiol, and their anti-depressant activity. Two compounds namely 5-{[1-(4-chlorophenyl)-3-(4-methoxy-phenyl)prop-2-en-1-ylidene]-amino}-5-benzylthio-1,3,4–thiadiazole (**41a**) and 5-{[1-(4-chlorophenyl)-3-(4-dimethyl-aminophenyl)-prop-2-en-1-ylidene]amino}-5-benzyl thio-1,3,4-thiadiazole

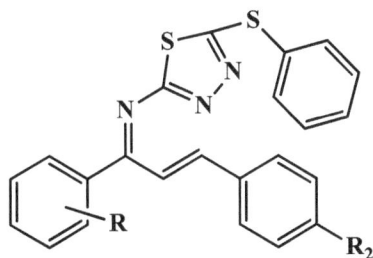

(**41b**) have shown significant anti-depressant activity, which decreased immobility time by 77.99% and 76.26% compared to the standard imipramine (82%). These compounds in the series have passed neurotoxicity tests also [116]. Biological activity of 2-substituted ethanamido-5- alkyl-1,3,4-thiadiazoles (**42**) as potential the CNS depressant, spasmolytic activity using standard drug Acetylcholine at concentration 12 to 32 mg/ml [117].

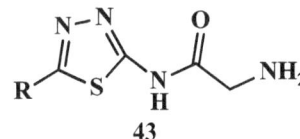

Compound 41a-b
41a R₁= OCH₃, R₂=Cl; 41b, R₁=(CH₃)₂N, R₂=Cl

Muscle Relaxant Activites

Anticonvulsant and muscle relaxant activities of substituted 1,3,4-oxidiazole, 1,3,4-thiadiazole and 1,2,4-triazole using standard drug Diazepam at concentration 10ml/kg [118].

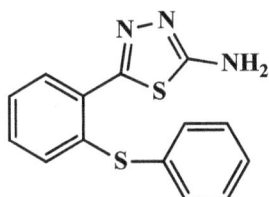

43

spp using standard drug Clotrimazole at cocentration 10μM [119].

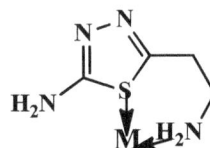

Metal Complexes

Biological activity of metal complexes of 5-(2-aminoethyl)-2-amino-1,3,4-thiadiazole as potential antifungal activity against *Aspergillus* and *Candida*

Antiprotozoal Activites

The   1-methyl-2-(1,3,4-thiadiazole-2-yl)-5-nitroimidazole and 1-methyl-2-(1,3,4-oxidiazoles-2-yl)-5-nitro-imidazole as potential antiprotozoal agents [120].

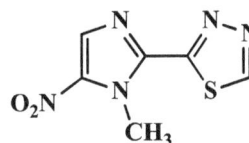

Diuretic Activites

Biological studies of Zn(II) complex of schiff base derived from 5-acetazolamido 1,3,4-thiadiazole-2-

sulphonamide as potential diuretic agents using standard drug Acetazolamide [121].

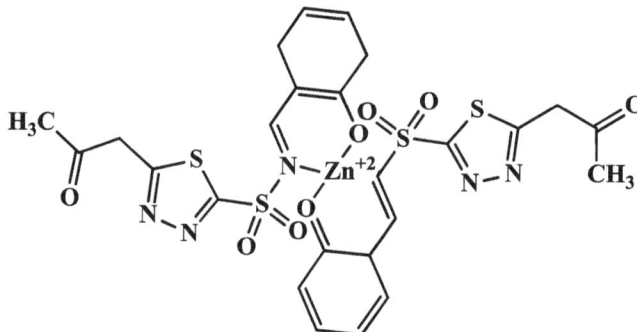

Carbonic anhydrase inhibitor activity

Docking studies of new 1,3,4-thiadiazole-2-thione derivatives with carbonic anhydrase inhibitory agents using standard drug Acetazolamide [122].

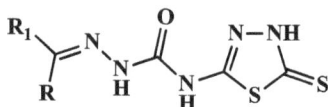

R=H, CH$_3$, C$_6$H$_5$, R$_1$=C$_6$H$_5$, 4-(OH)C$_6$H$_4$,

3(Br)C$_6$H$_4$, 4-(F)C$_6$H$_5$, C$_6$H$_5$, 3-pyridyl, 2-furyl,

Antidiabetic Activities

Biological evaluation of some 1,3,4-thiadiazoles as potential the anti-diabetic [123]. Biological activity of sulphanyl urea as potential anti-diabetic agent using standard drug Gliclazide at concentration 200mg/kg [124].

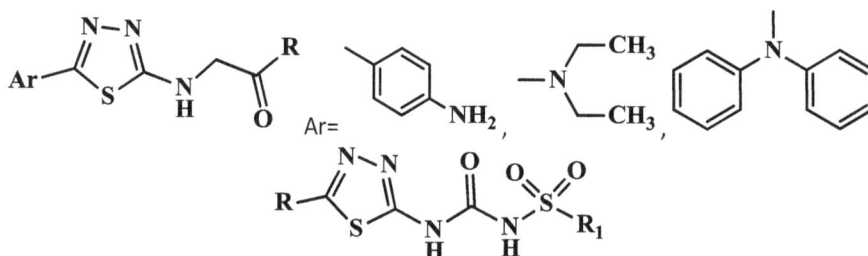

R= phenyl, methyl phenyl, chloro phenyl, methyl R$_1$= phenyl methyl

Antioxidant Activities

Biological activity of 3-alkyl/aryl-6-(1-chloro-3,4-dihydronaphth-2-yl)-5,6-dihydro-s-trizolo   (3,4-b)

(1,3,4) thiadiazoles as potential the antioxidant and antibacterial agent against *Escherichia coli* and *Staphylococcus aureus* usingstandard drug Sodium Nitroprusside at concentration 5µM [125].

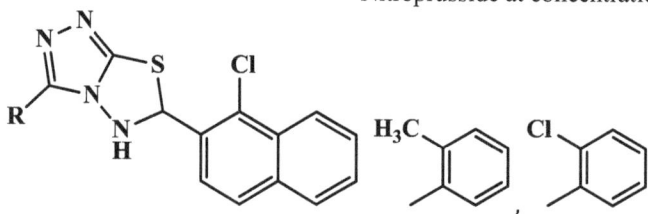

R = Phenyl, CH$_3$, C$_2$H$_5$, CH$_2$CH$_2$CH$_3$
R$_1$=H, OH, Cl, NO$_2$ R$_2$=OH, NO$_2$, Cl, R$_3$=OH, H, NO$_2$ R$_4$=H, OH, OCH$_3$

Acaricidal Activites

Acaricidal   activity   of   *N*-(1,3,4-thiadiazole-2-yl)cyclo-propane carboxamides against *Tetranychus urticae* [126].

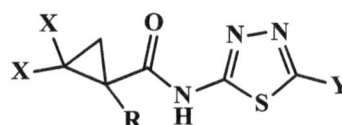

R=H, CH$_3$, C$_2$H$_5$, n-C$_3$H$_7$, i-C$_3$H$_7$;   X=H, Cl, Br;
Y=CF$_3$, CF$_2$CF$_3$, H, CH$_3$, C$_2$H$_5$

Discussion: The synthesis of 1,3,4-thiadiazoles that have been illustrates different approaches to the challenge of preparing these bioactive products and allows the synthesis of many novel chemical derivatives [127-132]. The 1,3,4-thiadiazoles are prepared by appropriate rearrangements, ring opening and substitution reaction. The area of the synthesis of 1,3,4-thiadiazole rings continues to grow, and the organic chemistry will provide more and better methods for the synthesis of this interesting heterocycle, allowing the discovery of new drug candidates more active, more specific and safer [133-140]. Thiadiazole are the most important classes of heterocyclic compounds and possess versatile type of biological activities.

Conclusion: The 1,3,4-thiadiazole have advantageous in the medicinal properties. Some 1,3,4- thiadiazole containing drugs having several pharmacological activity. Chemical properties of 1,3,4-thiadiazole have been reviewed in the last few years. This review provides a brief summary of the medicinal chemistry of 1,3,4-thiadiazole system and highlights some examples of 1,3,4-thiadiazole-containing drug substances in the current literature. A survey of representative literature procedures for the preparation of 1,3,4-thiadiazole is presented in sections by generalized synthetic methods.

References

1- Boschelli DH, Connor DT, Bornemeier DA, Dyer RD, Kennedy JA, Kuipers PJ, Okonkwo GC, Schrier DJ, Wright CD, *J. Med. Chem.,* 1993; 36(13): 1802-1810.

2- Ramprasad J, Nayak N, Dalimba U, Yogeeswari P, Sriram D. *Bioorg Med Chem Lett.* 2015; 25(19):4169-73.

3- Pal D, Tripathi R, Pandey DD, Mishra P. *J Adv Pharm Technol Res.* 2014; 5(4):196-201.

4- Guan P, Wang L, Hou X, Wan Y, Xu W, Tang W, Fang H. *Bioorg Med Chem.* 2014; 22(21):5766-75.

5- Gurjar AS, Andrisano V, Simone AD, Velingkar VS. Bioorg Chem. 2014; 57:90-8.

6- Chávez P, Ngov C, de Frémont P, Lévêque P, Leclerc N. J Org Chem. 2014; 79(21):10179-88.

7- Chengyuan Wang, Peiyang Gu, Benlin Hu and Qichun Zhang. *J. Mater. Chem. C*, 2015; 3, 10055-10065.

8- Junbo Li and Qichun Zhang, *ACS Appl. Mater. Interfaces*, 2015; 7 (51), 28049–28062.

9- Li Y, Geng J, Liu Y, Yu S, Zhao G. *ChemMedChem.* 2013; 8(1):27-41.

10- Yang H, Zhang W, Kong Q, Liu H, Sun R, Lin B, Zhang H, Xi Z. Food Chem Toxicol. 2013; 53:100-4.

11- Sharma M, Deekshith V, Semwal A, Sriram D, Yogeeswari P. *Pain Ther.* 2012; 1(1):3.

12- Haj Mohammad Ebrahim Tehrani K, Sardari S, Mashayekhi V, Esfahani Zadeh M, Azerang P, Kobarfard F. *Chem Pharm Bull* (Tokyo). 2013;61(2):160-6.

13- Mayhoub AS, Marler L, Kondratyuk TP, Park EJ, Pezzuto JM, Cushman M. *Bioorg Med Chem.* 2012; 20(24):7030-9.

14- Lin RY, Lee CP, Chen YC, Peng JD, Chu TC, Chou HH, Yang HM, Lin JT, Ho KC. *Chem Commun (Camb)*. 2012; 48(99):12071-3.

15- Chen M, Lin S, Li L, Zhu C, Wang X, Wang Y, Jiang B, Wang S, Li Y, Jiang J, Shi J. Org Lett. 2012; 14(22):5668-71.

16- Kavlakova M, Bakalova A, Momekov G, Ivanov D. *Arzneimittelforschung.* 2012; 62(12):599-602.

17- Aliabadi A, Eghbalian E, Kiani A. *Iran J Basic Med Sci.* 2013; 16(11):1133-8.

18- Datar PA, Deokule TA. *Mini Rev Med Chem.* 2014; 14(2):136-53.

19- Harish KP, Mohana KN, Mallesha L. *Bioorg Khim.* 2014; 40(1):108-16.

20- Chandrakantha B, Isloor AM, Shetty P, Fun HK, Hegde G. *Eur J Med Chem.* 2014; 71:316-23.

21- Wang C, Han Y, Wu A, Sólyom S, Niu L. *ACS Chem Neurosci.* 2014; 5(2):138-47.

22- Bandi KR, Singh AK, Upadhyay A. *Mater Sci Eng C Mater Biol Appl.* 2014; 34:149-57.

23- Jiang Q, Zhang Z, Lu J, Huang Y, Lu Z, Tan Y, Jiang Q. *Bioorg Med Chem.* 2013; 21(24):7735-41.

24- Asadipour A, Edraki N, Nakhjiri M, Yahya-Meymandi A, Alipour E, Saniee P, Siavoshi F, Shafiee A, Foroumadi A. *Iran J Pharm Res.* 2013 Summer;12(3):281-7.

25- Zhao HC, Shi YP, Liu YM, Li CW, Xuan LN, Wang P, Zhang K, Chen BQ. *Bioorg Med Chem Lett.* 2013; 23(24):6577-9.

26- Yang Y, Qiu F, Josephs JL, Humphreys WG, Shu YZ. *Chem Res Toxicol.* 2012; 25(12):2770-9.

27- Sharp SY, Roe SM, Kazlauskas E, Cikotienė I, Workman P, Matulis D, Prodromou C. *PLoS One.* 2012;7(9):e44642.

28- Al-Masoudi NA, Al-Soud YA, Al-Masoudi WA, *Nucleosides Nucleotides Nucleic Acids*, 2004; 23(11): 1739-1749. Ajay Kumar K, Jayaroopa P, Vasanth Kumar G, *Inter J Chem Tech Res*, 2012; 4(4): 1782-1791.

29- Wang C, Hu B, Wang J, Gao J, Li G, Xiong WW, Zou B, Suzuki M, Aratani N, Yamada H, Huo F, Lee PS, Zhang Q. Chem Asian J. 2015;10(1): 116-9.

30- Xiaoli Kan, Xiaobing Yang, Fangzhong Hu, Yang Wang, Ying Liu, Xiaomao Zou, Hengyu Li, Hao Li, Qichun Zhang.Tetrahedron Letters, 2015; 56(45), 6198-6201.

31- Almasirad A, Vousooghi N, Tabatabai SA, Shafiee AKA, *Acta. Chem. Slov.*, 2007, 54, 317-324.

32- Demirbas A, Sahin D, Demirbas N, Karaoglu SA Eur. J. Med. Chem., 2009, 44, 2896-2903.

33- Kadi AA, El-Brollosy NR, Al-Deeb OA, Habib EE, Ibrahim TM, El-Emam AA, Eur. J. Med. Chem., 2007, 42, 235-242.

34- Bekhit AA, Abdel-Aziem T, Bio. Org. Med. Chem., 2004, 12, 1935-1945.

35- Shucla HK, Desai NC, Astik RR, Thaker KA, J. Indian Chem. Soc., 1984, 61, 168-171.

36- Mullican MD, Wilson MW, Connor DT, Kostlan CR, Schrier DJ, Dyer RD, J. Med. Chem., 1993, 36, 1090-1099.

37- Gupta JK, Yadav RK, Dudhe R, Sharma PK. Inter. J. Pharm. Tech. Res., 2010, 2(2), 1493-1507.

38- Bak B, Nygard L, Pederson EJ, and Andersen RJ, J. Mol. Spectr., 1966, 19, 283.

39- Bak B, Cristensen D, Nygard LH, Lipschitz L, and Andersen J R, J. Mol. Spectr., 1962, 9, 225.

40- Zahradnik R, Kouteckey J, Collection Czech. Chem. Commun., 1961, 26, 156.

41- Smith JW "Electric Dipole Moments" 1955, 92-94 Butterworth, London and Washington, DC.

42- Ajay K, Kumar GV, Renuka N. Inter J PharmTech Res. 2013, 5(1), 239-248.

43- Kempegowda, Senthil Kumar GP, Dev Prakash and Tamiz Mani T. Der Pharma Chemica, 2011, 3(2): 330-341.

44- Gupta JK, Yadav RK, Dudhe R, Sharma PK. Inter J Pharm Tech Res, 2010, 2(2), 1493-1507.

45- Ajay Kumar K, Lokeshwari DM, Pavithra G, Vasanth Kumar G. Res J Pharm and Tech, 2012; 5(12): 1490- 1496.

46- Mohd Amir, Arun Kumar, Israr Ali, Khan SA, Ind. J. Chem., 2009; 48B: 1288-1293.

47- Singh AK, Bose S, Singh UP, jana S, Shukla R, Singh V, Lohani M, The Pharma Res., 2009; 2: 133-137.

48- Pattan SR, Kittur BS, Sastry BS, Jadhav SG, Thakur DK, Madamwar SA, Shindhe HV, Ind.J. Chem., 2011; 50B: 615-618.

49- Saqib M, Chatrapati KS, Kallur HJ, Hariprasanna RC, Waseem M, Durgad SA, Am. J. PharmTech. Res., 2012; 2(3): 699-706.

50- El-Sadek MM, Hassan SY, Abdelwahab HE, Yacout GA, Molecules, 2012; 17(7): 8378-8396.

51- Kaleta Z, Makowski BT, Soos T, Dembinski R, Org. Lett., 2006; 8: 1625-1628.

52- Linganna N, Lokanatha Rai KM, Synth. Commun., 2006; 28(24): 4611-4617.

53- Padmavathi V, Reddy GS, Mohan AVN, Mahesh K, Arkivoc, 2008; (xvii): 48-60.

54- Salunkhe NG, J. Curr. Chem. Pharm. Sci., 2012; 2(2): 100-106.

55- De SK, Chen V, Stebbins JL, Chen LH, Cellitti JF, Machleidt T, barile E, Riel-Mehan M, Dahl R, Yang L, Emadadi A, Murphy R, Pellecchia M, Bioorg. Med. Chem., 2010; 18(2): 590-596.

56- Bharadwaj S, Jain K, Parashar B, Gupta GD, Sharma VK, Asian J. Biochem. Pharma. Res., 2011; 1(1): 139-146.

57- Hassan AA, Mohamed NK, Shawkyand AM, Dietrich D, Arkivoc, 2003; (i): 118-128.

58- Kundapur U, Sarojini BK, Narayana B, Molbank, 2012; M778: 1-4.

59- Pattan SR, Kekare P, Dighe NS, Nirmal SA, Musmade DS, Parjane SK, Daithankar AV, J.Chem. Pharm. Res., 2009; 1(1): 191-198.

60- Desai, Bhavsar, Shah , Anil Saxena, Indian J. Chem., 2008, 47B, 579-589.

61- Valarmathy J, Samuelijoshua L, Rathinavel G, Senthilkumar L. Der Pharma Chemica, 2010, 2(2), 23-26.

62- Sunil D, Isloor A, Shetty P. Der Pharma Chemica, 2009, 1(2), 19-26.

63- N. Pokhodylo, I. Krupa, V. Matiychuk, M. Obushak, Visnyk Lviv. Univ. Ser. Chem., 2009, 50, 188-193.

64- Dhiman AM, Wadodkar KN. Indian J. Chem., 2001, 40B, 636-639.

65- Kidwai M, Bhushan KR. Indian J. Chem., 1998, 37B, 427-428.

66- Goerdeler J, Ohm J, and Tegtmeyer O, Chem. Ber., 1956, 89, 1534.

67- Goerdeler J, Galinke J, Chem. Ber., 1957, 90, 202.

68- Saikachi H and Kanaoka M, Zasshi Y, 1961, 81, 1333; Chem. Abstr., 1962, 56, 7304.

69- Saikachi H and Kanaoka M, Zasshi Y, 1962, 82, 683 ; Chem. Abstr., 1963, 58, 4543.

70- Sandstrom J, Arkiv Kemi., 1956, 9, 255.

71- Albert A, "Heterocyclic chemistry" Oxford University Press (Athlone), 1959, London and New York.

72- Ohta M, Hgiwara R, and Mizushima Y, J. Pharm. Soc. Japan, 1953, 73, 701.

73- Bak B, Cristensen CH, Hansen TS, Pederson EJ, and Nielson JT, Acta Chem. Scand., 1965, 19, 2434.

74- Song Y, Connor DT, Sercel AD, Sorenson RJ, Doubleday R, Unangst PC, Roth BD, Beylin VG, Gilbertson RB, Chan K, Schrier DJ, Guglietta A, Bornemeier DA, Dyer RD, J. Med. Chem., 1999, 42, 1161-1169.

75- Mathew V, Keshavayya J, Vaidya V P, Giles D, Eur. J. Med. Chem., 2007, 42, 823-840.

76- Chapleo CB, Myers M, Myers PL, Saville JF, Smith ACB, Stilling MR, Tulloch IF, Walter DS, Welbourne AP, J. Med. Chem., 1986, 29, 2273-2280.

77- Chapleo CB, Myers PL, Smith AC, Stilling MR, Tulloch IF, Walter DS, J. Med. Chem., 1988, 31, 7-11.

78- Turner S, Myers M, Gadie B, Nelson AJ, Pape R, Saville JF, Doxey JC, Berridge TL, J. Med. Chem. 1988, 31, 902-906.

79- Turner S, Myers M, Gadie B, Hale SA, Horsley A, Nelson AJ, Pape R, Saville JF, Doxey JC, Berridge TL, J. Med. Chem., 1988, 31, 906-913.

80- Cressier D, Prouillac C, Hernandez P, Amourette C, Diserbo M, Lion C, Rima G, Bio. Med. Chem., 2009, 17, 5275-5284.

81- Matysiak J, Nazulewicz A, Pelczynska M, Switalska M, Jaroszewicz I, Opolski A, Eur. J. Med. Chem., 2006, 41, 475-482.

82- Chou JY, Lai SY, Pan SL, Jow GM, Chern JW, Guh JH, Biochem. Pharmacol. 2003, 66, 115-124.

83- Radi M, Crespan E, Botta G, Falchi F, Maga G, Manetti F, Corradi V, Mancini M, Santucci MA, Schenone S, Botta M, Bio. Org. Med. Chem. Letters, 2008, 18, 1207-1211.

84- Swamy SN, Basappa, Priya BS, Prabhuswamy B, Doreswamy BH, Prasad JS, Rangappa KS, Eur. J. Med. Chem., 2006, 41, 531-538.

85- Kucukguzel SG, Kucukguzel I, Tatar E, Rollas S, Sahin F, Gulluce M, Clercq ED, Kabasakal L, Eur. J. Med. Chem., 2007, 42, 893-901.

86- Hilfiker MA, Wang N, Xiaoping H, Zhimin D, Pullen MA, Nord M, Nagilla R, Fries HE, Wu Charlene W, Sulpizio AC, Jaworski J, Morrow D, Edwards RM, Jian J, Bio. Org. Med. Chem. Letters, 2009, 19, 4292-4295.

87- Gadad AK, Palkar MB, Anand K, Noolvi MN, Boreddy TS, Wagwade J, Bio. Org. Med. Chem., 2008, 16, 276–283.

88- Goksen US, Kelekci NG, Goktas O, Koysal Y, Kılıc E, Isık S, Aktay G, Ozalp M, Bio. Org. Med. Chem., 2007, 15, 5738–5751.

89- Kalgutar A, Marnett A, Crews B, Remmel R, Marnett L, J. Med. Chem., 2000, 43, 2860-2870.

90- Duflos M, Nourrison M, Brelet J, Courant J, Le Baut G, Grimaud N, Petit J, Eur. J. Med. Chem., 2001, 36, 545-553.

91- Amir M, Kumar H, Javed SA, Eur. J. Med. Chem., 2008, 43, 2056-2066.

92- Schenone S, Brullo C, Bruno O, Bondavalli F, Ranise A, Filippelli W, Rinaldi B, Capuano A, Falcone G, Bio.Org. Med. Chem., 2006, 14, 1698- 1705.

93- Amir M, Shikha K, Eur. J. Med. Chem., 2004, 39, 535-545.

94- Tijen Onkol, M. Bilge Cakir, Fethi Sahin, Turk J. Chem., 2004, 28, 461-468.

95- Gadad A, Palkar M, Anand K, Malleshappa N, Noolvi, Thippeswamy S, Wagawade JB, Bioorg. Med. Chem., 2008, 16, 276-283.

96- Lamani RS, Shetty NS, Kamble RR, Khazi IAM, Eur. J. Med. Chem., 2009, 44, 2828–2833.

97- Padmavathi V, Reddy GS, Padmaja A, Kondaiah P, Ali-Shazia, Eur. J. Med. Chem, 2009, 44, 2106–2112.

98- Padmavathi V, Thriveni P, Reddy GS, Deepti D, Eur. J. Med. Chem., 2008, 43, 917-924.

99- Vijay Dabholkar, Faisal Ansari. Acta. Polo. Pham. Drug Res., 2008, 65(5), 521-526.

100- Zamani K, Faghihi K, Tofighi T, Shariatzadeh MR. Turk J. Chem., 2004, 28, 95-100.

101- Sankangoud R, Chatni N, Goudanavar P. Der Pharma Chemica, 2010, 2(2), 347-353.

102- Anand Kumar Dubey, K. Naresh Sangwan, Ind J. Chem., 1994, 33(B), 1043-1047.

103- Nizamuddin, Q. Bano, N. Tiwari, S. Giri, Ind. J. Chem., 1992, 31(B), 714-718.

104- Johnson Jogul, Bharati Badami, J. Serb. Chem. Soc., 2006, 71(8-9), 851-860.

105- Mahendra Shiradkar, Rajesh Kale, Ind. J. Chem., 2006, 45(B), 1009-1013.

106- Karakus S, Rollas S, Farmaco., 2002, 57, 577-581.

107- Oruc EE, Rollas S, Kandermirli F, Shvets N, Dimoglo AS, J. Med. Chem., 2004, 47, 6760-6767.

108- Foroumadi A, Kargar Z, Sakhteman A, Sharifzadeh Z, Feyzmohammadi R, Kazemi M, Shafiee A, Bio. Org. Med. Chem. Letters, 2006, 16, 1164–1167.

109- Matysiak J, Eur. J. Med. Chem., 2007, 42, 940-947.

110- Kemal Sancak, Yasemin Unver, E. R. Mustafa, Turk J. Chem., 2007, 31, 125-134.

111- Kaliappan Ilango, Parthiban Valentine, Eur. J. Chem., 2010, 1(1), 50-53.

112- S. Giri, Nizamuddin and U. C. Srivastava, Agric. Boil. Chem., 1983, 47(1), 103-105.

113- Mirzaei J, Siavoshi F, Emami S, Safari F, Khoshayand MR, Shafiee A, Foroumadi A, Eur. J. Med. Chem., 2008, 43, 1575-1580.

114- Foroumadi A, Rineh A, Emami S, Siavoshi F, Massarrat S, Safari F, Rajabalian S, Falahati M, Lotfali E, Shafiee A, Bio. Org. Med. Chem. Letters, 2008, 18, 3315-3320.

115- Stilings MR, Welbourn AP, Walter DS, J. Med. Chem. 1986, 29, 2280-2284.

116- Dogan HN, Duran A, Rollas S, Sener G, Uysal MK, Gulen D, Bio.Org. Med. Chem., 2002, 10, 2893-2898.

117- Yar MS, Akhter MW. Acta. Polo. Pharm. Drug Res., 2009, 66(4), 393-397.

118- Rajak H, Aggarwal N, Kashaw S, Kharya MD, Mishra P. J. Korean Chem. Soc., 2010, 54(1), 158-164.

119- Yusuf M, Khan R, Ahmed B, Bio.Org. Med. Chem., 2008, 16, 8029–8034.

120- Shakya A, Mishra P, Patnaik, Shukla R. Acta. Pharm. Turcica 2000, 2, 56-63.

121- Almasirad A, Vousooghi N, Tabatabai SA, Shafiee AKA, Acta. Chem. Slov., 2007, 54, 317-324.

122- Barboiu M, Cimpoesu M, Guran C, Supuran C. Metal Based Drug, 1996, 3(5), 227-232.

123- A. Shafiee, B. Yazdikarimy and S. Sadrai, J. Heterocyclic Chem., 1989, 26, 1341-1343.

124- Ghosh S, Malik S, Bharti N, Jain G. Asian J. Exp. Sci., 2009, 23(1), 189-192.

125- Hamid MA, Hafez AA, El Koussi N, Mahfouz N, Innocenti A, Supran C. Bioorg. Med. Chem., 2007, 15, 6975-6984.

126- Pattan S, Kekare P, Dighe N, Nirmal S, Musmade D, Parjane S, Daithankar A. *J. Chem. Pharm. Res.*, 2009, 1(1), 191-198.

127- Chengyuan Wang, Masataka Yamashita, Dr. Benlin Hu, Dr. Yi Zhou, Jiangxin Wang, Dr. Jin Wu, Prof. Dr. Fengwei Huo, Prof. Dr. Pooi See Lee, Prof. Dr. Naoki Aratani, Prof. Dr. Hiroko Yamada and Prof. Dr. Qichun Zhang. *Asian Journal of Organic Chemistry.* 2015; 4(7),589–677.

128- Chengyuan Wang, Takuya Okabe, Guankui Long, Daiki Kuzuhara, Yang Zhao, Naoki Aratani, Hiroko Yamada, *Qichun Zhang. Dyes and Pigments*, 2015; 122, 231-237.

129- Wang C, Wang J, Li PZ, Gao J, Tan SY, Xiong WW, Hu B, Lee PS, Zhao Y, Zhang Q. Chem Asian J. 2014; 9(3):779-83.

130- Ajay Kumar K, Lokeshwari DM, Pavithra G, Vasanth Kumar G. *Res J Pharm and Tech*, 2012; 5(12): 1490- 1496.

131- Ramprasad J, Nayak N, Dalimba U. *Eur J Med Chem.* 2015; 106:75-84.

132- Diaz JR, Camí GE, Liu-González M, Vega DR, Vullo D, Juárez A, Pedregosa JC, Supuran CT. *J Enzyme Inhib Med Chem.* 2015:1-9.

133- You JM, Han HS, Jeon S. *J Nanosci Nanotechnol.* 2015;15(8):5691-8.

134- Rezki N, Al-Yahyawi AM, Bardaweel SK, Al-Blewi FF, Aouad MR. *Molecules.* 2015; 20(9):16048-67.

135- Mhaidat NM, Al-Smadi M, Al-Momani F, Alzoubi KH, Mansi I, Al-Balas Q. *Drug Des Devel Ther.* 2015; 9:3645-52.

136- Abdou WM, Ganoub NA, Sabry E. Acta Pharm. 2014; 64(3):267-84.

137- Huong TT, Dung do TM, Oanh DT, Lan TT, Dung PT, Loi VD, Kim KR, Han BW, Yun J, Kang JS, Kim Y, Han SB, Nam NH. *Med Chem.* 2015; 11(3):296-304.

138- Akimov MG, Gretskaia NM, Karnoukhova VA, Serkov IV, Proshin AN, Shtratnikova VIu, Bezuglov VV. *Biomed Khim.* 2014; 60(4):473-8.

139- Ardestani SK, Poorrajab F, Razmi S, Foroumadi A, Ajdary S, Gharegozlou B, Behrouzi-Fardmoghadam M, Shafiee A. *Exp Parasitol.* 2012; 132(2):116-22.

140- Jie Luo, Wenqin Wu, Li-Wen Xu, Yuezhong Meng, Yixin Lu. *Tetrahedron Letters* 2013; 54 (21), *2623-2626.*

10

Application of Taguchi design to produce polyols based on castor oil derivatives with diethylene glycol

Thiago Alessandre da Silva, Luiz Pereira Ramos, Sônia Faria Zawadzki and Ronilson Vasconcelos Barbosa *

Laboratory of Synthetic Polymers: Federal University of Paraná, Department of Chemistry

CEP – 81.531-980 – PO Box 19032, Brazil

Abstract: Castor oil (CO) is one of the most valuable oils, because of its characteristics and potential use in synthesis. A simple way to modify the castor oil structure is the transesterification reaction with an alcohol. Through this reaction, a renewable polyol could be produced and applied in the polyurethane industry. In this study, the transesterification was done by KOH catalysis following a Taguchi experimental design. The chosen alcohol was diethylene glycol (DEG) and the product components were estimated by GPC. This chromatographic technique allowed establishment of the most favorable conditions to produce monosubstituted DEG or disubstituted DEG. In the conditions suggested by the Taguchi design, the condition that favors monosubstituted DEG is 9:1 DEG:CO molar ratios, 200°C for 4 h and 0.5% KOH. This condition leads to approximately 75% monosubstituted DEG and 11% disubstituted DEG in a mixture with the remaining acylglycerols. The hydroxyl value of this product is 407, a high value for a product with a relatively low molecular weight. These characteristics suggest that it can be used as a polyol for polyurethanes.

Keywords: Castor oil; polyol; transesterification; GPC.

Introduction

One of the principal chemical modifications of vegetable oils is the transesterification reaction. There is an extensive discussion about transesterification of oils employing methanol or ethanol to produce biodiesel, a renewable fuel to replace petrodiesel [1]. Not for fuels applications, vegetable oils are a good alternative option as starting materials to produce polyols [2]. The polyurethane industry uses large amounts of petrochemical polyols. To reduce the demand for petrochemicals, more environmentally friendly materials are being developed [3]. Desirable characteristics of these new products include high hydroxyl content, thermal stability, and low viscosity [4]. The crude oil has low reactivity because of its low hydroxyl content. Reactions are needed to introduce more hydroxyl groups and primary hydroxyl groups, such as the transesterification[5]. The polyol monomer used to produce polyester or polyurethane polymer is important for the mechanical properties of the resulting polymer. The selection of polyol can impart a significant effect in inducing hard or soft regions in polyurethanes[6]. Castor oil has been used industrially because of its chemical properties, and has found applications in the synthesis of a series of polyurethanes, such as polyurethane elastomers, adhesives, and coatings [7].

The main fatty acid constituent of castor oil is ricinoleic acid, which has a hydroxyl group at carbon-12 [8]. This fatty acid promotes hydrophobic interactions by the main carbon chain and hydrophilic liks by the esters groups [9]. Polyols from vegetable oils are considered excellent resources employed in industry because they are produced worldwide and are easy to process and low-cost [10]. Modification of castor oil by the transesterification reaction is a simple, low-cost procedure and the structures of the compounds produced are controlled by the reaction conditions used. Xu, using methyl ricinoleate and diethylene glycol at 190-200°C, prepared polyricinoleate diol and used it to prepare polyurethanes with control of the mechanical properties[11]. Many polyhydroxylated compounds are used in transesterification reactions with castor oil. Valero and Gonzalez reported the use of pentaerythritol as a reactant in the transesterification process with castor oil [12]. In this reaction, the catalyst used was PbO at a temperature of 200°C for 2.5 hours of reaction. Valero used PbO and pentaerythritol to prepare diols, and synthesized a network polymer [13]. To increase the thermal stability of polyurethanes, Cakić was reacted castor oil with the product of the glycolysis of PET, introducing aromayic unities to the polyol [14].

Corresponding author: Ronilson Vasconcelos Barbosa
E-mail address: ronilson@ufpr.br

The design of experiment is a mathematical tool which function is to provide consistent and feasible analysis of results and predict scenarios and/or interactions. The Taguchi design of experiment is based on a selection of a quality in one matrix and optimizes the conditions factors to accomplish the quality [15]. The design itself is composed by a table where the factors are arranged by an orthogonal array. The number of runs is related to the number of factors, a typical four factors experiment with three levels can be optimized with nine runs [16]. This leads to an economy of time and materials involved in each factor with the benefits of revealing useful information with a reduced number of experiments [17]. To optimize the transesterification of castor oil with methanol, Ramezani studied the effect of reaction temperature, mixing intensity, alcohol to oil ratio and catalyst concentration as factors [16].

The aim of this study was to obtain a new polyol derived from castor oil and diethylene glycol. The study included an attempt to control the reaction products by the use of Taguchi experimental design, and determine the composition of the reaction products by Gel Permeation Chromatography (GPC)/Nuclear Magnetic Ressonance (NMR).

Experimental Section

The castor oil (CO) employed in the transesterification reactions was purchased from Farmanilquima (Curitiba, Brazil). Diethylene glycol (99%), potassium hydroxide (99.5%), and disodium sulfate from Vetec (Rio de Janeiro, Brazil) were used as the reagent, catalyst, and drying agent, respectively. Dichloromethane (99%) from Synth (Diadema, Brazil) was used as the extraction solvent,

and tetrahydrofuran (HPLC grade, Vetec) was used as the mobile phase for GPC analysis.

Transesterification of castor oil

In a transesterification reaction, some parameters can influence the formation of the products. In this study, the diethylene glycol (DEG):CO molar ratio, temperature, reaction time and catalyst concentration were evaluated as variables (factors). The statistical model adopted to evaluate those parameters together in three levels was the robust Taguchi design. Table 1 provides details of the nine runs needed to perform the Taguchi design, with the exact values for each variable in each run.

All reaction runs were conducted in a 125 mL round-bottom flask equipped with a magnetic stirrer. Heating was maintained via an oil bath at the specified temperature. In the first step, the amount of castor oil, 50 g (0.05 mol), was weighed. In the second step, the catalyst (KOH) was dissolved in the DEG for alkoxy formation, provided by heating. Then, this solution was added to the weighed castor oil in a round-bottom flask and the reaction started. After the reaction, the flask was cooled to room temperature and the contents transferred to a separatory funnel. A first wash was done with water to remove glycerol and unreacted DEG, followed by solubilization of the ester products in dichloromethane. The product was again washed with water. The ester products were obtained by rota-evaporation of the solvent and retained for analysis. Each run from the Taguchi design was analyzed by GPC, with no other purification procedure, to investigate the profile of the main components.

Table 1. Taguchi design for castor oil transesterification

Run	molar ratio (DEG:CO)	Temperature (°C)	Reaction Time (h)	Catalyst KOH (w:w %)*
1	3:1	150	2	0.1
2	3:1	175	4	0.3
3	3:1	200	6	0.5
4	6:1	150	4	0.5
5	6:1	175	6	0.1
6	6:1	200	2	0.3
7	9:1	150	6	0.3
8	9:1	175	2	0.5
9	9:1	200	4	0.1

*Based on castor oil

In general, the GPC showed the contents of monosubstituted and disubstituted ester of DEG, 2-(2-hydroxyethoxy)ethyl (9Z,12R)-12-hydroxyoctadec-9-enoate and oxybis(ethane-2,1-diyl) (9Z,12R,9'Z,12'R)bis(12-hydroxyoctadec-9-enoate), that could be obtained in the reaction, and unreacted triacylglycerol. After analyzing the results from the GPC combined with the Taguchi design, two new runs were performed in the optimum conditions suggested by the design, focusing on

monosubstituted DEG and disubstituted DEG. The optimum conditions to produce monosubstituted DEG (MS-DEG) were: DEG:CO molar ratio 9:1, 200°C for 4 h with 0.5% catalyst; and to produce disubstituted DEG (DS-DEG) were: DEG:CO molar ratio 3:1, 175°C for 18 h with 0.3% catalyst.

An internal standard was made by the hydrolysis of the castor oil, 50g of castor oil, 15g of KOH dissolved in 50g of water and heated at 80°C/2h. The fatty acids salts were neutralized until pH 5 with concentrated HCl, and the top layer dried in rotary-evaporator. The mixture of free fatty acids was esterified with DEG employing 2%w:w manganese acetate as the catalyst at 175°C for 2 h. The product was washed with water and dried in rotary-evaporator.

Analytical procedures

The ester conversion was monitored by GPC with a methodology similar to Schoenfelder, in a Waters 1515 chromatograph, with a Waters 2487 refractive index detector and two columns in series: Supelco TSK 1000 (<1,500Da) and Waters Styragel 100 (<1500Da). THF was used as the mobile phase at

The products were characterized by FTIR spectra recorded in a Varian 660-IR spectrometer, with 64 scans at 4 cm^{-1} resolution from 400 to 4000 cm^{-1} and KBr discs as the support for film formation. The hydroxyl values were determined following ASTM D 1957-86, with the scale modified, using half quantities. For the optimum conditions, ^{1}H NMR spectroscopy was done in a Bruker DPX200 operating at 4.7 T and 200 MHz. Samples were prepared in deuterated chloroform containing 0.1% TMS.

0.8 mL/min and 40°C column temperature [18]. The relative concentrations of the ester products were determined by normalized areas of peaks. To identify the monosubstituted ester, disubstituted ester, and triacylglycerol, standards of monolein, diolein, and triolein were employed. The peak for the monosubstituted ester comprises the monosubstituted DEG plus monoacylglycerol, and the peak for the disubstituted ester comprises the disubstituted DEG plus diacylglycerol. To distinguish the co-eluted products, NMR spectroscopy was used for the products obtained under the optimized conditions.

Results and Discussion

Figure 1 shows a general scheme for product formation when castor oil reacts with DEG. In the first step, the catalyst (KOH) dissolves in DEG, leading to an alkoxide salt after heating, as proposed by Gok [19]. The reaction of the DEG with the metal hydroxide produces water, which can be eliminated by evaporating since the DEG boiling point is 245°C. This alkoxide salt can react again with another catalyst molecule and generate a di-alkoxide salt. These alkoxides salts react with the triacylglycerol (TAG) once or twice, forming the monosubstituted or disubstituted DEG.

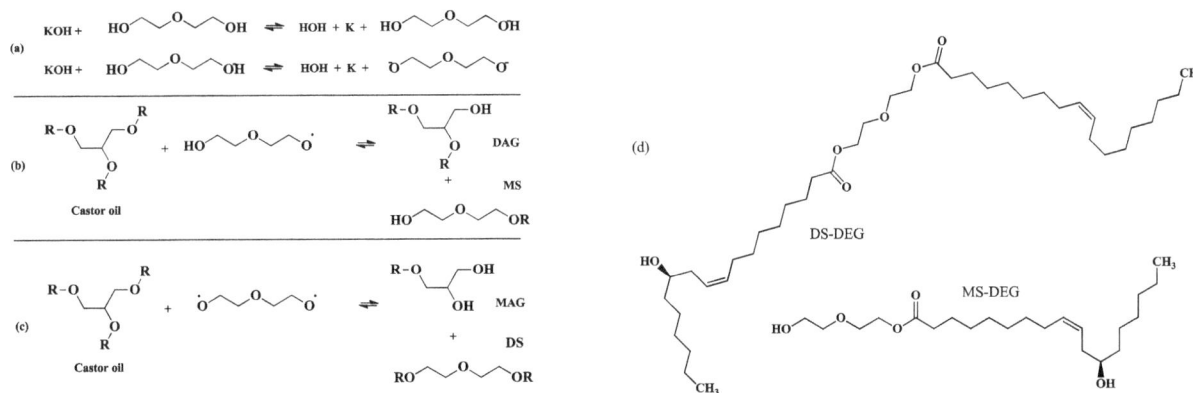

Figure 1. Scheme of alkoxy generation [a], monosubstituted DEG (MS) [b] and disubstituted DEG (DS) [c] formation with the exact structures [d]. R= ricinoleic acid structure, DAG= diacylglycerol and MAG= monoacylglycerol.

The Taguchi design is a helpful tool to evaluate 4 parameters with few runs, affording economies of time and reagents. The mass yield (mass of the material obtained after the rota-evaporation step in % relative to initial CO mass) and reaction conversion (% of products, DS and MS related to initial 100% TAG) are described in Table 2. In general, the transesterification reaction showed a good mass recovery of ester (mass yield), except runs 4 and 7,

in which mass recovery was compromised by emulsification during the water wash step. The values of mass yield above 100% are explained by the molar quantities of the compounds in the transesterification of castor oil with DEG. One mol of castor oil has approximately 933.4g and considering 100% reaction conversion to mono ester of DEG, three mols of MS product generate 1159.8g; which results in 124% of mass yield.

Table 2. Mass yield and composition of esters produced in transesterification with DEG

run	DEG:CO ratio	T (°C)	Time (h)	KOH (w:w%)	Mass Yield (%)	Composition (%)			Reaction conversion (%)
						TAG	DS	MS	DS+MS
1	3:1	150	2	0.1	99.23	82.03	13.47	4.50	17.97
2	3:1	175	4	0.3	101.08	7.57	32.32	60.12	92.44
3	3:1	200	6	0.5	107.46	3.42	36.79	59.79	96.58
4	6:1	150	4	0.5	42.06	4.55	20.12	75.33	95.45
5	6:1	175	6	0.1	113.87	17.11	26.93	55.96	82.89
6	6:1	200	2	0.3	113.55	0.72	19.11	80.16	99.27
7	9:1	150	6	0.3	63.42	9.28	20.01	70.71	90.72
8	9:1	175	2	0.5	104.18	0.83	12.23	86.94	99.17
9	9:1	200	4	0.1	117.41	2.28	15.54	82.19	97.73

Direct evidence of the reaction conversion is the remaining TAG content in each condition. Runs 1 and 7, using the lowest temperature combined with low and medium catalyst concentrations, yielded products with a high TAG content. On the other hand, in run 6, with high temperature and medium catalyst concentration, only 0.72% TAG remained, indicating a high conversion. Cavalcante obtained 97.89% conversion by employing a response surface reacting castor oil and ethanol catalyzed by KOH[20]. The optimized condition in ethanolysis was 11:1 ethanol:castor oil ratio, 90 min and 1.75% KOH. With a similar objective of producing a new polyol, Valero reacted castor oil with pentaerythritol at 210°C for 2 h with 0.05% litharge catalyst and 10% mass of pentaerythritol. The conversion obtained was 21.2% monosubstituted pentaerythritol, 8.5% disubstituted pentaerythritol, with 20.8% DAG, 40.5% MAG and 9% other components remaining[13]. Amiza obtained approximately 40% MAG and 50% DAG from the transesterification of palm oil employing 0.5 KOH as catalyst [14] In this study, except for run 1, the conditions led to a good

conversion reaction, higher than 90%, considering the sum of MS- and DS-DEG determined by GPC. Also, run 8, with a high molar ratio and 175°C, gave approximately 87% yield only in MS-DEG. Clearly, modulating the independent factors changes the quantity of MS- or DS-DEG obtained.

Considering the amount of MS-DEG as a dependent variable, Figure 2 (a) shows the relationship of the Taguchi analysis. This analysis suggests that a high DEG:CO ratio, temperature, and catalyst concentration, and a medium time favored the production of MS-DEG. The principal factor that influenced the MS-DEG was the initial amount of DEG employed; for this set of experiments, the correlation matrix between the quantities of MS-DEG and DEG was 0.68. On the other hand, the Taguchi analysis with DS-DEG as the dependent variable (Figure 2, b) suggested that a low DEG:CO ratio, medium temperature and catalyst concentration, and a long time produced a larger quantity of DS-DEG. In this case the factor with the most impact on DS-DEG was time, with a correlation matrix of 0.69.

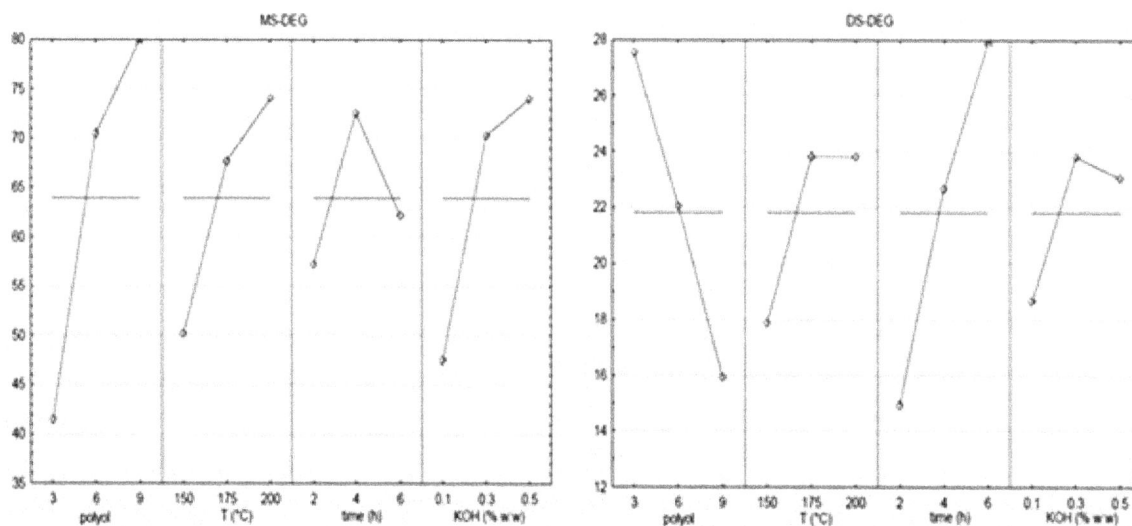

Figure 2. Factor analysis from the Taguchi design.

The FTIR spectroscopy showed a characteristic profile of aliphatic compounds for both the castor oil and transesterification products. This is evident in Figure 3, attributing to castor oil the bands of C-H stretch from the CH_2/CH_3 groups at 2926 and 2956 cm^{-1}, the methyl bend at 1458 and 1377 cm^{-1} (asymmetric and symmetric), and the methylene bend at 1463 cm^{-1}. The ethylene C=C group found in the ricinoleic acid structure (C9-C10) was characterized by C-H stretching and bending at 3007 and 1417 cm^{-1}. Also, the hydroxyl group present on C12 is demonstrated by O-H stretching at 3419 cm^{-1} (associated) and bending at 1352 cm^{-1}, and the C-O stretch from secondary alcohol at 1097 cm^{-1}. The ester group is present at 1745 cm^{-} corresponding to C=O stretch and O-C stretch at 1240 and 1164 cm-^{1}. The spectra obtained from the different runs were similar to each other and also to the castor oil. The bands mentioned above for the experimental runs had very close values to castor oil. The main importance of the FTIR analysis is to emphasize the additional broadening and intensification at O-H stretch 3397 cm^{-1}, presence of primary hydroxyl groups at 1058 cm^{-1} from C-O stretch, and 1126 cm^{-1} corresponding to the ether linkage from DEG [21].

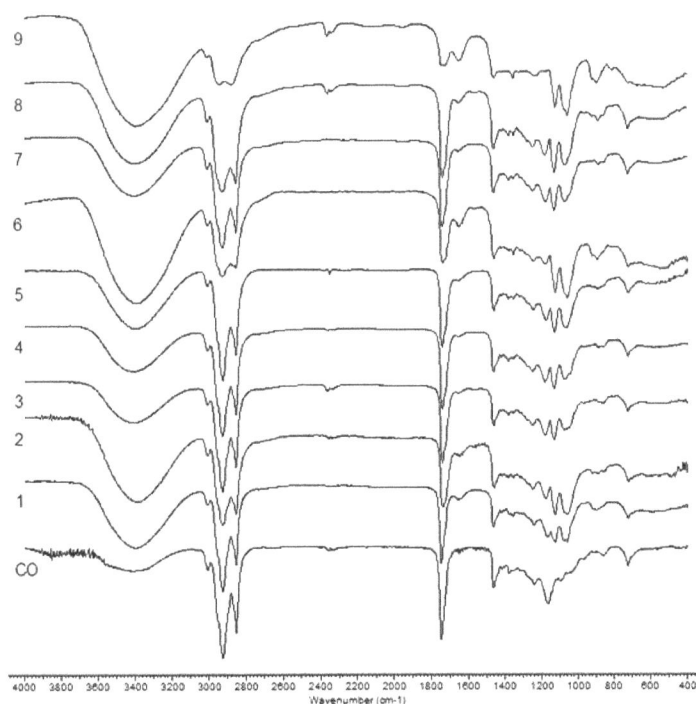

Figure 3. FTIR spectra of the castor oil (CO) and runs under different conditions.

The hydroxyl value determined for the castor oil is similar to those reported by other authors, such as Teramoto in which cator oil showed 161mg KOH/g oil[22]. The oil employed had a hydroxyl value of 159, and the transesterification procedure led to products with higher hydroxyl contents (Table 3). The highest content was obtained with run 9 (OH 492) and run 6 (OH 408); the lowest OH number was obtained in run 4 (227), which is much higher than the castor oil OH value. These high hydroxyl values are important for the production of polyol monomers. In the present study, the products had a relatively low molecular weight. This can influence the hydroxyl content; Xu synthesized a polyol with a moderate molecular weight and a hydroxyl content of 43.5[11]. The starting material of Xu was methyl ricinoleate, with a hydroxyl content of 180. The reaction with DEG was done in two stages: 200°C for 105 min, followed by vacuum at the same temperature for 90 min. The key to obtain the product was the molar ratio employed, 7.7:1 methyl ricinoleate:DEG. Some of the runs shown had a similar hydroxyl content to the product obtained by Valero, 310 (OH number) with 29% of the esters derived from pentaerythritol [13].

Table 3. Hydroxyl content of the transesterification products from castor oil and DEG

Run	Hydroxyl	StdDev	RSD
Castor	159.10	2.27	1.43
1	331.62	19.83	5.98
2	308.35	6.08	1.97
3	263.31	1.22	0.46
4	227.04	0.67	0.29
5	291.39	3.77	1.29
6	408.02	6.86	1.68
7	321.34	9.43	2.93
8	350.60	12.69	3.62
9	492.91	14.31	2.90

*Relative standard deviation

Based on the Taguchi analysis (Figure 2), two new runs were proposed, one focusing on the production of MS-DEG and the other on DS-DEG. The conditions suggested for a maximum MS-DEG yield were a 9:1 DEG:CO ratio, 200°C for 4 h, employing 0.5% catalyst. The suggested conditions to produce DS-DEG were a 3:1 DEG:CO ratio, 175°C for 18 h, with 0.3% catalyst. The main characteristics of the products are described in Table 4.

Table 4. Compositions of products from conditions suggested by the Taguchi design

run	DEG:CO	T (°C)	Time (h)	KOH	Mass Yield (%)	Composition (%)			
						oligomers	TAG	DS	MS
MS-DEG	9:1	200	4	0.5	122.1	-	0.26	14.67	85.07
DS-DEG	3:1	175	18	0.3	130.7	5.47	5.47	32.9	58.44

In general, the results obtained using the suggested conditions were close to those obtained in the experiments done with the Taguchi experimental design. This indicates that the levels for each factor in the Taguchi design must be chosen with care, so that the Taguchi design is not developed with levels that depart too much from the best value. Both conditions showed contents of MS-DEG and

DS-DEG that were similar to some conditions obtained in the experimental design. The difference in the suggested conditions is the remaining quantity of TAG in MS-DEG, which is quite low. The hydroxyl content of MS-DEG was 407.70 ±12.57 (RSD 3.08%) and for DS-DEG was 425.63 ± 18.08 (RSD 4.25%). Again, these values are interesting and

important when choosing a candidate polyol monomer.

The approximately 5% TAG remaining in DS-DEG is intriguing. With a longer reaction time, 5% TAG may represent a reversion in the reaction or the beginning of oligomerization, since ricinoleic acid had a hydroxyl group at C12

There could be an error inherent in the GPC quantification method used.

In this technique the compounds are separated by their molecular weight, and compounds with similar molecular weights may overlap. Another technique was employed to estimate the quantity of acylglycerols and DEG esters. Figure 4 shows the NMR spectra of the products from the suggested conditions.

To determine the best signal to estimate the DEG ester products, a standard compound was synthesized. The free fatty acids from castor oil were esterified with DEG employing manganese acetate as the catalyst at 175°C for 2 h. The goal of this reaction was to synthesize MS-DEG; and its quantity determined by HPLC was above 98%. In the NMR spectrum generated by the synthesized standard, it is clear that the CH_2 group attached to an oxygen from the ester linkage shows a distinct triplet near 4.25 ppm. The other CH_2 groups from the DEG structure generate a group of poorly resolved signals in the range of 3.5-3.75 ppm, restricting their use in quantitative analysis. Compton showed that the 1H NMR spectrum is a useful tool to quantify MAG and DAG[23]. The remaining acylglycerols in the DEG ester products, such as MAG and DAG, showed signals near 3.9, 4.05 and 4.09 ppm, while TAG showed a signal near 5.23 ppm. It is expected that MAG exists as glycerol substituted at C1 or C2. The C1 substituted glycerol is responsible for the signal at 3.9 ppm, and the C2 substituted glycerol signal produces the signal at 4.09 ppm [15].

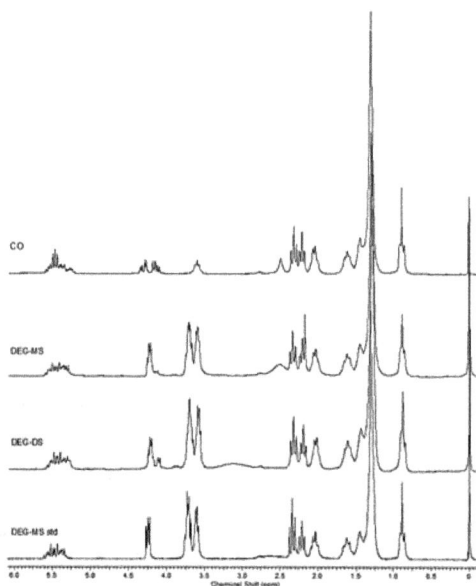

Figure 4. NMR spectra of suggested conditions from Taguchi design.

Table 5. Estimated ester composition (%)

	DEGE	TAG	DAG	MAG*
MS-DEG	84.14	1.99	3.15	10.72
DS-DEG	64.61	1.7	2.52	31.17

*sum of MAG C1 and C2

The use of the NMR technique to estimate components in the ester mixture has advantages and disadvantages. The advantages include the ability to detect each possible compound present in the mixture, rapidity, and economy in use of reagents. In the NMR spectra, the signals from each compound group appear in a different region. The GPC analysis employed cannot adequately separate MS-DEG from MAG, or DS-DEG from DAG. By NMR, there is a significant quantity of MAG in MS-DEG (Table 5), which was not revealed by the chromatogram. Also, the real conversion for sample MS-DEG was 82.37%, counting both MS- and DS-DEG compounds. One of the disadvantages of NMR in this case is the overlap of the MS- and DS-DEG signals that could be used to quantify components in the mixture. The signal employed is the CH2 group from DEG attached to the O atom of the ester group. This CH2 group has the same neighbor in the MS- and DS-DEG structures, preventing them from being distinguished.

Conclusion

The transesterification of castor oil can be accomplished by employing KOH as a catalyst. This can reduce the cost and toxicity of the process, e.g., compared to lead catalysts. The Taguchi design for the transesterification of castor oil with DEG demonstrates the importance of temperature in this process. To obtain good conversion, higher than 80%, high temperatures are needed. The desired properties of the products, such as more monosubstituted DEG or more disubstituted DEG is linked to the conditions of the reaction. The condition that favors monosubstituted DEG is 9:1 DEG:CO molar ratios, 200°C for 4 h and 0.5% KOH.

The results demonstrated that the GPC can be a helpful technique to monitor and estimate the transesterification products and reaction conversion. Although the MAG is not precisely determined by GPC, this technique is important to monitor the reaction conversion, the change of the oil in to products. The characteristics of each polyol are important for selecting a specific product to use as a new polyol for the polyurethane industry.

Acknowledgments

The authors thank to supporting agency FINEP and Ioto International Ltda in developing this work.

References

1 - A.S. César, M.O. Batalha. Energy Policy, **2010**, 38, 4031-4039.

2 - A.S. Carlsson. Biochimie, **2009**, 91-6, 665-670.

3 - J. Salimon, N. Salih, E. Yousif. Eur. J. Lipid. Sci. Technol., **2010**, 112, 519-530

4 - V. Sharma, P.P. Kundu. Progress in Polymer Sci., **2006**, 31, 606-632.

5 - M.Z. Arniza, S.S. Hoong, Z. Idris, S.K. Yeong, H.A. Hassan, A.K. Din, Y.M. Choo. J. Am. Oil. Chem. Soc., **2015**, 92, 243–255.

6 - Z.S. Petrovic. Contemporary Materials, **2010**, I-1, 39-50.

7- S. OPREA. Polymer Bulletin, 2012, 65, 8

8- J. Lin, A. Arcinas, L. A. Harden. Lipids, 2009, 44, 359–365.

9 - M.F. Valero, L.E. Díaz. Quim. Nova, **2014**, 37-9, 1441-1445.

10 - D.S. Ogunniyi. Bioresource Technology, **2006**, 97, 1086-1091.

11 – Y. Xu, Z. Petrovic, S. Das, G.L. Wilkes. Polymer, **2008**, 49, 4248-4258.

12 -M. Valero, A. Gonzalez. J. Elastomer and Plastics, **2010**, 42, 255-265.

13 – M. Valero, J.E. Pulido, A. Ramirez, D.C. Camargo, D. Navas. Polímeros, **2011**, 21-4, 293-298.

14 - S.M. Cakić, I.S. Ristić, M.M. Cincović, D.T. Stojiljković, C.J. János, C.J. Miroslav, J.V. Stamenković. Progress in Organic Coatings, **2015**, 78, 357-368.

15 - T. Wang, C. Huang. Eur. J. of Operational Research, 2007, 176, 1052-1065.

16 - K. Ramezani, S. Rowshanzamir, M.H. Eikani. Energy, **2010**, 35, 4142-4148.

17 - M. Hvalec, A. Goršek, P. Glavič. Acta Chim. Slov., **2004**, 51, 245–256.

18 - W. Schoenfelder. Eur. J. Lipid Sci, **2003**, 105, 1, 45-48.

19 - H.Y.F. Gok, J. Shen, S. Emami, M.J.T. Reaney. J. Am. Oil. Chem., **2012**, 90, 291-298.

20 - K.S.B Cavalcante, M.N.C.Penha, K.K.M.Mendonça, H.C.Louzeiro, A.C.S.Vasconcelos, A.P. Maciel, A.G.de Souza, F.C. Silva. Fuel, **2010**, 89, 1179-1186.

21 - R. Silverstein, F.X. Webster, D. Kiemle. Spectrometric Identification of Organic Compounds. 7th edition. John Wiley & Sons: New York, **2005.**

22 - N. Teramoto, Y. Saitoh, A. Takahashi, M. Shibata. J. Appl. Polymer Sci., **2010**, 115, 6, 3199-3204.

23 - D.L. Compton, K.E. Vermillion, J.A. Laszlo. J. Am. Oil. Chem. Soc., **2007**, 84, 343-348.

11

Recovery of hexavalent chromium from water using photoactive TiO$_2$-montmorillonite under sunlight

Ridha Djellabi [1,*], Mohamed Fouzi Ghorab [1], Claudia Letizia Bianchi [2], Giuseppina Cerrato [3] and Sara Morandi [3]

[1] Laboratory of Water Treatment and Valorization of Industrial Wastes, Chemistry department, Faculty of Sciences, Badji-Mokhtar University, BP12 2300, Annaba, Algeria

[2] Università degli Studi di Milano, Dip. Chimica & INSTM-UdR Milano, Milano, Italy

[3] Università degli Studi di Torino, Dipartimento di Chimica & NIS Interdepartmental Centre & Consorzio INSTM-UdR Torino, Italy

Abstract: Hexavalent chromium was removed from water under sunlight using a synthesized TiO$_2$-montmorillonite (TiO$_2$-M) employing tartaric acid as a hole scavenger. Cr(VI) species was then reduced to Cr(III) species by electrons arising from TiO$_2$ particles. After that, the produced Cr(III) species was transferred to montmorillonite due to electrostatic attractions leading to set free TiO$_2$ particles for a further Cr(VI) species reduction. Furthermore, produced Cr(III), after Cr(VI) reduction, does not penetrate into the solution. The results indicate that no dark adsorption of Cr(VI) species on TiO$_2$-M is present, however, the reduction of Cr(VI) species under sunlight increased strongly as a function of tartaric acid concentration up to 60 ppm, for which the extent of reduction is maximum within 3 h. On the other hand, the reduction extent of Cr(VI) species is maximum with an initial concentration of Cr(VI) species lower than 30 ppm by the use of 0.2 g/L of TiO$_2$-M. Nevertheless, the increase of the Cr(VI) initial concentration led to increase the amount of Cr(VI) species reduced (capacity of reduction) until a Cr(VI) concentration of 75 and 100 ppm, for which it remained constant at around 221 mg/g. For comparison, the increase of Cr(VI) species concentration in the case of the commercial TiO$_2$ P25 under the same conditions exhibited its deactivation when the reduced amount decreased from 198.1 to 157.6 mg/g as the concentration increased from 75 to 100 ppm.

Keywords: TiO$_2$-montmorillonite; Deactivation; Hexavalent chromium; Reduction; Water; Sunlight.

Introduction

Chromium is a common heavy metal contaminant in manufacturing regions because of its environmental risk [1, 2]. It primarily exists in aqueous effluents both in the form of anion of Cr(VI) (chromate) and cation Cr(III) species. Cr(VI) species is very toxic to most organisms and has been classified as carcinogenic and mutagenic, whereas Cr(III) species is less toxic and readily precipitates as Cr(OH)$_3$ [3, 4]. Furthermore, Cr(VI) species is far more mobile than Cr(III) species and more difficult to remove from water. For these reasons, the maximum concentration limit of Cr(VI) species allowed by the European Union for discharging into industrial and civil wastewaters is 0.2 mg/L [5]. The World Health Organization (WHO) prescribed 0.5 mg/L as a maximum level of Cr(VI) species for drinking water [6]. As, Cr(VI) species exhibits a higher toxicity and mobility than Cr(III), the remediation of chromium polluted wastewaters by its reduction from Cr(VI) to Cr(III) species is an effective approach frequently chosen. Among the various reduction methods, such as chemical, microbial and electrochemical [7], photocatalytic reduction of Cr(VI) species has been mainly used in recent years [8-11]. This process consists of the irradiation with UV light of a semiconductor, like for instance TiO$_2$: it results in excited electron-hole pairs that can be applied as a redox system for the degradation/reduction of pollutants in water, including the possible use of solar irradiation and operation at ambient conditions [12, 13]. In the context of metal reduction, this reaction is highly dependent on the metal redox potentials relative to the TiO$_2$ conduction band (CB) edge. The energy of the semiconductor band gap must be more negative than the energy of the metallic couple (M^{n+}/M$^{(n-z)+}$) [14, 15]. Furthermore, in order to ensure the reduction of metals by the photogenerated electrons, the promotion by additives (A) as a holes scavenger, usually in the form of organic compounds, is required [16, 17].

*Corresponding author : Ridha Djellabi
Email adress : ridha.djellabi@yahoo.com

Hence, it is necessary to choose an appropriate sacrificial electron donor. In our previous work [18], we studied the effect of a wide kind of hole scavenger molecules on the photoreduction of Cr(VI) using TiO_2 under sunlight.

TiO_2 is usually used in photocatalytic reduction of Cr(VI) species due to its excellent chemical and optical properties [19-21]. The efficiency rate of the photocatalytical reduction of Cr(VI) species by TiO_2 depends on Cr(VI) species concentration in the TiO_2 vicinity, as well as both the number of electrons produced and the role played by the organic additive as holes/radicals scavenger. By assuming a constant surface area of pure TiO_2 and using an efficient holes scavenger molecule with a sufficient concentration, the amount of Cr(III) species produced and deposited on the TiO_2 surface increases with the increase of the initial concentration of Cr(VI) species up to a maximum. In this condition, the TiO_2 surface is totally covered by Cr(III) species which will be inactive to reduce more Cr(VI) species. Furthermore, a large amount of Cr(III) species could be set free from the TiO_2 surface after saturation of the solution. Because Cr(III) species also represents a high hazard for both environment and human health at a certain concentration and also can be oxidized to Cr(VI) species in wastewaters, it must be removed by another treatment, like for instance ion-exchange, membrane filtration or by simple chemical precipitation, all these processes leading to increase the global treatment costs. Recently, in the context of a better and environmental-friendly chromium removal, a combination of supported materials containing TiO_2 particles has been demonstrated the feasibility of Cr(VI) species reduction with the subsequent adsorption of Cr(III) species on the supported material. Wen Liu *et al.* have studied the synergy of photocatalysis and adsorption for simultaneous removal of both Cr(VI) and Cr(III) species with TiO_2 and titanate nanotubes (TNTs) [22]. The nanotubes play two important roles: (i) the decrease of electron/hole pairs recombination and (ii) the transfer of the Cr(III) species produced from the TiO_2 surface to the nanotubes material. In this work, we used the TiO_2-montmorillonite composite for the reduction of chromium species.

Montmorillonite, a natural clay silicate, possesses a 2:1 layer structure: the negative charge between layers, caused by substitution, is neutralized by different hydrated cations adsorbed to the sheets surface. A characteristic feature of montmorillonite clay is the extensive surface for the adsorption of water and cations; therefore its cation-exchange capacity is very high.

These proprieties give to TiO_2-pillared montmorillonite material a very special advantage for the removal of chromium species from water by photocatalysis if compared to commercial TiO_2 materials. After the reduction of Cr(VI) species by the TiO_2 particles fixed on the montmorillonite support, the Cr(III) species produced is directly attracted and adsorbed by the negative charges of the montmorillonite itself, thus leading to: (i) set free TiO_2 sites for a further Cr(VI) species reduction; (ii) avoid the penetration of Cr(III) species into the solution.

The present work demonstrates the efficiency of the TiO_2-montmorillonite system for the reduction and recovery of Cr(VI) species under natural sunlight.

Materials and methods

Materials

The montmorillonite used in this study is a natural sodium-exchanged bentonite (Na-M) from the Roussel deposit in Maghnia (Algeria) and was used without any further treatment or purification. Its cation-exchange capacity was determined by the methylene blue method and is 89.30 mmol/100 g. Potassium dichromate (Sigma-Aldrich, ≥99.0%) solution was prepared using double-distilled water. Adjustment of the pH solution was achieved with H_2SO_4 (Sigma-Aldrich) and monitored by a pH meter (HANNA HI 9812-5). Tartaric acid (Sigma-Aldrich, ACS reagent, ≥ 99.5%) was used as a holes scavenger.

Synthesis and characterization of of titania-montmorillonite

Titania-montmorillonite (TiO_2-M) was prepared by impregnation Na-M with $TiCl_4$ (Aldrich, 99.99%). The weight ratio of Ti/montmorillonite during the preparation was 10% (g/g). The experimental synthesis procedure and the characterization of the materials by different methods (FTIR, TG-TDA, BET, XRD and SEM-EDX) are reported in our previous paper [23]. In the present paper, we added the results of TiO_2-M XPS analysis performed using M-probe Apparatus (Surface Science Instruments) equipped with a monochromatic source of Al Kα (1486.6 eV).

Photocatalytic tests

The photocatalytic reduction of Cr(VI) species was performed using a static batch reactor, consisting of Pyrex beakers open to air under natural sunlight at sea level (in front of the Chemistry Department without any obstacle) on sunny days (at Annaba University) and were started at 10:00 am for a duration of 4 h. The Cr(VI) solution (250 mL) in the presence of the photocatalyst and tartaric acid was exposed to natural sunlight under a constant stirring. In this study, we used a pH value of 2.2 based in the results of our previous study [18]. The reduction of Cr(VI) needs an acidic medium in order to ensure a high difference between the energy level of the conduction band (E_{CB}) of TiO_2 and the redox Cr(VI) potential (E_0 Cr(VI)/Cr(III)). The use of tartaric acid as a hole scavenger does not change the pH of the solution since the working pH was fixed after adding of tartaric acid and the solution pH was controlled and was keeping at 2.2 along the experiments by adding dropping of NaOH or H_2SO_4 if it changed. During the experiments, samples (4 mL) were collected at selected time intervals. The photocatalyst was removed by filtration (0.45 μm, Whatman) and the residual concentration of Cr(VI) species was determined at a wavelength of 540 nm via the 1,5-diphenylcarbazide (DPC) method [25]. The intensity of sunlight radiation at 365 nm was measured using a VLX-3W radiometer (Vilber Loumart, France) with a cell diameter of 1 cm^2: it always was in the 16 to 18 W/m^2 range during the experimental period. The extent of water evaporation during the solar photocatalysis experimental was in average of 6.0% in volume after 4 h of irradiation and was considered to be within the experimental errors. Due to the inherent non reproducibility of solar radiations and in order to minimize the experimental errors, the study of the effects of each parameter was performed simultaneously on a set of parallel experiments.

The extent of Cr(VI) reduction species was calculated using Eq.(1). The kinetics of the reduction process was investigated using a pseudo-first-order reaction Eq.(2). The amount of Cr(VI) species removed (mg/g) from the solution was estimated using Eq. (3).

$$\text{Extent of reduction} = \frac{(C_0 - C_t)}{C_0} \times 100 \qquad (1)$$

$$\ln \frac{C}{C_0} = -k.t \qquad (2)$$

$$Q_{removed} = \frac{(C_0 - C_t) \times V}{m \times 1000} \qquad (3)$$

Where C_0 and C_t represent the Cr(VI) concentrations (mg/L) before and after the treatment, t is the irradiation time (min); k is the apparent rate constant; V is the solution volume (mL); m is the mass of the photocatalyst (g).

Results and discussion

Characterization of the material
As reported in our previous papers [23, 24], the main results of material characterization can be summarized in the following points:

- FTIR spectra in the 4000-1200 cm^{-1} range, for both sample (Na-M and TiO$_2$-M) showed different bands like that of OH groups, SiO$_2$-like matrix and water. However, Na-M exhibits a sharp and complex spectral component at 1490 cm^{-1} which is totally absent in TiO$_2$-M.

- The specific surface area has been changed after the introduction of TiO$_2$ from 49 m^2/g of the starting material (Na-M) to 51 m^2/g of TiO$_2$-M. The change of the pore volume is more evident (from 0.107 to 0.144 cm^3/g): for Na-M and TiO$_2$-M samples, respectively.

- TG/DTA curves of Na-M and TiO$_2$-M samples in the 25-500°C temperature range reported a similar weight loss for both materials, most likely due to the evaporation of physisorbed/adsorbed water.

- XRD pattern of TiO$_2$-M exhibits the crystalline features of titania, in which only the anatase polymorph is evidenced. The average crystallite size of anatase was estimated of 15-20 nm.

- SEM image of TiO$_2$-M composite showed a clear change of the montmorillonite morphology, compared to that of Na-M, by the addition of TiO$_2$ species.

- EDX results showed that the content of Ti element in TiO$_2$-M is 48.6 wt%.

Fig.1 presents the XPS survey spectra of TiO$_2$-M sample. The chemical composition derived from XPS analysis is shown in **Table 1**. Sample depicts mainly Ti and O elements and C at 284.8 eV as well. Adventitious C is always present and is explained by the environmental contamination. High resolution Ti2p and O1s is reported in **Fig.2**. One single Ti species is present due to Ti(IV) [26]. On the contrary, O1s spectrum can be fitted with three different peaks attributed to lattice oxygen in TiO$_2$ at lower BE, OH species and O-C band in montmorillonite at higher BE.

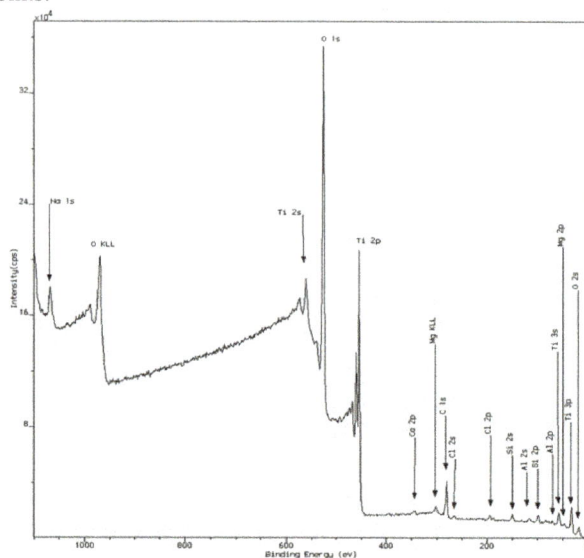

Figure 1. XPS survey spectra of TiO$_2$-M sample.

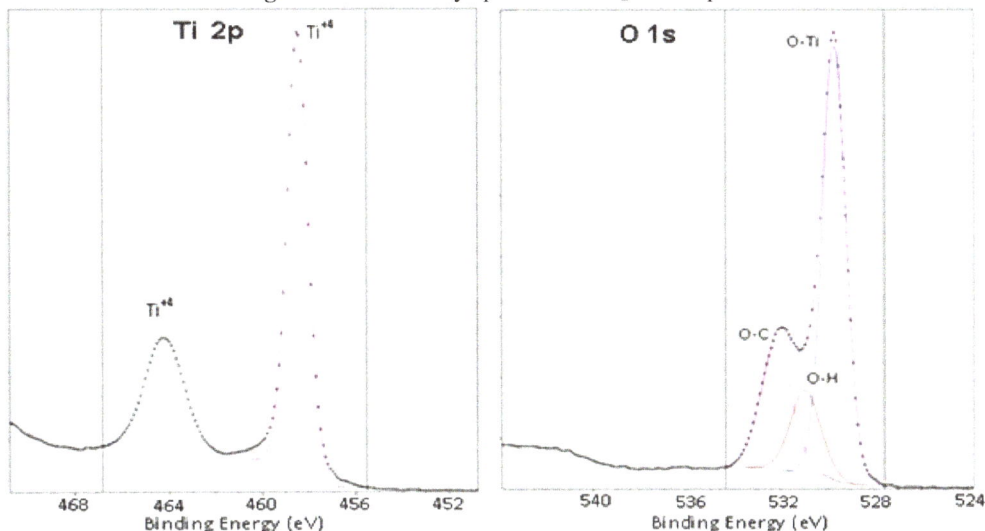

Figure 2. High resolution XPS core level spectra of Ti 2p and O 1s.

Table 1. Measured mass ratios of TiO$_2$-M sample.

Element	Na 1s	O 1s	Ti 2p	N 1s	Ca 2p	C 1s	Cl 2p	Si 2p	Al 2p	Mg 2p
Mass ratio (%)	2.05	39.78	37.22	0.24	0.53	9.41	1.30	3.74	2.47	3.25

Effect of tartaric acid concentration on Cr(VI) adsorption and reduction

One of the most important keys to reduce metal cations by photocatalysis is the presence of an efficient holes scavenger molecule. In this work, we used tartaric acid as scavenger molecule based on the results of our previous study [18], in which it was demonstrated to be the most efficient agent for the reduction of Cr(VI) species among the various tested molecules. In order to show the photocatalytic reduction of hexavalent chromium, the effect of tartaric acid concentration on the adsorption and reduction of the Cr(VI) at 30 ppm under sunlight has been studied at the 0-120 ppm range. The results of Cr adsorption and photoreduction are reported in **Figs 3** and **4,** respectively. The direct photolysis of Cr(VI) species both in the absence and in the presence of tartaric acid was negligible: as reported in **Fig.3**, it is clear that the Cr(VI) species adsorption onto the TiO$_2$-M surface is negligible. Natural clays, such as montmorillonite, are not usually efficient to adsorb anions because of their intrinsic negative charge. However, a few anionic Cr(VI) species can result on the montmorillonite surface through either physical sorption or electrostatic binding mechanisms. It is worth to note that the presence of tartaric acid increases slightly the adsorption of chromate ions, may be due to the chelating effect caused by tartaric acid molecules that can fix Cr(VI) ions on the montmorillonite surface. However, the photo-reduction of Cr(VI) species increases strongly with the increase of tartaric acid concentration **(Fig.4)** up to a concentration of 60 ppm where the reduction extent is maximum within 3 h.

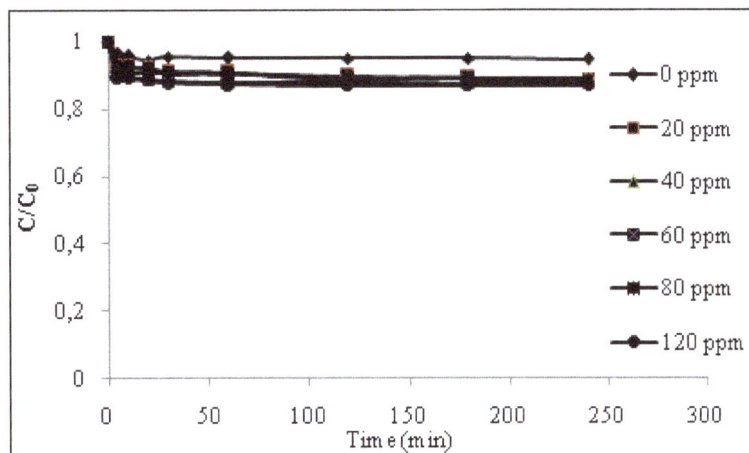

Figure 3. Effect of tartaric acid concentration on the dark adsorption of Cr(VI) on TiO$_2$-M. Conditions: [Cr(VI)] : 30 ppm, [TiO$_2$-M] : 0.2 g/L, pH : 2.2.

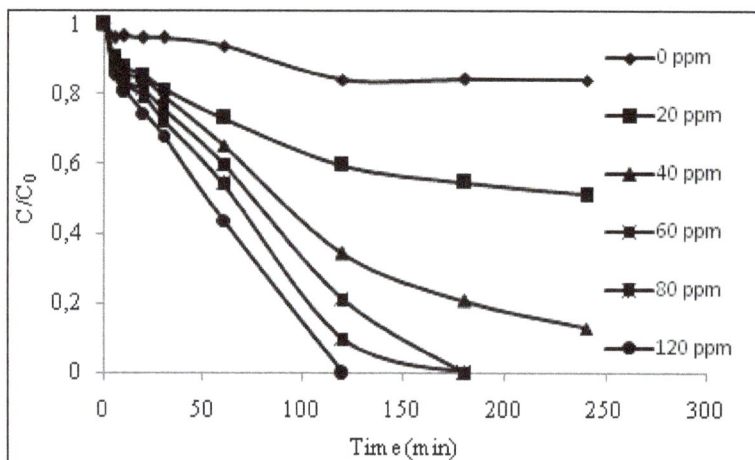

Figure 4. Effect of tartaric acid concentration on reduction of Cr(VI) using TiO_2-M under sunlight. Conditions: [Cr(VI)] : 30 ppm, [TiO_2-M] : 0.2 g/L, pH : 2.2.

This observed enhancement confirms the importance of the presence of a hole scavenger molecule that can improve the photochemical quantum which, in turn, facilitates the separation of pairs (electron/hole) by scavenging the photogenerated holes (**Eq.6**), thus (i) decreasing the chance of charges (electron/hole) recombination and (ii) setting free more electrons in the conduction band available to reduce Cr(VI) ions (**Eq.8**). Since experiments were achieved at low pH, only $Cr_2O_7^{2-}$ and CrO_4^{2-} species can exist. The net reaction of both species is a three electron-reduction of Cr(VI) to Cr(III) which are shown in Eqs. (8) and (9), respectively. Assuming that one electron transfer step could be occurred in Cr(VI)reduction process, Cr(V) and Cr(IV) species could be formed as intermediates in the first and second stages Eq. (10). Furthermore, the reaction between tartaric acid molecules and $^•$OH which possess a high oxidizing potential (**Eq.7**), limits the oxidation of the produced Cr(III) species to Cr(VI) species (**Eq.11**). As discussed before, in the present case, the removal of chromium from water by the TiO_2-montmorillonite composite passes through two important steps: (i) negatively charged species ($Cr_2O_7^{2-}$, $HCrO_4^-$), which exhibit a very low adsorption behavior on the composite surface, is reduced to a positive species (Cr(III)) by electrons coming from the conduction band of TiO_2 particles deposited on the montmorillonite material. It is worth noting that binding between Cr(VI) and TiO_2 could take place. It is worth noting that the TiO_2 particles fixed on the montmorilonite can exhibit a positive charge at low pH which increases the attraction of chromium negatively charged species ($Cr_2O_7^{2-}$, CrO_4^{2-}). In general, if a pure commercial TiO_2 is employed, the produced Cr(III) ions still deposited on the TiO_2 surface lead to its deactivation. Hence after saturation, Cr(VI) ions cannot reach the TiO_2 surface. Furthermore, the desorption of Cr(III) species from the surface to the solution in the case of pure TiO_2 is strongly possible. (ii) The second important step caused by the montmorillonite support is the transfer of Cr(III) cations from the TiO_2 particles to the negative inter-layer pores where binding can take place between Cr(III) and the negative charge of montmorillonite sheets. The fixation of Cr(III) on the montmorillonite takes place through ion exchange, physical and in some cases chemical adsorption mechanisms. On the other hand, a part of Cr(III) could be immobilized onto TiO_2 particles fixed on the momtmorillonite surface directly after Cr(VI) reduction through physical adsorption. This positive effect avoids the penetration of Cr(III) species in the solution and ensures a continuous contact of Cr(VI) species with photocatalytic TiO_2 sites. It is important to note that, for a high concentration of tartaric acid, the reduction rate of Cr(VI) species was not depleted, due to the high adsorption capacity of this material.

The main steps of the effect of tartaric acid on the reduction of Cr(VI) are reported in the following reactions and described in **Fig.5**.

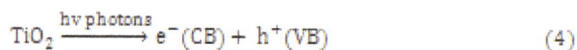

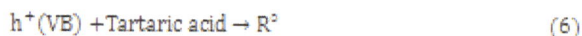

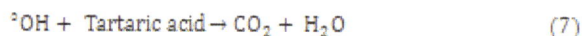

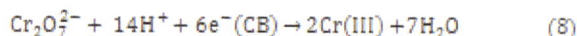

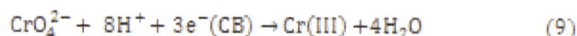

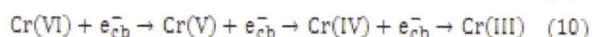

$$TiO_2 \xrightarrow{hv\,photons} e^-(CB) + h^+(VB) \tag{4}$$

$$h^+(VB) + H_2O \rightarrow {}^°OH + H^+ \tag{5}$$

$$h^+(VB) + Tartaric\;acid \rightarrow R^° \tag{6}$$

$${}^°OH + Tartaric\;acid \rightarrow CO_2 + H_2O \tag{7}$$

$$Cr_2O_7^{2-} + 14H^+ + 6e^-(CB) \rightarrow 2Cr(III) + 7H_2O \tag{8}$$

$$CrO_4^{2-} + 8H^+ + 3e^-(CB) \rightarrow Cr(III) + 4H_2O \tag{9}$$

$$Cr(VI) + e_{cb}^- \rightarrow Cr(V) + e_{cb}^- \rightarrow Cr(IV) + e_{cb}^- \rightarrow Cr(III) \tag{10}$$

$$Cr(III) + 3\,{}^°OH \rightarrow Cr(VI) + 3OH^- \tag{11}$$

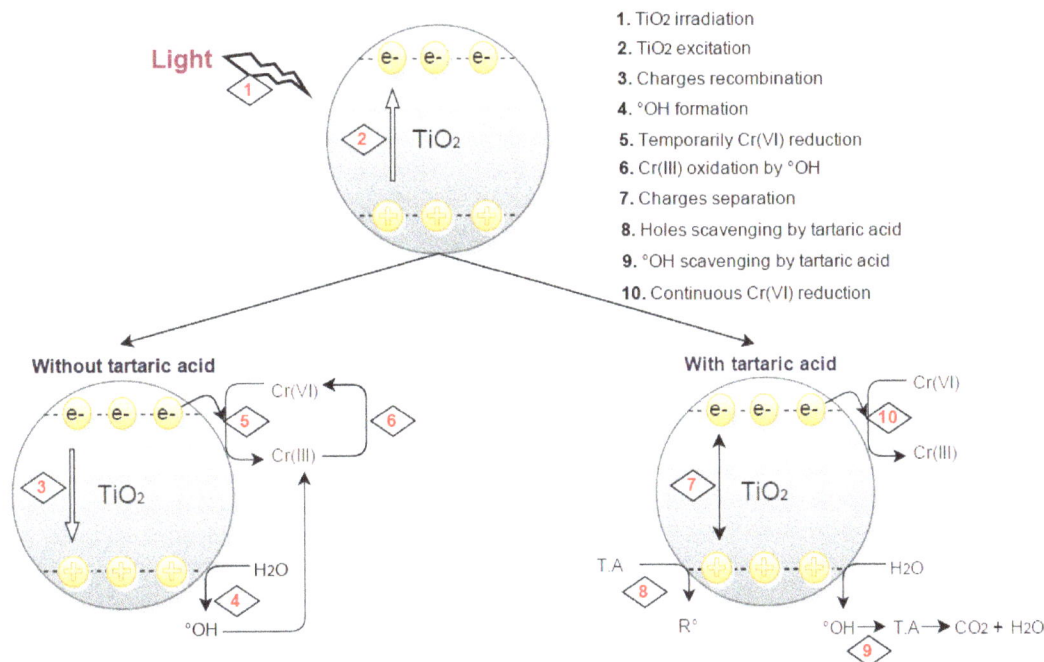

1. TiO2 irradiation
2. TiO2 excitation
3. Charges recombination
4. °OH formation
5. Temporarily Cr(VI) reduction
6. Cr(III) oxidation by °OH
7. Charges separation
8. Holes scavenging by tartaric acid
9. °OH scavenging by tartaric acid
10. Continuous Cr(VI) reduction

Figure 5. Scheme reporting the main steps of tartaric acid effects on the photoreduction of Cr(VI) species.

Effect of Cr(VI) concentration on the photoreduction efficiency

As it is known, there is a relationship between the amount of reduced ions and the amount of the used photocatalyst. For a fixed amount of photocatalyst, the amount of reduced ions increases proportionally to the increase of the initial concentration of ions in the solution up to a maximum. If we evaluate the efficiency of the photocatalyst by using the amount of reduced ions, it is important to show the effect of initial concentration of Cr(VI) on the reduced amount. **Table 2** shows the effect of Cr(VI) species concentration on both the extent of reduction and the amount of Cr(VI) species reduced using a TiO_2-M mass of 0.2 g/L.

Table 2. Effect of Cr(VI) concentration on the photoreduction efficiency. [TiO_2-M]: 0.2 g/L, pH: 2.2, [tartaric acid]: 60 mg/L

Cr(VI) concentration	20 ppm	30 ppm	50 ppm	75 ppm	100 ppm
Extent of reduction (%)	100	100	73.7	58.9	44.2
Q removed (mg/g)	100	150	184.25	220.75	221
k_{app} (min^{-1})	0.0188	0.0109	0.0052	0.0345	0.0021

The data indicate that the reduction extent of Cr(VI) species is maximum with an initial concentration Cr(VI) species lower than 30 ppm, whereas for higher values, the extent of reduction decreases. This feature may be attributed to an insufficient number of electrons produced by the photocatalyst and able to reduce all Cr(VI) ions present in the solution at high concentration, as their production is constant for a given amount of both photocatalyst and a selected irradiation time. However, the amount of reduced Cr(VI) species increases proportionally with the increase of Cr(VI) species concentration up to 75 ppm. For higher values, it remains constant at the value found for 75 ppm. When concentration of Cr(VI) species increases in the solution, more Cr(VI) ions will be in contact with the photocatalyst surface leading to the reduction of a larger amount of Cr(VI) species. For values higher than 75 ppm, the reduced amount remains constant (~ 220 mg/g) because of two possible reasons: (i) the activated surface of TiO_2-M mass (0.2 g/L) is unable to reduce more Cr(VI) species (a saturation step has been reached); (ii) the amount of tartaric acid molecules used (60 ppm) is consumed and this leads to a stop of the reduction process. In both cases, the amount of both reduced Cr(VI) species by this material is relatively high. On the other hand, as the reduced amount is still regular after saturation, we can state that the photocatalyst deactivation at concentration of 100 mg/L is not observed.

For comparison purposes, the results of the effect of Cr(VI) species initial concentration on the reduced amount of Cr(VI) species by a pure commercial TiO_2 (P25 by Evonik) under the same conditions are presented in **Fig.6**.

It is clear that the reduced amount increases with the increase of the initial concentration up to 198.1 mg/g at 75 ppm. Nevertheless, it strongly decreases to 157.6 mg/g at 100 ppm. This decrease, not observed with TiO$_2$-M, may be explained by the deactivation of TiO$_2$ particles at high concentration. The Cr(III) species produced from the reduction of

Cr(VI) species still deposited on the TiO$_2$ surface limit the access of more Cr(VI) ions to the surface.

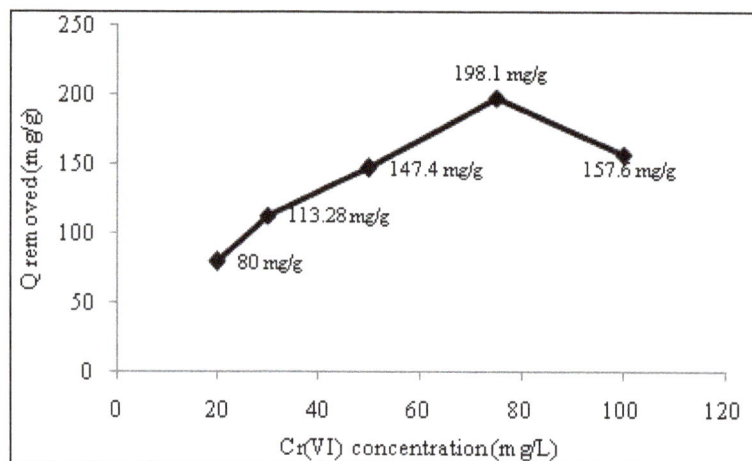

Figure 6. Effect of Cr(VI) concentration on the reduced Cr(VI) amount using commercial TiO$_2$P25 under sunlight. Conditions: [tartaric acid]: 60 ppm, [TiO$_2$P25]: 0.2 g/L, pH: 2.2.

The pathways of Cr(VI) photoreduction by commercial TiO$_2$ and TiO$_2$-M can be summarized in **Fig. 7.**

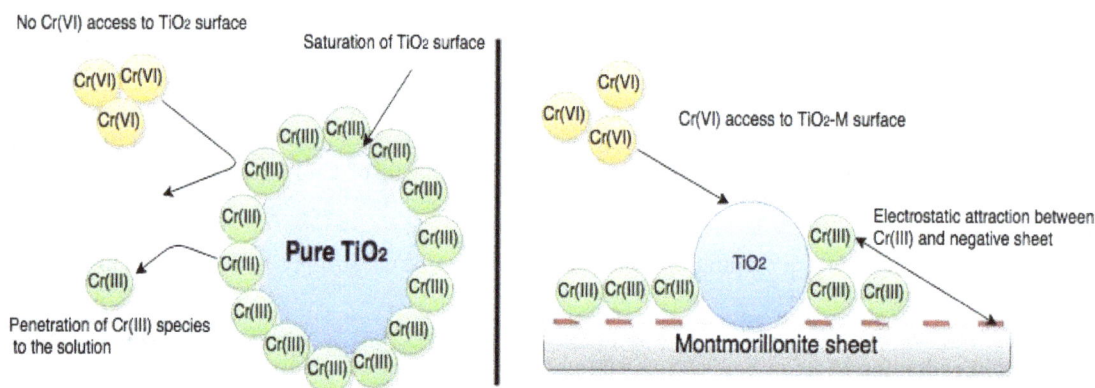

Figure 7. Pathways of Cr(VI) photoreduction by commercial TiO$_2$ and TiO$_2$-M under sunlight.

Conclusions

In the present study, we reported the high efficiency of the synthesized TiO$_2$-montmorillonite composite under sunlight for reducing and recovering chromium from water. Firstly, the characterization of this material by XPS analysis demonstrates the presence of only TiO$_2$ oxide in the Ti(IV) form in TiO$_2$-M. The use of tartaric acid as a hole scavenger shows high efficient for reducing Cr(VI) where the reduction process is proportional with the increase of its concentration. For a constant TiO$_2$-M mass, the amount of reduced Cr(VI) increases with the raise of the initial concentration of chromium until a constant value at about 220 mg/g at 75 and 100 ppm of initial Cr(VI). The non-deactivation of TiO$_2$-M at high Cr(VI) concentration can be explained by the special behavior of this composite where the produced Cr(III) is transferred from the TiO$_2$ surface to the negative interlayer of the montmorillonite support. Otherwise, with a pure

TiO$_2$ sample, like the commercial P25, a high Cr(VI) concentration leads to decrease the quantity of reduced Cr(VI) because of its saturation by Cr(III) deposited into the surface.

References

1 - D. Blowes, Tracking hexavalent Cr in groundwater, Science **2002**, 295, 2024-2025.

2 - M.M. Benjamin, Water Chemistry, McGraw-Hill, New York, **2002**.

3 - K. Selvaraj, S. Manonmani, S. Pattabhi, Removal of hexavalent chromium using distillery sludge, Bioresour. Technol **2003**, 89, 207-211.

4 - P. Miretzkya, A. Fernandez Cirelli, Cr(VI) and Cr(III) removal from aqueous solution by raw and modified lignocellulosic materials: a review, J. Hazard. Mater **2010**, 180, 1-19.

5 - F. Di Natale, A. Erto, A. Lanciaa, D. Musmarra. Equilibrium and dynamic study on hexavalent

chromium adsorption onto activated carbon. J. Hazard. Mater **2015**, 281, 47-55

6 - J. Hu, I.M.C. Lo, G. Chen, Comparative study of various magnetic nanoparticles for Cr(VI) removal, Sep. Purif. Technol **2007**, 56, 249-256.

7 - E.C. Barrera-Díaz, L.L. Violeta, B. Bilyeu. A review of chemical, electrochemical and biological methods for aqueous Cr(VI) reduction. J. Hazard. Mater **2012**, 223-224, 1- 12.

8 - Y. Liu, L. Wang, F. Wu, N. Deng, Enhancement of heterogeneous Cr(VI) reduction using clay minerals in the presence of organic carboxylic acids under UV irradiation, Desalin. Water Treat **2013**, 51, 7194-7200.

9 - T. Iwata, M. Ishikawa, R. Ichino, M. Okido, Photocatalytic reduction of Cr(VI) on TiO_2 film formed by anodizing, Surf. Coat. Technol **2003**, 169, 703-706.

10 - S. Chakrabarti, B. Chaudhuria, S. Bhattacharjeea, A.K. Rayb, B.K. Dutta, Photoreduction of hexavalent chromium in aqueous solution in the presence of zinc oxide as semiconductor catalyst, Chem. Eng. J **2009**, 153, 86-93.

11 - J. Yoon, E. Shim, S. Bae, H. Joo, Application of immobilized nanotubular TiO_2 electrode for photocatalytic hydrogen evolution: Reduction of hexavalent chromium (Cr(VI)) in water, J. Hazard. Mater **2009**, 161, 1069-1074.

12 - J.K. Yang, S.M. Lee, Removal of Cr(VI) and humid acid by using TiO_2 photocatalysis. Chemosphere **2006**, 63, 1677-1684

13 - J.M. Herrmann Heterogeneous photocatalysis: fundamentals and applications to the removal of various types of aqueous pollutants. Catal. Today **1999**, 53, 115-129.

14 - K. Kabra, R. Chaudhary, R.L. Sawhney, Treatment of hazardous organic and inorganic compounds through aqueous-phase photocatalysis: A review, Ind. Eng. Chem. Res. **2004**, 43, 7683-7696.

15 - N. Wang, Y. Xu, L. Zhu, X. Shen, H. Tang, Reconsideration to the deactivation of TiO_2 catalyst during simultaneous photocatalytic reduction of Cr(VI) and oxidation of salicylic acid, J. Photochem. Photobiol., A **2009**, 201, 121-127.

16 - M.I. Litter, Heterogeneous photocatalysis transition metal ions in photocatalytic systems, Appl. Catal., B **1999**, 23, 89-114.

17 - T. Tan, D. Beydoun, R. Amal, Effects of organic hole scavengers on the photocatalytic reduction of selenium anions, J. Photochem. Photobiol., A **2003**, 159, 273-280.

18 - R. Djellabi, M.F. Ghorab. Photoreduction of toxic chromium using TiO_2-immobilized under natural sunlight: effects of some hole scavengers and process parameters. Desalin. Water Treat **2014**, 1-8.

19 - G.Cappelletti, C.L. Bianchi, S. Ardizzone, Nano-titania assisted photoreduction of Cr(VI) the role of the different TiO_2 polymorphs. Appl. Catal. B: Environ **2008**, 78 (34), 193-201.

20 - W. Quanping, Z. Jun, Q. Guohui, W. Chengyang, T. Xinli, X. Song, Photocatalytic reduction of Cr(VI) with TiO_2 film under visible light. Appl. Catal. B: Environ **2013**, 142-143, 142-148.

21 - G. Colón, M.C. Hidalgo, J.A. Navío, Photocatalytic deactivation of commercial TiO_2 samples during simultaneous photoreduction of Cr(VI) and photooxidation of salicylic acid. J. Photochem. Photobiol., A **2001**, 138, 79-85

22 - W. Liu, J. Ni, X. Yin, Synergy of photocatalysis and adsorption for simultaneous removal of Cr(VI) and Cr(III) with TiO_2 and titanate nanotubes. Water Res. **2014**, 53, 12-25.

23 - R. Djellabi; M.F. Ghorab; G. Cerrato; S. Morandi; S. Gatto; V. Oldani; A. Di Michele; C.L. Bianchi, Photoactive TiO_2-montmorillonite composite for degradation of organic dyes in water, J. Photochem. Photobiol., A. **2014**, 295, 57-63.

24 - R. Djellabi, M.F. Ghorab, C.L. Bianchi, G. Cerrato, S. Morandi. Removal of crystal violet and hexavlent chromium using TiO_2-Montmorillonite under sunlight: effect of TiO_2 content. JCEPT., **2016**, 7: 276.

25 - J. Kotas, Z. Stasicka, Chromium occurrence in the environment and methods of its speciation. Environ. Pollut. **2000**, 107 (3), 263-283.

26 - G. Iucci, M. Dettin, C. Battocchio, R. Gambaretto, C. Di Bello, G. Polzonetti, Novel immobilizations of an adhesion peptide on the TiO_2 surface: An XPS investigation. Mater. Sci. Eng., C **2007**, 27, 1201-1206.

12

Structural and optical characterization of poly (aniline) (PANi) films obtained by spin-coating method

Zina Barhoumi [1,*], **Noureddine Amdouni** [1], **Jean François Stumbé** [2] and **Amalendu Pal** [3]

[1] Unité Physico-Chimie Des Matériaux Solides, Département de Chimie, Faculté des Sciences de Tunis, Campus Universitaire, 2092 El Manar Tunisie

[2] Laboratoire de Photochimie et d'Ingénierie Macromoléculaires, Université de Haute Alsace, 3rue A.werner 68093 Mulhouse cedex, France

[3] Department of Chemistry, Kurukshetra University, Kurukshetra 136 119, India

Abstract: The structural and optical properties of the PANi films were investigated in terms of different solvent content by using X-ray diffraction (XRD), RAMAN spectroscopy, Scanning electron microscopy (SEM), Photoluminescence (PL) and optical absorption spectroscopy. The XRD results revealed that the crystallinity through films increased as the molar ratio of dimethylacetamide (DMAc and dimethylformamide (DMF) increased to 0.6:0.4, but then decreased when the molar ratio was increased from 0.6:0.4 to 0.8:0.2. It is shown that even small clusters in PANi films can explain the observed RAMAN scattering. The results of the optical experiments showed that films prepared from different molar ratio of DMAc and DMF showed good near-infrared characteristic in the range of 800-2000 nm. The optical band gap value from absorption coefficient data is found to be 3.57 eV. The SEM also showed changes in the morphology of PANi films.

Keywords: Conducting polymer; Spin-Coating; Raman spectroscopy and scattering; Optical properties.

Introduction

In recent years, developments of inorganic-organic hybrid materials on nanometer scale have been receiving significant attention due to a wide range of potential applications and high absorption in the visible part of the spectrum and high mobility of the charge carriers. Polyaniline (PANi) has been one of the most thoroughly investigated conducting polymers due to its environmental stability, high density of charge storage capability, and low cost and simple synthesis [1-3]. These polymers have extensive applications such as anticorrosive [4-6]; sensors and biosensors [7-9], as actuators [10-11], in advanced biomedical materials [12], in electronic device [13], in transistors [14] and hole injection layers for flexible light emitting diodes [15,16]. However, such industrial applications are limited for PANi because of its poor processability and intractable nature. A p-dopable (i.e., electron deficient) material, PANi exhibits reversible electrochemical and physical properties controlled by its oxidation and protonation state properties that have been exploited for the development of several technological applications including optoelectronic device. PANi can be easily polymerized by direct chemical oxidation of aniline or electrochemical polymerization [17].

But, in the preparation of PANi dispersions, surfactants have been rarely employed as stabilizer and micelles as reaction medium. Han et al. prepared PANi nanoparticles in anionic sodium dodecylsulfate (SDS) micelles with ammonium peroxydisulfate (APS) as an oxidant. More structured PANI particles showed a higher conductivity[18]. Li and co-workers successfully prepared stable nano-polyaniline waterborne latex by emulsion polymerization of aniline in micellar solution of sodium dodecylbenzene sulfonate (SDBS), dodecylbenzene sulfonic acid (DBSA), or sodium dodecylsulfate (SDS) as both emulsifier and dopant. The uniform PANi-SDS/polyvinyl alcohol (PVA) composite films were prepared by casting a direct mixture of the PANi latex and PVA aqueous solution and good properties including film formability, anti-corrosivity, and electrical conductivity[19]. Sainz et al. obtained soluble polyaniline-multi-wall-carbon nanotube (CNT-PANi) composites. It was found that the presence of multi-wall-carbon during the polymerization of aniline induces the formation of a planar conformation of PANi[20]. Studies of Ma and co-workers [21] have suggested that the use single-stranded DNA offers better solubilization of the CNTs prior to the water-soluble nanocomposite preparation (PANi-CNTs).

Corresponding authors: Zina Barhoumi
E-mail address: zinabarhoumi112@gmail.com

This PANi-CNTs composite in functionalized forms exhibited good enhanced conductivity, thermal stability and luminescent behaviors. Therefore, conformational changes lead to favorable interactions between emeraldine base (EB) and CNTs and are responsible for both enhancement effects and the processability of the composite. The effect of the solvent type and casting temperature on the electrical properties of plastdoped polyaniline films was investigated by Bohli et al. the films prepared from PANi in dichloroacetic acid (DCAA) after heating treatment at 298 K showed the highest electrical conductivities[22]. In all these views, we endeavored to study the structural, optical properties of polyaniline films and to cater the needs of optical materials in near future.

Experimental

Materials

Aniline monomer C_6H_7N was purchased from Sigma-Aldrich and bi-distilled under vacuum before use. Ammonium persulfate $((NH_4)_2S_2O_8,$ APS) was obtained from Fluka. N, N- dimethylacetamide and dimethylformamide were obtained from Sigma Aldrich. The chemicals were used as received without further purification.

Preparation of PANi: DMAc/DMF composites film-forming solution

Polyaniline emeraldine base (PANi-EB), which was synthesized was dissolved into variable ratio DMAc/DMF mixtures to prepare a PANi (3 wt %) solution. Stock solutions were stirred using a magnetic stirrer for 1 h. A Speed line Technologies P-6000 spin coater was used to spin cast polymer solutions for 30 s onto the substrates. All films were held under vacuum overnight prior to measurements to remove any residual solvents. The dried polymer films, in which thickness values around 100 nm, were analyzed.

Characterization

The XRD spectra of the PANi films were measured on X'Pert PRO Alpha-1-ray diffractometer with CuKα (1.5405 Å), and it was operated at 40 KV voltage and 50 mA anode current. The XRD patterns were recorded in the range of 3-65° with a scan speed 0.5°/min. Raman spectra of films were carried out at room temperature in the range of 100-2000cm^{-1} using FT-Raman Bruker SENTERRA which is having Nd: YAG laser source (488nm).The Photoluminescence (excitation and emission) spectra of the PANi films were recorded on Cary Eclipse spectrophoto-fluorometer using the corresponding excitation wavelength with 2.5 nm slits. Optical properties of the samples have been analyzed using a VARIAN spectrophotometer at room temperature over a large spectral range. The wavelength range for both the reflectance and transmittance measurements was ranged between 300-2200 nm. Scanning electron microscopy experiments were made with Philips Field Effect Gun XL-30, at an accelerating voltage of 5 kV. Image treatments were performed using the digital instructions software.

Results and discussion

X-ray diffraction

Structural properties of PANi films were characterized by X- ray diffraction (XRD). XRD pattern of different films is shown in Figure 1.

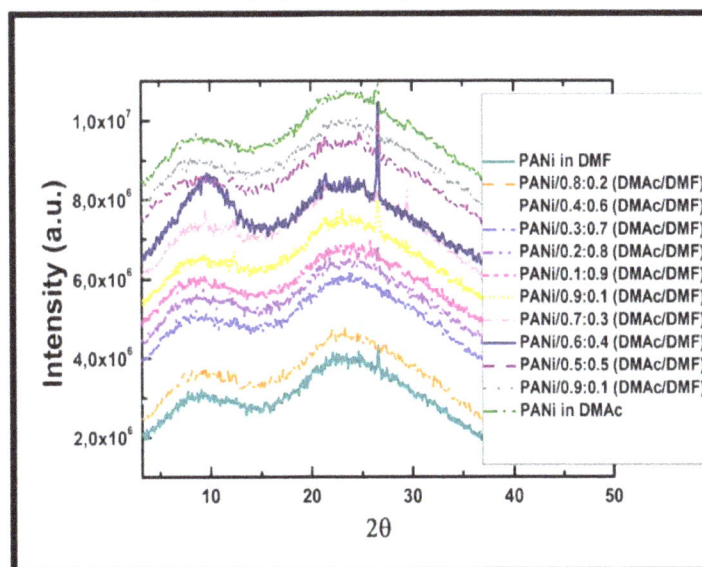

Figure 1. XRD patterns of the PANi films at different molar ratio of DMAc and DMF after heating treatment at 35℃.

As is evident in PANI, broad diffraction peaks occur between10° and 30° due to the parallel and perpendicular periodicity of the polymer (PANI) chain. The plot of PANi:0.5:0.5(DMAc/DMF) and

PANi :0.6:0.4(DMAc/DMF) in Fig.1, exhibited significant peak at diffraction angle of 27° none of which are present in it PANi:0.4:0.6 (DMAc/DMF or PANi:0.8:0.2 (DMAc/DMF). The peak observed at

2θ ~30 ° indicates the increase of crystalline structure in the PANi films[23]. T. Rajavardhana Rao et al. observed similar results for low molecular weight PVAEG and PVAEG-cu (II) systems. The diffraction peaks was sharp in Cu^{2+} complex of PVAEG films comparing with the pure PVAEG films and the crystalline phase increase after addition of Cu^{2+} ions [24]. In case of PANi: 0.6:0.4(DMAc/DMF), diffraction peaks are shifting towards higher angles from 2θ=9.7° to 10.1°. The peaks are shifting toward higher angles indicate probably the formation of a stereo-complex between polymer and solvents on the glass substrates. The crystalline structure of this polymer may be regarded as an amorphous matrix in which small crystallites are randomly distributed.

Analysis of X-ray line profile is a simple and powerful method to estimate crystallite size. Among the available methods, Scherrer analysis is a simplified method where size average was estimated [25]. The crystallite size D, which is about 13 nm, was calculated from the Scherrer equation (Eq.1):

$$D = 0.9 \, \lambda/\beta \, \cos\theta \qquad \textbf{Eq.1}$$

Where λ is the wavelength of the X-rays, θ is the diffraction angle, and β is the corrected full width, at half-maximum of the peak. The estimation of the particle size is tested further by evaluating it with Meulenkamp equation [26]. According to Meulenkamp the wavelength λ½ at which the absorption is 50% of that excitonic peak is directly related to the size of the particle via the fitted expression. The values of

particle size evaluated from Meulenkamp equation is in fair agreement with the values obtained from Scherrer method Eq.2.

$$1240 = \lambda_{1/2}(3.301 + \frac{294}{D^2} + \frac{1.09}{D}) \qquad \textbf{Eq. 2}$$

Raman spectra

The Raman spectra of films prepared in the presence of DMAc/DMF reflect the differences in the molecular structure corresponding to polyaniline. The band corresponding to the C-C stretching vibration of the benzenoid ring in the spectrum of the sample prepared in DMAc (Fig. 2-c) is shifted to 1680 cm^{-1} in comparison to the spectrum of a sample prepared in the presence of 0.6 DMAc: 0.4 DMF mixtures. Only a small shoulder at 1593 cm^{-1}, assigned to the C=C stretching of the quinoid ring vibrations [27], is observed in the spectrum. The maximum of the band located at 1490 cm^{-1} and attributed to the C=N stretching vibration in quinonoid units in the spectrum is detected. The maximum of the band attributable to $C - N^{+\cdot}$ vibrations is shifted from 1346 to 1338 cm^{-1} and its intensity is lowered. The red shift of these bands corresponds most probably to the crosslinked and phenazine-like structures present in the sample. The position of the band at 1339 cm^{-1} reveals the presence of shorter and/or crosslinked structures or both. The peak of C-N stretching vibrations of various benzenoid, quinonoid or polaronic forms observed at 1225 cm^{-1} [28].

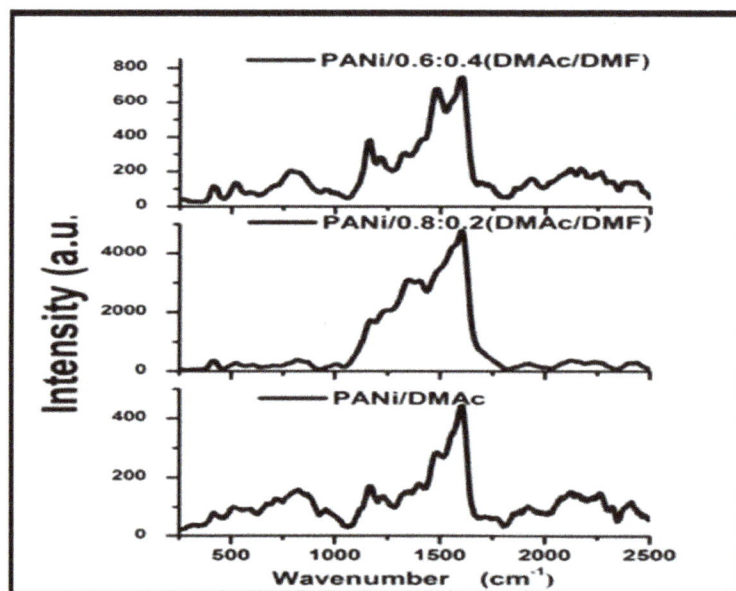

Figure 2. Resonance Raman spectra of polyaniline films: (a) PANi/0.6:0.4 (DMAc/DMF); (b) PANi/0.8:0.2 (DMAc/DMF) and (c) PANi/DMAc. 488 nm. Baseline corrected.

The shape of the spectrum of PANi film prepared from a 0.6:0.4 molar ratio of DMAc and DMF (Fig. 2a) is similar to the spectrum of PANi film prepared in 0.5 DMAc/0.5 DMF mixtures. The two most intensive bands are situated at 1600 and 1505 cm^{-1}. The two bands at 1262 and 1223 cm^{-1} of

various C-N stretching vibrations and the peak at 1171 cm^{-1} of C-H bending vibration of the semi-quinonoid rings are present in the spectrum. Bands at 815 cm^{-1} with a shoulder at 875 cm^{-1}, attributed to benzene ring deformation [29]. The characteristic peaks at 1667 and 1033 cm^{-1} are attributed to the N-H bending and C-N stretching modes in the amines, respectively, which are due to the presence of stereo complex resulting in hydrogen bonding between DMAc or DMF molecules and amine (NH) and imine (=N) sites on the PANi chains. The changes in the spectra of the PANi base derived from the PANi film prepared from a 0.7:0.3 molar ratio of DMAc and DMF (Fig. 3a) are considerably smaller comparing to those of the PANi: 0.8:0.2 (DMAc/DMF). The decrease of the shoulder at 1610 cm^{-1} is observed. The band at 1496 cm^{-1} is shifted to 1480 cm^{-1} and its intensity increases. A small peak at 1379 cm^{-1} with a band at 1339 cm^{-1} corresponding to $C - N^{+}$ vibrations of charge delocalization, moved to higher wavelength and decreased in intensity. A weak band at 1266 cm^{-1} are also present in the spectrum as well as the band at 1165 cm^{-1} of the C-H

bending vibration of quinonoid ring. In the region ~1350 cm^{-1}, the bands are related to the vibration of the protonated stretching vibration of the $C - N^{+}$ group and the changes in the intensities and positions of these bands reflect the conformational changes, leading to conversion to the PANi of the aromatic rings. The course of deprotonation is confirmed by the decrease in the intensity of the band at 809 cm^{-1}. The decrease of the band at 1262 cm^{-1} can be linked to a decrease in the number of quinonoid and semi-quinonoid rings in the structure. The marked decrease of the band at 1170 cm^{-1}, which is corresponding to the deformation vibration of the C-H groups on the quinonoid rings, confirms the diminishing number of them [28].

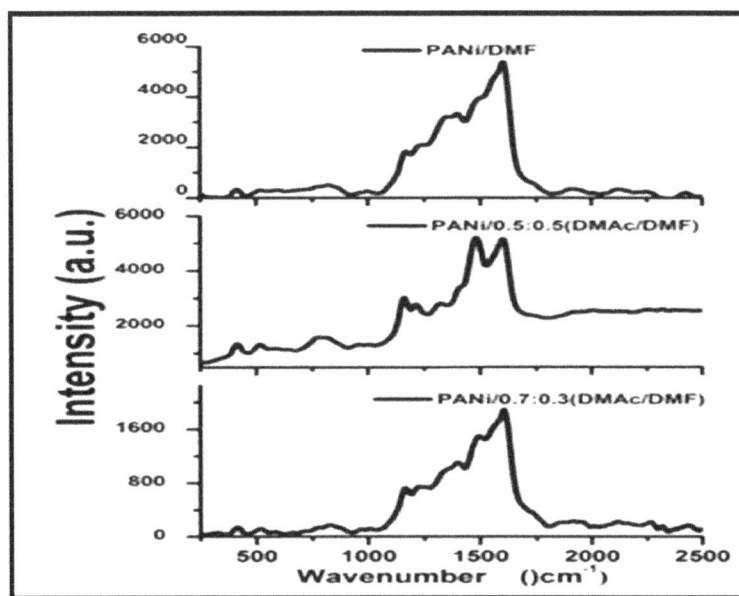

Figure 3. Resonance Raman spectra of polyaniline films: (a) PANi/DMF; (b) PANi/0.5:0.5 (DMAc/DMF) and (c) PANi/0.7:0.3 (DMAc/DMF). 488 nm. Baseline corrected.

Optical studies

The energy gap defaults to the difference between the frontier orbitals, the HOMO and LUMO. In organic conjugated systems, HOMO is typically an occupied pi bonding orbital and LUMO is typically an unoccupied anti bonding orbital. Since an organic LUMO-HOMO excitation generates a highly bound exciton instead of free electron and a hole. The optical properties of the PANi films were investigated by optical transmittance and reflectance measurements. Figures 4-5 show the optical reflectance and transmittance

spectra of PANi films .prepared at different molar ratio of compounds DMAc and DMF in the wavelength region from 300 to 2200 nm. It can be observed that PANi films exhibit high transparency in near-infrared region. The average transmittance of films is over 80% in near-infrared region from 800 to 2200 nm. One of the possible reasons for the transparency of films in the near-infrared could be the formation of a more planar conformation of polyaniline leading to favorable interaction between constituents.

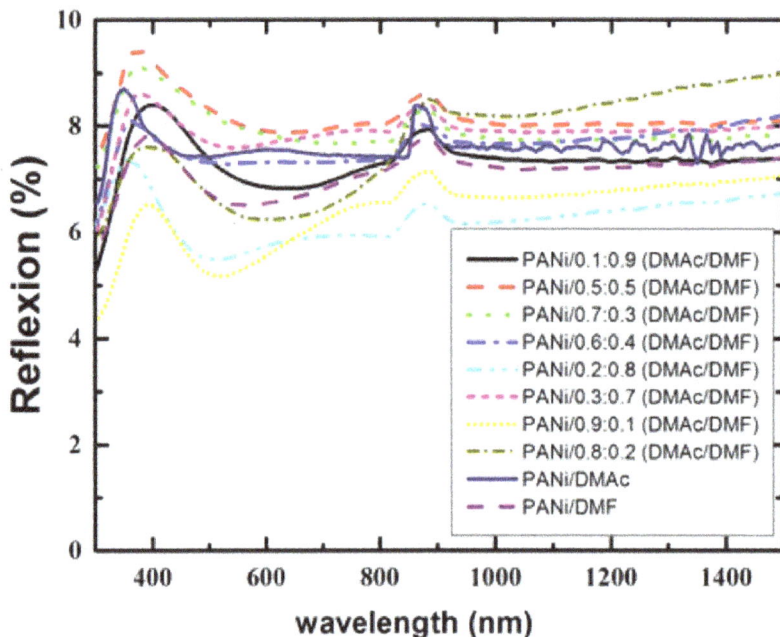

Fig.4: Optical reflectance spectra of films with the molar ratio of DMAc and DMF =0.1:0.9,0.2:0.8,0.3:0.7,0.5:0.5,0.6:0.4,0.7:0.3,0.8:0.2 and 0.9:0.1 after heating treatment at 35°C.

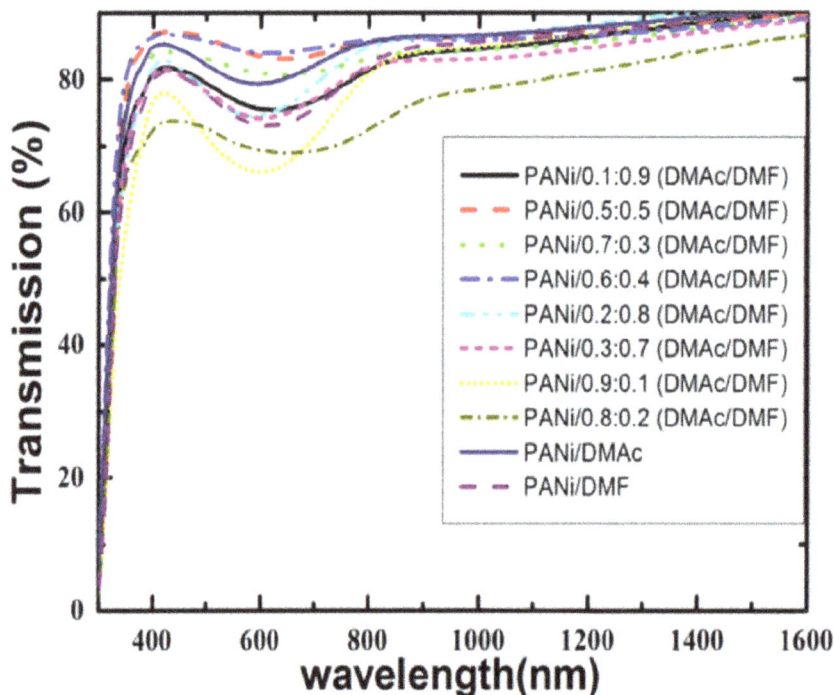

Figure 5. Optical transmittance of films with the molar ratio of DMAc and DMF = 0.1:0.9,0.2:0.8,0.3:0.7,0.5:0.5,0.6:0.4,0.7:0.3,0.8:0.2 and 0.9:0.1 after heating treatment at 35°C

The study of the fundamental absorption edge in the UV region is a useful method for the investigation of optical transitions and electronic band structure in crystalline and non-crystalline materials. The optical absorption spectra can be used for the calculation of the value of energy band gap (Eg) for polymer sample.

The absorption coefficient (α) near the band edge varies with the photon energy (hv) as the equation Eq.3:

$$\alpha^n(E) = \beta(E - E_g)$$

Eq. 3

Where E is the energy of the incidence photon, Eg is the optical band gap energy, β is a constant known as the disorder parameter which is dependent on composition and independent to photon energy. Parameter n is the power coefficient with the value that is determined by the type of possible electronic transitions, i.e.n=1/2, 3/2, 2 or 1/3 for direct allowed, direct forbidden, indirect allowed and indirect forbidden respectively. The direct band gap energy can be determined from the extrapolation of the linear section of the curves to x-axis, as described by Mott & Davis concept. The optical band gaps of the PANi films obtained from Figure 6-8 are around 3.57 eV. Thamilselvan et al. reported optical and magnetic characterizations of the interaction between polyaniline and copper oxide $Cu_2O/PANI$ Nanocomposite [30]. In their opinion, the increase of band gap of Cu_2O compared with its bulk band due to the presence of bioorganic phases on Cu_2O due to the formation of polaron and bipolaron charge carrier in the nanocomposite. The substitutions in the ring affect the tensional angle; hence dihedral/torsional angle between adjacent aromatic rings is increased [31]. It caused the increase in twisting of torsional angle which increases the band gap as compared with the band gap of polyaniline.

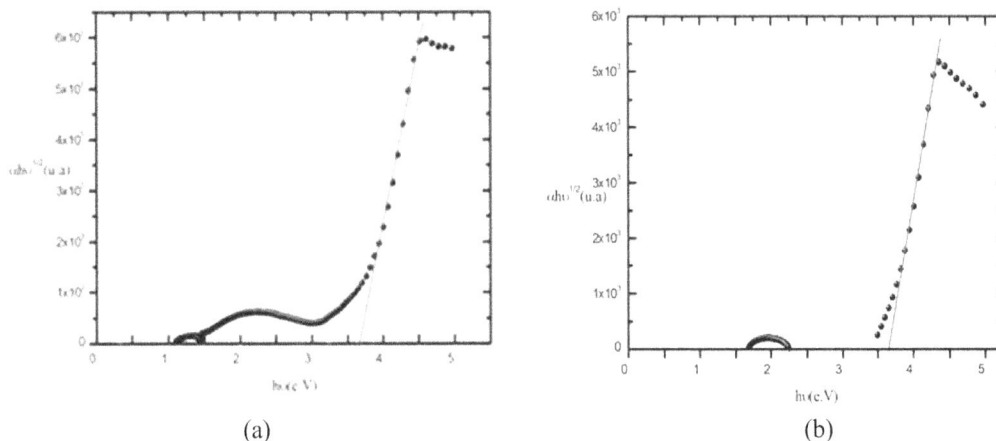

(a)   (b)

Figure 6. Plot of $(\alpha h\nu)^{1/2}$ versus photon energy (hν) of the PANi films: (a) PANi/DMAc and (b) PANi/DMF

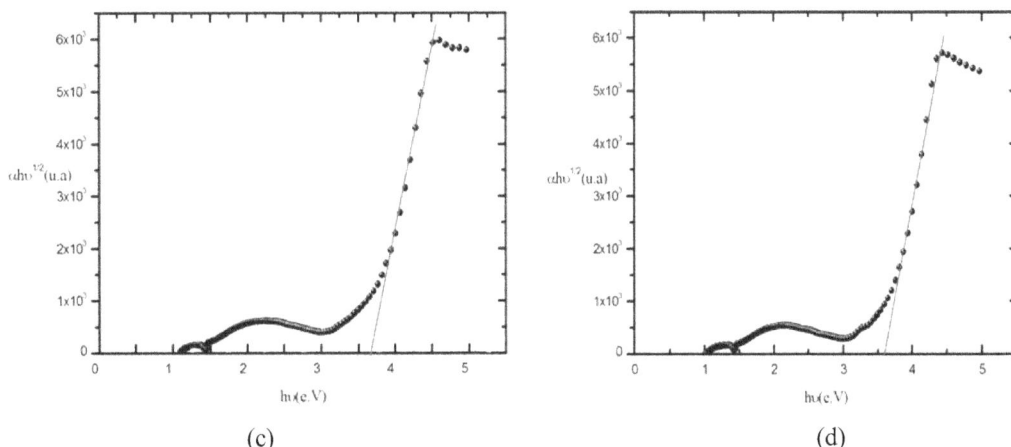

(c)   (d)

Figure 7. Plot of $(\alpha h\nu)^{1/2}$ versus photon energy (hν) of the PANi films: (c) PANi/0.5:0.5 (DMAc/DMF) and (d) PANi/0.6:0.4 (DMAc/DMF).

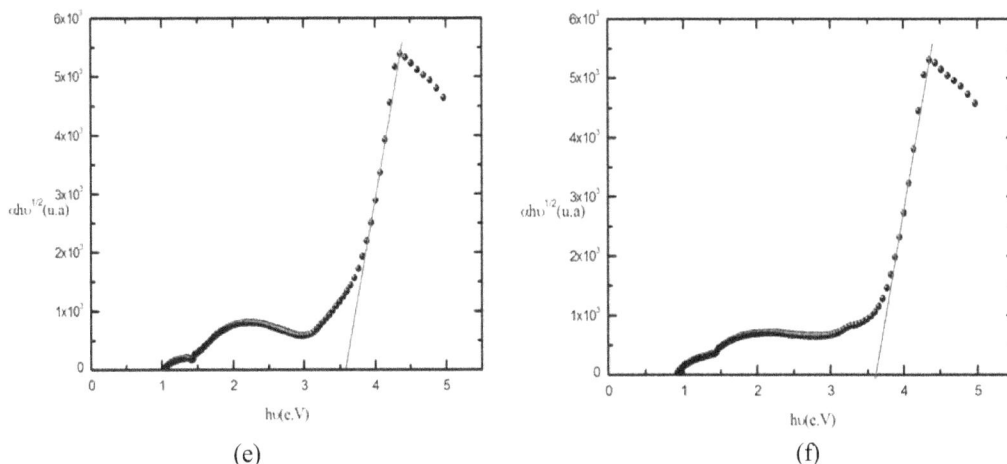

(e)   (f)

Figure 8. Plot of $(\alpha h\nu)^{1/2}$ versus photon energy (hν) of the PANi films: (e) PANi/0.7:0.3 (DMAc/DMF) and (f) PANi/0.8:0.2 (DMAc/DMF).

Fluorescence spectral

Figure 9 illustrates the photoluminescence (PL) spectra obtained from the PANi films. Its PL increase with a change of DMAc/DMF contents. Musa et al. showed that the PANi polymer had two broader emission peaks with wavelength at 420 and 470 nm respectively for EB and ES [32].

PL spectra of PANi: DMAc/DMF cast films of a series of molar ratio (0.7:0.3; 0.8:0.2 and 0.9:0.1) were investigated under laser excitation around 300 nm. The PL peaks of 480 nm are broadened and shifted to lower wavelength (~380 nm). The PL of PANi: 0.6:0.4 (DMAc/DMF and PANi: 0.5:0.5 (DMAc/DMF) thin films exhibited two peaks at 420 and 480 nm. The PANi: 0.6:0.4 (DMAc/DMF) film has a somewhat stronger peak at 480 nm that of the PANi: 0.3:0.7 (DMAc/DMF) or PANi: 0.5:0.5 (DMAc/DMF) films. The thin film with polymer cast from 0.6:0.4 and 0.5:0.5 molar ratio of (DMAc/DMF), exhibited the highest PL increase

without any peak shift. We suggest the difference in PL peak position for polymer cast films should be attributed to the different dispersion states of solvent molecules in polymer matrices due to the differences in the interaction and miscibility between DMAc/DMF and each polymer. PANi appeared to be well dispersed in 0.6:0.4(DMAc/DMF) solvents but some particles of aggregated polymer molecules are formed in PANi :0.7:0.3 or PANi: 0.8:0.2 thin film. We refer to the earlier reports of Zhang et al. where PL increases with an increase of rare earth ions [33]. The emission peaks of polyaniline doped with Fe^{3+} and Al^{3+} ions at the excitation wavelength of 370 nm are located at 477 nm. The Fe-PANi or AL-PANi exhibited a very weak PL peak around 410 nm, while their main emission peak shifted to higher wavelength with fluorescent intensity increase. In their opinion, the progressive increase of fluorescent intensities led eventually to the complexation between Fe^{3+} and Al^{3+} ion with the nitrogen atoms in the polymeric chains.

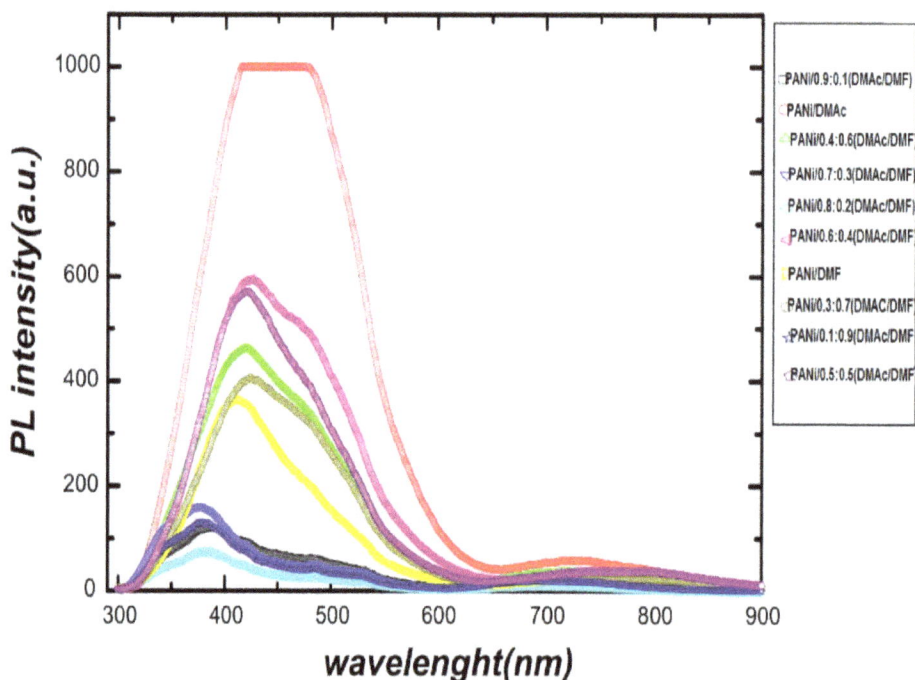

Figure 9. Photoluminescence spectra of the PANi films, under 300 nm excitation.

The same observation was also made for the PANi films .An increase in the rigidity of its structure observed an increase in the fluorescence intensity .Since formation of complex PANi/0.6(DMAc/0.4(DMF) has already contributed towards the rigidity of the structure. With the

decreasing of fluorescent intensity (as shown in Figure 9), a cross-linking structure is eventually formed can be explained by the formation of cross-linking structure by 0.8 DMAc/0.2 DMF or 0.7DMAc/0.3 DMF interaction with adjacent polyaniline chains.

SEM studies

Morphologies of the PANi films with different

DMAc/DMF contents are shown in Fig. 10-12.

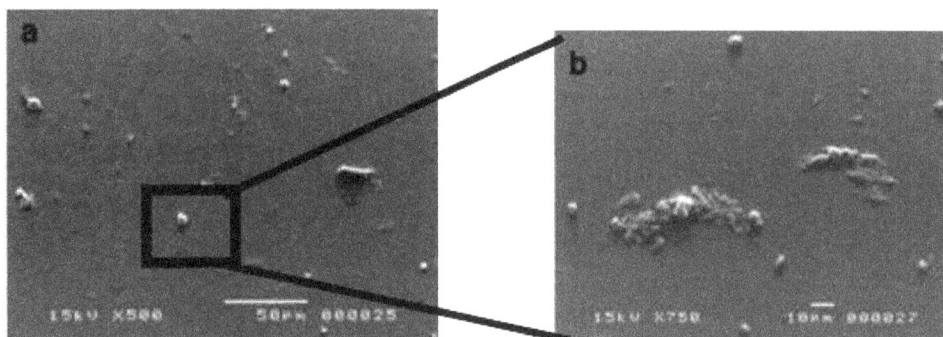

Figure 10. SEM image of PANi /0.6:0.4 (DMAc/DMF) film

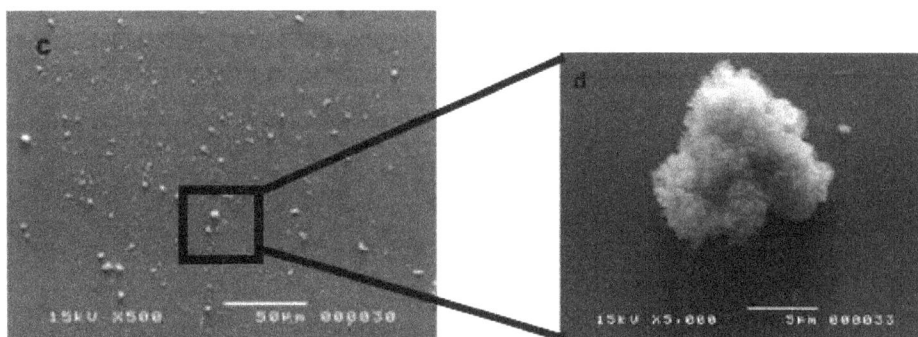

Figure 11. SEM image of PANi / 0.8:0.2 (DMAc/DMF) film

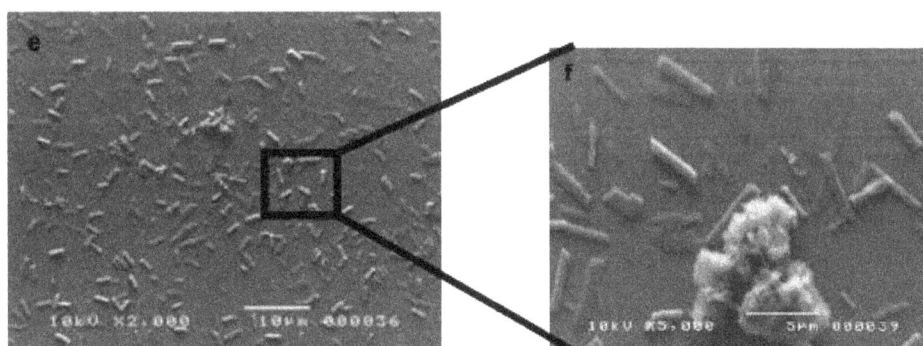

Figure 12. SEM image of PANi/0.7:0.3 (DMAc/DMF) film

All the studied systems reveal flaky shaped structure, in which the size of flakes increases as certain DMAc/DMF is added into PANi. It can be seen that the size of final cluster of DMAc/DMF covered by PANi with which 0.8:0.2 (DMAc/DMF) nanorods content due to the possibility of agglomeration of particles, and PANI has declined.The SEM images help us draw a conclusion that the DMAc/DMF casting solvent has a strong effect on the morphology of PANi, since PANi has various structures such as granules, nanofibers, nanotubes, nano- spheres, microspheres and flakes [34]. According to the SEM micrographs in Fig.10, PANi and 6DMAc/4DMF crystal have formed a nanocomposite in which the nanorods are embedded in the polymer matrix.

Conclusion

The polyaniline cast films have been prepared in DMAc/DMF solutions. Structural analysis indicated that the interactions of PANi with DMAc/DMF solvents occur at the nitrogen atom of imine groups on polymeric chain backbones. Two DMAc/DMF compositions can significantly increase the fluorescent intensity of polyaniline due to a steady plane rigid structure for PANi chains formed by the complex of nitrogen atoms on polymer chains with DMAc or DMF molecules. The result indicates that DMAc/DMF mixtures can sensitize the fluorescence of PANi by improving the structure of polymeric chains. Addition of DMAc/DMF into polyaniline matrix increases the direct band gap (3.57eV) of the films when compared to polyaniline.

This gap increases with increases the highlight of this work is the simple production of highly transparent thin films with electrical and optical properties appropriate for optoelectronic applications.

References

1- A.G. MacDiarmid, Angew. Chem. Int.**2001**, 40, 2581-2590.

2- C.D. Dimitrakopoulos, P.R.L. Malenfant, Adv. Mater. **2002**, 14, 99-117.

3- S.Virji, R.B. Kaner, B.H. Weiller, Chem. Mater. **2005**, 17, 1256-1260.

4- S. Venkatachalam, P.V. Prabhakaran, Synth .Met. **1998**, 97, 141-146.

5- Z.M.Tahir, E.C. Alocilja, and D.L. Grooms, Sensor. **2007**, 7, 1123-1140.

6- L. Liang, J. Liu, C.F. Windisch, G.J. Exarhos, and Y. Lin, Angew.Chem. Int. **2002**, 41, 3665-3668.

7- L. F. He, Y. Jia, F. L. Meng, M. Q. Li, and J. H. Liu, Mater. Sci. Eng. **2009**, 163, 76-81.

8- Y. H. Lin, X. L. Cui, Chem. Commun. **2005**, 2226-2228.

9- M. Q. Li, L. H. Jing, Electrochim. Acta. **2007**, 52, 3250-3257.

10- V. Mottaghitalab, B. B. Xi, G. M. Spinks, and G. G. Wallace, Synth.Met. **2006**, 156, 796-803.

11- G. M. Spinks, V. Mottaghitalab, M. Bahrami-Saniani, P. G. Whitten, and G. G. Wallace, Adv. Mater. **2006**, 18, 637-640.

12- S. Ben-Valid, H. Dumortier, M. Decossas, R. Sfez, M. Meneghetti, A. Bianco, and S. Yitzchaik, J. Mater. Chem. **2010**, 20, 2408-2417.

13- P. C. Ramamurthy, A. M. Malshe , W. R. Harrell, R. V. Gregory, K. McGuire, and A. M. Rao, Solid State. Elect. **2004**, 48, 2019-2024.

14- J. H. Lim, N. Phiboolsirichit, S. Mubeen, M. A. Deshusses, A. Mulchandani, and N. V. Myung, Nanotech. 2010, 21, 075502.

15- B. L. He, B. Dong, W. Wang, and H. L. Li, Mater. Chem. Phys. **2009**, 114, 371-375.

16- S. R. Sivakkumar , W. J. Kim, J.A. Choi, D. R. MacFarlane, M. Forsyth, and D.W. Kim, J.Power. So. **2007**, 171, 1062-1068.

17- S. Bhadra, D. Khastgir, N. Singha and J. Lee, Progress in Polym. Sci. **2009**, 34, 783-810.

18- S.G. Oh, B.J. Kim, M.G. Han, and S.S. Im, Synth. Met. **2001**, 122, 297-304.

19- X.G. Li, M.R. Huanga, J.F. Zeng, and M.F. Zhu, Colloids and Surfaces A: Physicochem. Eng. **2004**, 248, 111-120.

20- R. Sainz, A. M. Benito, M. T. Martinez, J. F. Galindo, J. Sotres, A. M. Baro, B. Corraze, O. Chauvet, A. B. Dalton, R. H. Baughman, and W. K. Maser, Nanotech. **2005**, 16, 150-154.

21- Y. F. Ma, S. R. Ali, L. Wang, P. L. Chiu, R. Mendelsohn and H.X. He, J. American Chem. Soc. **2006**, 128, 12064-12065.

22- N. Bohli, F. Gmati, A. B. Mohamed, V. Vigneras and J. L. Miane, J. Phys. D: Appl. Phys. **2009**, 42, 205404.

23- D. Prasanna, H. S. Jayanna, A.R. Lamani, M. L. Dinesha, C. S Naveen, and C. Shankaramurthy, J. Chin. Phys. Lett. **2011**, 28, 7701.

24- T. Rajavardhana Rao, I. Omkaram , K-V. Brahmam, Ch. Linga Raju, J. Mol. Struct. 2013, 1036, 94-101

25- H. Klug, L. Alexander, Eds.; X-ray Diffraction Procedures, Wiley: New York, **1962**, pp.125.

26- E. A. Meulenkamp, J.Phys. Chem. B. **1998**, 102, 5566-5572.

27- M. Cochet, G. Louarn, S. Quillard, M-I. Boyer, J-P. Buisson, S. Lefrant, J. Raman Spectrosc. **2001**, 11, 1029-39.

28- I. Sedenkova, M. Trchova, J. Stejskal, Polym. Degrad. Stability, **2008**, 93 , 2147-2157

29- A. Uygun, L. Oksuz, A. G. Yavuz, A. Guleç, and S. Sen , Current App.Phys. **2011**, 11, 250-254.

30- K. Gopalakrishnan, C. Ramesh, M. Elango, and M. Thamilselvan, Mat. Sci. **2014**, ID 567927.

31- V. A. Khati, S. B. Kondawar, and V. A. Tabhane, Analytical & Bioanalytica Electrochem. **2011**, 3, 614-624.

32- F. S. Mehamod, R. Daik and A. Musa, Malaysia. J. Chem. **2002**, 4, 35-40.

33- Z. Jiali, W. Hao, Y. Shimei , W. Shaohui, Y. Shaoming , J. App. Polym. Sci. **2012**, 125, 2494-2501.

34- J .Stejskal, I .Sapurina, M .Trchova, Progress In Polym. Sci, **2010**, 35, 1420-1481.

13

Treatment and valorization of olive mill wastewaters

Nabila Slimani Alaoui *, Anas El Laghdach, Mostafa Stitou and Aniss Bakkali

Laboratory of Water ,Studies and Environmental Analysis, Department of chemistry, Faculty of Sciences, University Abdelmalek Essaâdi, B.P. 2121, Mhannech II, 93002 Tetouan, Morocco.

Abstract: This study aims to evaluate the effectiveness of the physicochemical process with lime and ferric chloride in removing the pollution generated by the olive mill wastewaters (OMW) .The characterization of the samples has shown that they are acidic, with a black color and a strong organic load due to the presence of phenolic compounds. The combination of the lime and the ferric chloride allows the removal of 87% of the total suspended solid (TSs), 58% of chemical oxygen demand (COD) and 75% of Phenolic compounds. After purification the treated OMW were valorized as wash water or used for irrigation of green spaces and the generated sludge were dried and used as burning material.

Keywords: OMW, Characterization, Physicochemical Treatment, Valorization.

Introduction

The olive tree is the main fruit planted in Morocco, it occupies more than 58% of the national tree area[1], the production is estimated at 1.500.000 tons of olives on an area of 933.475 hectares, a figure that the kingdom aims doubling in 2020[2]. In terms of production Morocco ranks the fifth place in the world after Spain, Italy, Tunisia, and Greece [3].

Since ancient times the oil was traditionally obtained by pressure until the introduction of the centrifugation in olive mills. In the late 80s in Spain, after a severe drought and a big increase in production, the extraction of olive oil was obtained by the continuous extraction system including a vertical and horizontal centrifugation which separates the olive mixture in three-phase: oil, pomace and a black liquid effluent called olive mill wastewaters (OMW) or in two phases: oil and wet pomace. The two-phase system uses a small amount of water and therefore a lower dissolution of phenolic compounds remains in the oil, but who also make it bitter [4-5].

The olive mill wastewaters generated from olive oil extraction causes a major environmental issue when they are directly discharged into rivers, they greatly change the quality of surface waters and produce serious environmental damages and disrupt the operation of treatment plants when they come through the sewers. The annual production of this effluent exceeds 30 million m^3 per year; pressing 1Kg of olives can generate 0.5-1L of OMW with modern production modes [6-7].

The problem of disposing of OMW was apprehended in different ways by the Mediterranean olive-growing countries, which have adopted point solutions to solve this problem. The difficulty of treatment of these effluents is due to their poor biodegradability due to the presence of phenolic compounds[8-9], lipids and organic acids.
Taking into account the composition and the toxicity degree of the OMW, simple treatments are not sufficient to ensure their purification. Some biological (with aerobic and anaerobic microorganisms) [10, 11], chemical [12] and physical [13] treatments have been applied and have proved to be effective for reduction of the organic content of OMW. Many solutions of valorization were adopted to prevent the environmental pollution. The OMW is proposed as a renewable resource to produce valuable photochemical compounds, to extract natural chemicals or to be used as fertilizers[14,15].

This work aims to provide an effective treatment valid for the different olive extraction systems, able to reduce or even eliminate the organic matter contained in the OMW. The treatment proposed is the physicochemical treatment using the lime and the ferric chloride followed by a decolourization and desinfection with chlorine. The capacity of removing the phenolic compounds, COD and total suspended solids will be evaluated

The capacity of removing the phenolic compounds, COD and total suspended solids will be evaluated, in addition to the valorization of the treated OMW and recycling the sludge generated.

*Corresponding author: Nabila Slimani Alaoui
Email adress : *slimanialaouinabila@gmail.com*

Experimental Section
Sampling

The samples of OMW studied were obtained from a three phase olive oil extraction unit located in the city of Meknes. The OMW were collected every day from the storage tank which gathers all the OMW produced in the industry, namely: Washing water, OMW generated after extraction and the rainwater. Samples were analyzed in the laboratory of the unit.

Physical-chemical analysis of OMW

The color measurement is done using a rotating disk apparatus (*Lovibond Comparator 2000+*); the color of the OMW was compared to the discs with different colors from white to brown. Once the colors is close the corresponding value given in Hazen is taken by taking into account the dilutions.

The pH of the samples was continuously measured using a pH meter (*pH 1100) Eutech instruments pH/ mV/°C*.
Fats were extracted from the OMW by Hexane; the mixture was evaporated in a rotary evaporator (*Nahita, Rotary Evaporator 9200*) at 80°C. Fats were calculated by weight difference.

The total solid (TS) was determined by weighting the samples before and after drying at 105°C for 24h.
The total suspended solid (TSs) was determined after filtering a sample through a filter (0.45μm) using vacuum pump and drying the filtrate obtained at 105°C for 24h.

The mineral matter (MM) was measured after calcination of the OMW in an oven at 600°C for 4h.
The biological oxygen demand (BOD_5) is the amount of oxygen required to oxidize biologically the organic matter contained in the OMW .The samples of OMW were diluted and introduced in DOB_5 bottles (*OxiTop IS6*) containing a bar magnetic, they were then placed in a thermo cupboard at 20°C. After 5 days of incubation the value of BOD_5 was measured taking into account the dilution factor.

The biochemical oxygen demand is an oxidation of organic matter by excess of potassium dichromate. COD of the OMW was calculated from the volume of salt Mohr used for titration of Mohr's salt.

Polyphenols are a family of organic molecules widely present in the OMW they were assayed by colorimetric method using tannic acid as a standard at 725 nm.

All parameters were analyzed in triplicate, and determined following standard methods [16-17].

Treatment by Coagulation-Flocculation

The Physico-chemical treatment by coagulation-flocculation is based on the addition of a reagent to the mixture to be treated, in order to destabilize the fine suspended matter by reducing electrostatic repulsion forces[18-19].

In this work the physicochemical treatment by coagulation flocculation is based on the use of lime and ferric chloride.

Samples of OMW were allowed to stand for 1hour until the suspended matter has settled.
In a series of 100 mL beakers, raw OMW were used without adjusting pH (pH = 4.6) and we followed this protocol:

- Increasing doses of lime (Purity>92%) were added (1.5-7.5)g/L on raw OMW to adjust the acidic pH (pH=4.6) to a pH widely basic (pH = 10) ;

- The mixture was agitated with rapid magnetic stirring for 3 min at 180rpm followed by slow mix for 30 min at 30 rpm;

- The ferric chloride ($FeCl_3$, 41%) was then added (0.5-1.5)g/L on the supernatants recovered after treatment with lime and the mixture was agitated again by rapid magnetic stirring 3 min at 180 rpm followed by a slow mix for 30min at 30rpm;

- The mixture must be decanted at least 5 hours to separate the liquid phase from the particulate phase;

- After decantation the mixture was filtered over a filter paper (100 mm) using a vacuum pump.

Reuse of treated OMW

When we need to reuse the treated OMW, 200ml/L of bleach can be added the filtrate obtained. It allowed the disinfection and the decolourization of the OMW.

Sludge Treatment

The physicochemical treatment generates sludge during the various processing steps. The sludge were collected after clarification and dried in an oven for 4h at 105°C.

Results and Discussion

Characterization of the OMW

The characteristics of the studied raw OMW are summarized in the following Table:

Table 1. Characteristics of OMW produced in a three phase continuous extraction unit

Parameters	Values
pH	4.5-5
Conductivity (mS/cm)	14
Color	Brown-Black
Fats	0.6
Total suspended solid (TSs) (g/L)	14.8
Total solid (TS)(g/L)	23.9
Mineral matter (MM)(g/L)	4.96
BOD$_5$ (g O$_2$ /L)	45
COD (g O$_2$/L)	64
Total phenols (g/L)	2.1

The characterization of the studied samples shows an acidic pH. This low value indicates that the use of this effluent for irrigation can damages soil [20], except calcareous soils where the equilibrium can be established (e.g. of the city of Meknes). The conductivity (14mS/cm) is high compared to other studies, this is essentially due to the significant addition of salt to preserve the olives [21-22]. The OMW are brown to brown-blackish, the color intensifies with the storage time and the oxidation of phenolic compounds [23]. In our case these molecules are present in low content in comparison with results

reported by S. Azabou et al [24]. The OMW are rich in TSs which may be due to bad separation of OMW from the olive pomace. In our case the content of fats is much lower compared to that reported in the literature [25], the OMW studied were collected from a storage tank which allowed the removing of the excess of suspended matter and fats.

The ratio BOD$_5$/COD indicates the biodegradability of waste water. Industrial wastewaters which have a ratio higher than 0.3 are easily biodegradable [26], in our case the biological treatment can be applied but it will be limited by the presence of phenolic compounds which inhibits the microbial activity [27].

Treatment of the OMW by Coagulation-Flocculation

The OMW are unstable in the same day which requires the use of a treatment highly adaptable to changes to the effluent.

After a simple decantation, 15% of TSs was eliminated without affecting the pH. The addition of the lime (7.5g/L) allowed the adjustment of the acidic pH (4.6) to a basic pH (10). Burnt lime (CaO) is often used for water treatment for its avaibility and effectiveness The neutralization by lime leads the transformation of phenols to phenates with the formation of C$_6$H$_5$O$^-$ ions [28] (Fig 1) , the phenolic compounds obtained lose much of their antibacterial effect and biological activity can therefore start [29-30].The percentage of elimination of polyphenols with lime attain 68%, similar results were reported by E. S. Aktas and al where they shows that the precipitation with lime allowed the elimination of 65% of Polyphenols and only 28% of volatile phenols, the phenolic substances could be removed totally or partially and some of them were not affected [31].

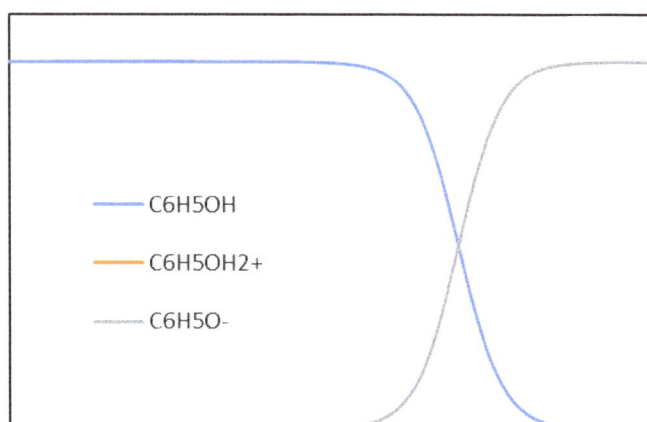

Figure 1 . Different forms of phenol in different pH

Lime allowed the removal of 57% of TSs and 41% of COD. The treatment with lime removes suspended matter, grease and oils which causes the

reduction of the COD and forms large aggregations of floc [32]. Figure 2 represents the evolution of the different parameters of OMW after treatment with lime.

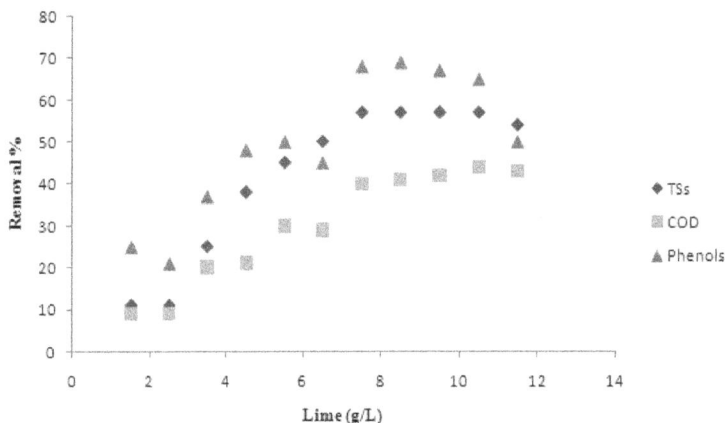

Figure 2. Evolution of TSs, COD and Phenols after treatment with lime

Flocculants are always used at low dosage (0.5 to 2g/L) to complete the initial coagulation of colloids provided by liming. The addition of ferric chloride as a flocculent allows the removal of organic matter present in the OMW [33]. It provides a good flocculation of the particles contained in the OMW and promotes the removing of phosphates coming from fertilizers or agricultural activities [34]. Ferric chloride allows also the elimination of fats which remains in the OMW after decantation and liming. The Percentage of removal of TSs, COD and polyphenols is respectively 87%, 58% and 75%. The ferric chloride allows the passage of the dissolved form of pollutants to an insoluble particulate form which can be easily retained by simple decantation and gives settleable floc [30, 35]. Figure 3 summarizes the evolution of the characteristics of OMW after treatment with lime and ferric chloride.

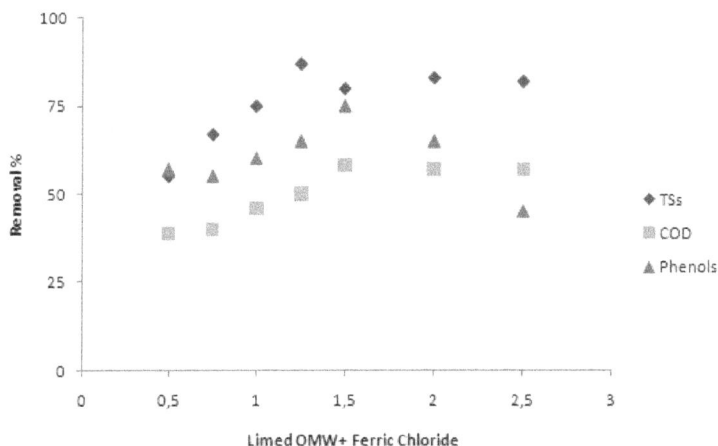

Figure 3. Evolution of TSs, COD and Phenols after treatment with lime and ferric chloride

The clarification of the treated OMW for at least 5 hours is necessary it allowed the separation of the liquid phase of the colloids and hydroxide floc which agglomerate after treatment with ferric chloride [36]. Figure 4 summarizes the steps performed for the removal of organic matter from raw OMW obtained from the three phase extraction.

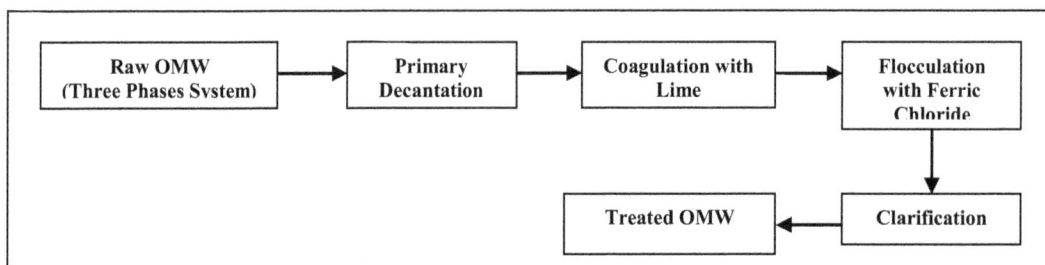

Figure 4. Steps performed for the removal of organic matter from raw OMW

Reuse of treated OM

The chlorination (Post-chlorination) is an optional step which can be added when we want to reuse the treated OMW as water wash .Chlorine ensures the disinfection and decolourization of the effluent, the neutralization of the pH (pH=7) and the removal of organic impurities which are not retained after clarification [35]. The percentage of COD and decolourization reach respectively 65% and 71%. The treated OMW remain on their same condition when they are exposed to light or stored which explains that the majority of phenolic compounds, especially responsible of coloration, were removed.

In our case after chlorination and secondary decantation we proposed the reuse of the treated OMW as wash water or for irrigation.

Some tests were performed on raw OMW to evaluate the effectiveness of the physicochemical treatment by coagulation-flocculation followed by chlorination. Figure 5 includes the treatments used only for decolourization of OMW.

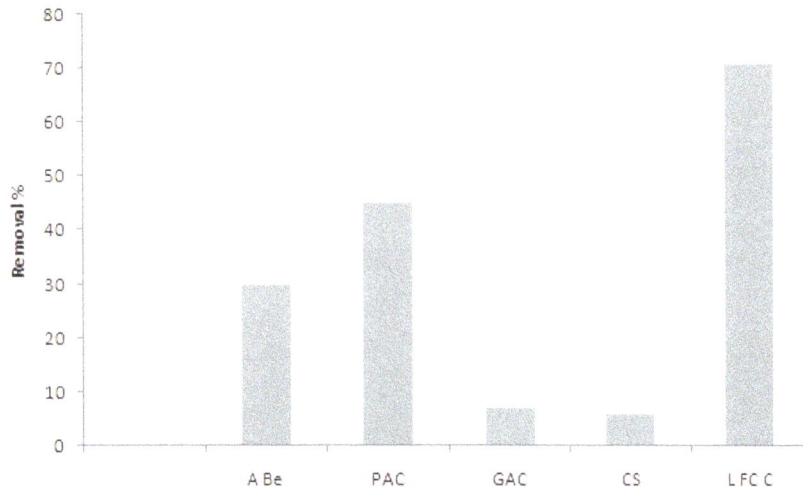

Figure 5. Treatments tested for Discoloration of the raw OMW
A Be: Adsorption with Activated Bentonite, PAC: Powder Activated Carbon, GAC: Granular Activated Carbon, CS: charcoal stick, LFC C: Lime+Ferric C +Chlorine

In our study the physicochemical treatment by coagulation-flocculation using lime and ferric chloride followed by chlorination after clarification gave the best discoloration of raw OMW with a removal percentage reaching 71%.

Sludge treatment

The disadvantage of the physico-chemical treatment by coagulation-flocculation is the production of large amounts of sludge and the difficulty of regeneration of reagents, which increase the cost of processing and generates another source of pollution.

In our study we solved the problem of sludge at the factory as follows:

• The primary decantation produces a fresh sludge, it represents 20% of the volume (depends on the retention time) and it usually mixed with the wet pomace.

• The sludge generated from the treatment with lime (limed sludge) is conform to legislation; they are often highly valued by farmers due to their economic interest[32]. This sludge can neutralize the acidic sludge generated after treatment with ferric chloride.

• The sludge generated after clarification are acid too on which the burnt lime or limed sludge can be added to adjust the pH. After drying in for 4 hours, the dried sludge is recovered in the form of coal and used for combustion.

Picture 1. Dried sludge

Conclusion

Samples of raw OMW were collected from a three-phase olive mill to determine the pollutant responsible of their toxicity. The characterization shows that they are acidic, rich in organic matter and phenolic compounds which limits their biodegradation. The treatment of the OMW by coagulation flocculation with lime and ferric chloride allowed a destabilization of colloidal particles and a transformation of the phenolic compounds, which facilitated the agglomeration of the hydroxide floc by simple decantation. Chlorination must be used when a reuse of the treated OMW is planned; it ensures the disinfection and decolourization of the effluent. The sludge generated after clarification was dried and used as burning material.

References

1- B. Boulouha, L'amélioration génétique de l'olivier au Maroc programme prometteur pour le développement de l'oléiculture nationale ; Bulletin trimestriel de l'INRA : Marrakech, **2007**, Vol.2, N°01. http://www.inra.org.ma/docs/119052009163829. pdf.

2- Ministère d'agriculture et de la pêche maritime, Veille économique-Secteur oléicole, Maroc, **2013**, Note stratégique N°95.

3- I. El Mouhtadi, M. Agouzzal, F. Guy, L'olivier au Maroc. OCL : Oléagineux, Corps Gras, Lipides, **2014**, 21(2), pp.1-3. http://www.ocl-journal.org/articles/ocl/pdf/2014/02/ocl130034. pdf.

4- H. Chimi, Technologies d'extraction de l'huile d'olive et gestion de sa qualité ; Bulletin mensuel d'information et de liaison de PNTTA : Maroc, **2006**, N° 141, pp.1-4.

5- Agro-pôle Olivier et Agence d'Exécution du Projet CFC/IOOC/04, Les bonnes pratiques d'épandage des margines et du compost sur les terres agricoles: Cas de l'olivier: Meknes, **2004**, pp.1-17.

6- C. Pintucci et al, Fresh olive mill waste deprived from polyphenols as feedstock for hydrogen photo-production by means Rhodopseudomonas palustris 42OL, Renew. Energy, **2013**, 51, 358-363.

7- N. Benyahia and K. Zein, Analyse des problèmes de l'industrie de l'huile d'olive et solutions récemment développées ; Sustainable Business Associates : Lausanne **2003**, pp.1-8.

8- I. Ntaikou et al, Exploitation of olive oil mill wastewater for combined biohydrogen and biopolymers production, Bioresour. Technol, **2009**,100, 3724-3730. doi: 10.1016/j.biortech.2008.12.001

9- K. Baransi, Y.Dubowski, I.Sabbah, Synergetic effect between photocatalytic degradation and adsorption processes on the removal of phenolic compounds from olive mill wastewater, Water Res, **2012**, 46, pp.789-798. http://www.ildesal.org.il/pdf/gwri_abstracts/2012 1/14.pdf.

10- S.E. Garrido Hoyos, L. Martinez Nieto, F. Camacho Rubio, Ramos Cormenzana, Kinetics of aerobic treatment of olive-mill wastewater (OMW) with Aspergillus terreus, Process Biochem, **2002**, 37, pp.1169-1176. doi: 10.1016/S0032-9592(01)00332-6

11- M.R. Gonçalves, J.C. Costa, I.P. Marques, M.M. Alves, Strategies for lipids and phenolics degradation in the anaerobic treatment of olive mill wastewater, Water Res, **2012**, 46, pp.1684-1692. doi: 10.1016/j.watres.2011.12.046

12- H.El Hajjouji et al, Photochemical UV/TiO2 treatment of olive mill wastewater (OMW), Bioresour.Technol, **2008**, 99, pp.7264-7269. https://core.ac.uk/download/files/437/12041824. pdf.

13- M.Achak, N.Ouazzani,L.Mandi, Traitement des margines d'une huilerie moderne par infiltration-percolation sur un filtre à sable,Rev. Des Sci. L'eau, **2009**, 22(3), pp. 421-433.

14- A. Nefzaoui, Valorisation des sous-produits de l'olivier, Options méditerranéens. **1991**, 16, pp. 101-108.

15- A.Ouatmane, V.Dorazio, M. Hafidi, J.C.Revel, N.Senesi, Elemental and spectroscopic characterization of humic acids fractionated by gel permeation chromatography, Agronomie, **2000**, 20, pp.491-504.

16- J. Rodier, L'analyse de l'eau : Eaux naturelles, eaux résiduaires, eaux de mer ; Dunod, ISBN: 2-04-010045-8, **1978**,160-193.

17- J. Rodier, L. Bernard and M. Nicole, L'Analyse de l'eau: Eaux Naturelles, Eaux Résiduaires, Eau de Mer. 8th ed.by Dunod, DL, Paris, **2005**, ISBN: 2100496360, pp: 1383.

18- A. Jaouani, M. Vanthournhout, M.J. Penninckx, Olive oil mill wastewater purification by combination of coagulation-flocculation and biological treatments,Environmental Technology, 2005, 26, pp. 633-641

19- P. Mouchet, Traitement des eaux avant utilisation, Substances dissoutes, Techniques de l'ingénieur traité Environnement : Strasbourg, **2000**, ISSN : 1282-9080.

20- A. Nefzaoui, Contribution à la rentabilité de l'oléiculture par une valorisation optimale des sous-produits, Options méditerranéens. **1991**, pp. 153-173.

21- M. Achak, A. Hafidi, N. Ouazzani, S.Sayadi , L. Mandi ,Low Cost Biosorbent "Banana Peel" for the Removal of Phenolic Compounds from Olive Mill Wastewater: Kinetic and Equilibrium Studies, J. Hazard. Mater, 2009, 166, pp 117-125.

22- F.Aouidi, H.Gannoun, N.Ben Othman, L.Ayed, M.Hamdi, Improvement of fermentative decolorization of olive mill wastewater byLactobacillus paracasei by cheese whey's addition, Process Biochem, 2009, 44, pp. 597- 601.

Despite these satisfactory results physico-chemical treatment by coagulation flocculation remains an expensive treatment, responsible for another source of pollution, which can be applied as a pre-treatment or secondary treatment.

23- S.Khoufi, F.Feki, S.Sayadi, Detoxification of olive mill wastewater by electrocoagulation and sedimentation processes. J. Hazard. Mater, **2007**, 142, pp.58- 67.

24- S.Azabou, W.Najjar, A.Gargoubi, A.Ghorbel, S.Sayadi, Catalytic wet peroxide photo-oxidation of phenolic olive oil mill wastewater contaminants. Part II.Degradation and detoxification of low-molecular mass phenolic compounds in model and real effluent, Appl. Cat. B: Environ, **2007**, 77, pp.166-174.

25- A. Ben Sassi, A. Boularbah, A. Jaouad, G.Walker, A. Boussaid, A comparison of olive oil mill wastewaters (OMW) from three different processes in Morocco, Process Biochem, 41,**2006**,pp.74-78

26- J. Boeglin, Propriétés des eaux naturelles. Techniques de l'Ingénieur, traité Environnement: Nancy, **2001**, ISSN : 1776-0135, G1250.

27- F.Caponino, M.T.Bilancia, A.Pasqualone, E.Sikorska, T.Gomes, Influence of the exposure to light on extra virgin olive oil quality during storage, Eur. Food. Res. Technol, **2005**, 221, pp. 92-98.

28- J.J.Macheix, A. Fleuriet and J.A. Billo, Fruit phenolics, Boca Raton Florida: CRC Press Inc, **1990**, pp. 378.

29- F. Medeci, C.Merli, E. Spagnoli, Anaerobic digestion of olive mill wastewater: a new process. H.In Ferranti, M.P.Ferrero, H.Vaveau, Eds, Anaerobic digestion and carbohydrates hydrolysis of waste;Elsevier : London, **1985**, pp. 385-398.

30- A. Yaakoubi, A. Chahlaoui, M. Rahmani, M. Elyachioui, Y. Oulhote, Effet de l'épandage des margines sur la microflore du sol, Agro solutions, **2010**, 20 (1), pp. 35-43. http://www.irda.qc.ca/assets/documents/Publicati ons/documents/yaakoubi-et-al 2009_article_effet_epandage_margines_microflo re_sol.pdf.

31- E. S. Aktas, S. Imre and L. Ersoy, Characterization and lime treatment of olive mill wastewater, Wat. Res, **2001**, 35(9), pp. 2336-2340. PII: S0043-1354(00)00490-5

32- G. Lolos, A. Skordilis & G. Parissakis, Polluting characteristics and lime precipitation of olive mill wastewater. Journal of Environmental Science & Health Part A, **1994**, 29, 7, pp. 1349 -1356. http://dx.doi.org/10.1080/10934529409376115

33- P. Mouchet, Traitement des eaux avant utilisation : Matières particulaires, Traité Environnement, **2000**, G1, G1172.

34- L. Sigg, P.Behra, W. Stumm, Chimie des milieux aquatiques - Chimie des eaux naturelles et des interfaces dans l'environnement, 5ème ed Dunod : Paris, **2000**, pp. 547. http://www.unitheque.com/UploadFile/Documen tPDF/C/H/IKMG-9782100588015.pdf.

35- J. Boeglin, Traitements physico-chimiques de la pollution soluble, Techniques de l'Ingénieur, traité Environnement: Nancy, **2001**, G1271.

36- J. Boeglin, Pollution industrielle de l'eau, Stratégie et méthodologie, Techniques de l'Ingénieur, traité Environnement: Nancy, **1998**, G1220.

14

Click chemistry approach to a series of calcitriol analogues with heterocyclic side chains

Zoila Gándara [1], Pedro-Lois Suárez [1], Alioune Fall [2], Massène Sène [2], Ousmane Diouf [2], Mohamed Gaye [2], Generosa Gómez [1,*] and Yagamare Fall [1,*]

[1] Departamento de Química Orgánica, Facultad de Química and Instituto de Investigación Biomedica (IBI), University of Vigo, Campus Lagoas de Marcosende, 36310 Vigo, Spain
[2] Laboratoire de Chimie de Coordination Organique (LCCO), Département de Chimie, Faculté des Sciences et Techniques: Université Cheikh Anta Diop de Dakar, Sénégal

Abstract: We report a straightforward synthesis of a series of novel calcitriol analogues from vitamin D_2 with some modification of the procedures described by Calverley and Choudhry. This approach allows the large scale synthesis of a late-stage intermediate common to all the analogues of the series. This intermediate was successfully employed to synthesize a huge number of calcitriol analogues using a "click" chemistry approach.

Keywords: Calcitriol; Vitamin D_2; triazole; azaanalogue; "Click" chemistry.

Introduction

1,25-Dihydroxyvitamin D_3 (**1**, calcitriol) (Fig.1), the hormonally active metabolite of vitamin D_3 (**2**), acts as a regulator in calcium and phosphate homeostasis[1]. Next to these classical activities, calcitriol has been shown to inhibit cellular proliferation and to induce cellular differentiation[2]. However the therapeutic utility of **1** is hampered by the effective doses leading to calcemic side effects and this has stimulated the search for analogues having a relatively weak systemic effect on calcium metabolism while maintainig potent regulatory effects on cell differentiation and proliferation.

As part of our ongoing program on the synthesis of vitamin D analogues modified at the side chain,[3] we envisaged the synthesis of various calcitriol analogues having heteroatoms on their side chain. The rational that could explain this choice was: we have already synthesized Aza-vitamin D analogues[3r] and the biological activity of some of these derivatives was later studied showing that they had less calcemic effect than calcitriol. The strategy we used so far involved construction of the triene unit on the CD fragment following the introduction of the side chain. This strategy is inconvenient if a lengthy series of analogues with modified side chains are to be prepared for systematic biological evaluation.

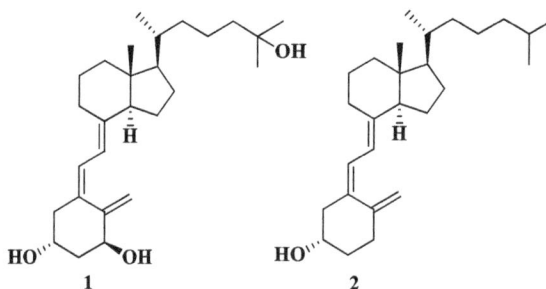

Figure 1. Structures of 1,25-Dihydroxyvitamin D_3 (**1**) and vitamin D_3 (**2**).

Results and Discussion

We examined the possibility of preparing a series of analogues modified at the side chain from a common intermediate, in which the labile triene system was already present. The use of this strategy involved the concept of the triene system protection to allow chemical modification of the vitamin D side chain. This concept received relatively little attention.[4] Among these approaches, the one using the preparation and subsequent thermolysis of the sulfur dioxide adducts of vitamin D_2[4b,c] seemed to us more appropriate for a large scale synthesis of a late-stage intermediate such as **10** (Scheme 1).

Corresponding author : Generosa Gómez, Yagamare Fall
Email adress : ggomez@uvigo.es, yagamare@uvigo.es

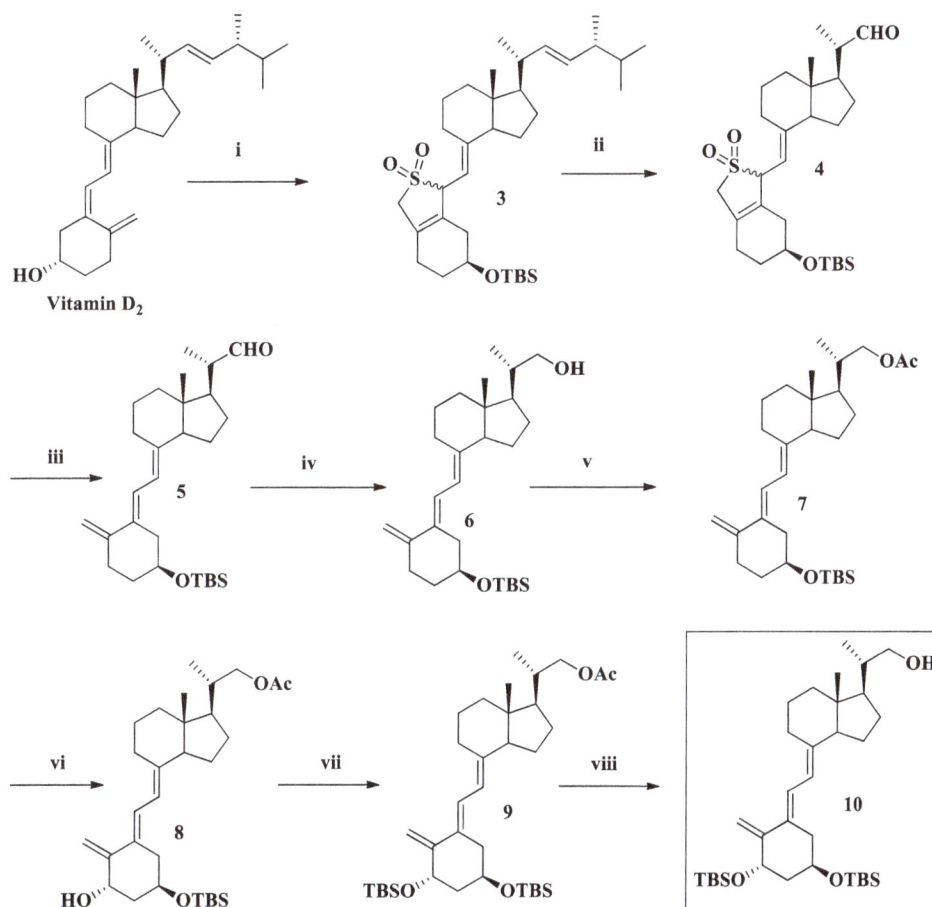

Scheme 1. Synthesis of intermediate **10** from vitamin D_2. *Reagents and conditions*: (i) a) SO_2, CH_2Cl_2, -25 °C to -10° C; b) TBSCl, cat. DMAP, CH_2Cl_2, -5° C to rt (97%, 2 steps); (ii) a) O_3, CH_2Cl_2, MeOH, -78° C; b) PPh_3, 0° C to 25° C; (iii) $NaHCO_3$, EtOH, reflux; (iv) $NaBH_4$, MeOH, 0° C (85% from **3**); (v) Ac_2O, Et_3N, DMAP (96%); (vi) a) SeO_2, MeOH, reflux; b) **7**, NMO, 50° C (60 %); (vii) TBSCl, imidazole, DMAP, CH_2Cl_2 (88%); (viii) K_2CO_3, MeOH (94%).

Accordingly vitamin D_2 was converted to its SO_2-adducts **3** in 97% yield by dissolving in liquid sulfur dioxide and subsequent silylation. The ozonolysis of **3** resulted to be extremely troublesome and after much experimentation the best reaction conditions could be established. The results are summarized in Table 1.

Table 1. Ozonolysis of **3**.

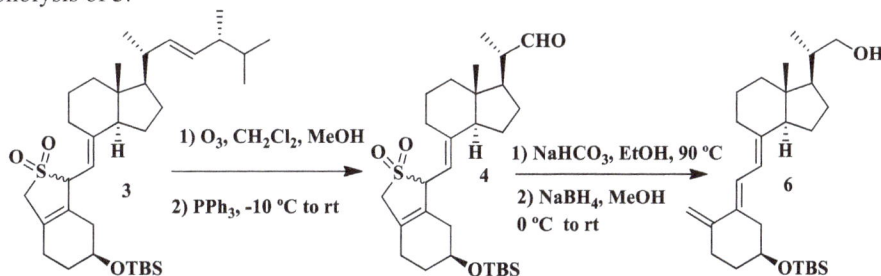

Entry	3 quantity (g)	F(x100L /h) F = gas flow	I(A) I = current	Reaction time	Yield % of 6 (from 3)
1	1	1.5	0.6	10 mn	0
2	1	1.5	0.3	10 mn	0
3	1	1.0	0.1	15 mn	30
4	1	1.0	0.03	1 h	85
5	5	1.0	0.03	5 h	85

The optimized reaction conditions for running the ozonolysis of **3** were as described in entries 4 and 5.

The time necessary for the ozonolysis to be completed was substrate dependant (1 h to ozonolyze 1 g of substrate **3**). Aldehyde **4** was unstable and was immediately converted to alcohol **6** by thermal chelotropic extrusion of sulfur dioxide (SO_2) in the presence of sodium bicarbonate ($NaHCO_3$) followed by sodium borohydride reduction of the intermediate aldehyde **5**. Alcohol **6** was obtained in 85% overall yield (3 steps).

Reaction of alcohol **6** with acetic anhydride gave 96% yield of acetate **7** which was hydroxylated at C-1 with selenium dioxide in the presence of 4-methylmorpholine *N*-oxide (NMO) to afford allylic alcohol **8** in 60% yield. The latter was silylated, giving 88% yield of acetate **9**. Reaction of the latter with potassium carbonate (K_2CO_3) in methanol afforded key intermediate **10** with the 5-(*E*) triene system.

The overall yield of intermediate **10** from vitamin D_2 was 39%. Worth mentioning that the synthesis of

10 could be carried out in multigrams quantities and the compound could be stored in the fridge during months without alteration. Compound **10** is more stable than its 5-(*Z*) isomer and can be further elaborated in order to introduce the desired side chain.

The advantages of this present approach compared to Calverley and Choudhry's approaches are: 1) Only one SO_2 triene protection is needed, hence one SO_2 extrusion. 2) The ozonolysis of the side chain is carried out before the C-1 hydroxylation. 3) For the synthesis of **10** from vitamin D_2, we found an overall yield of 39% which is a bit better then 37% calculated using Calverley's procedure.

We anticipated that intermediate **10** could lead to calcitriol analogues **15**, **16**, and **17** using a "Click" chemistry approach[5] between azide **11** and commercially available alkynes **12**, **13**, and **14**. Our retrosynthetic basis is outlined in Scheme 2.

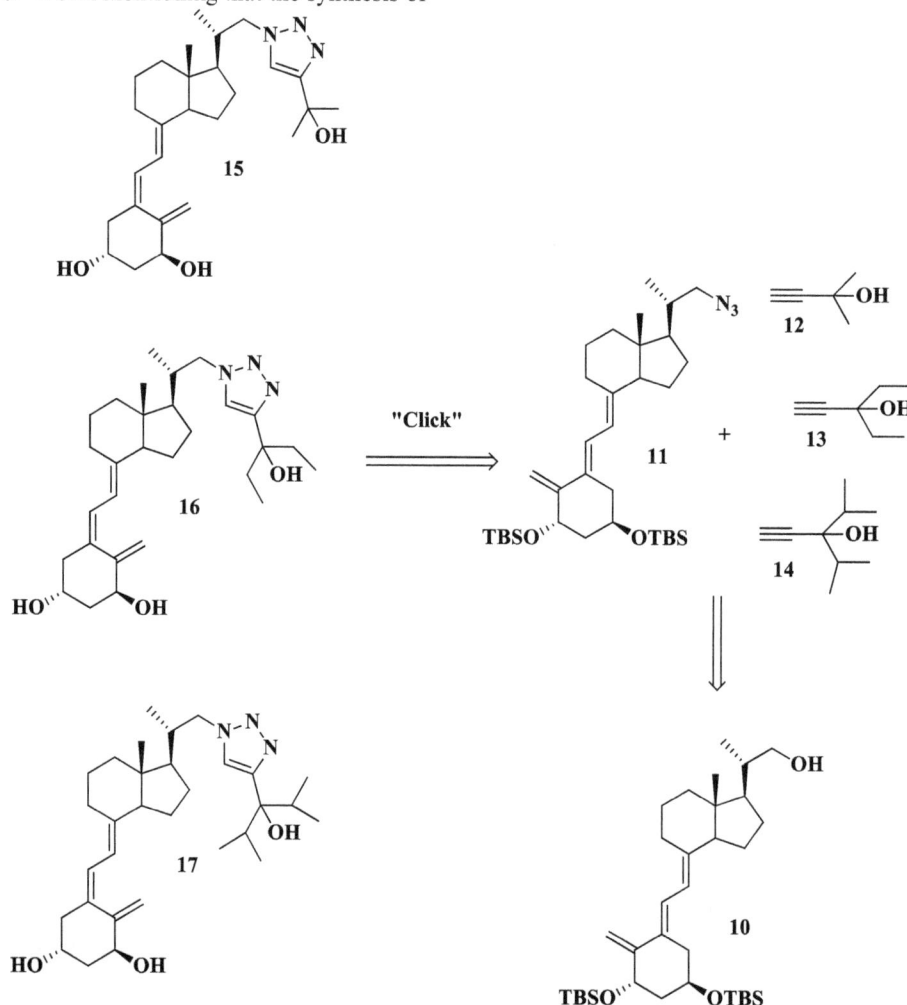

Scheme 2. Retrosynthetic analysis of analogues **15, 16** and **17**

Accordingly compounds **15-17** were prepared as outlined in Scheme 3.

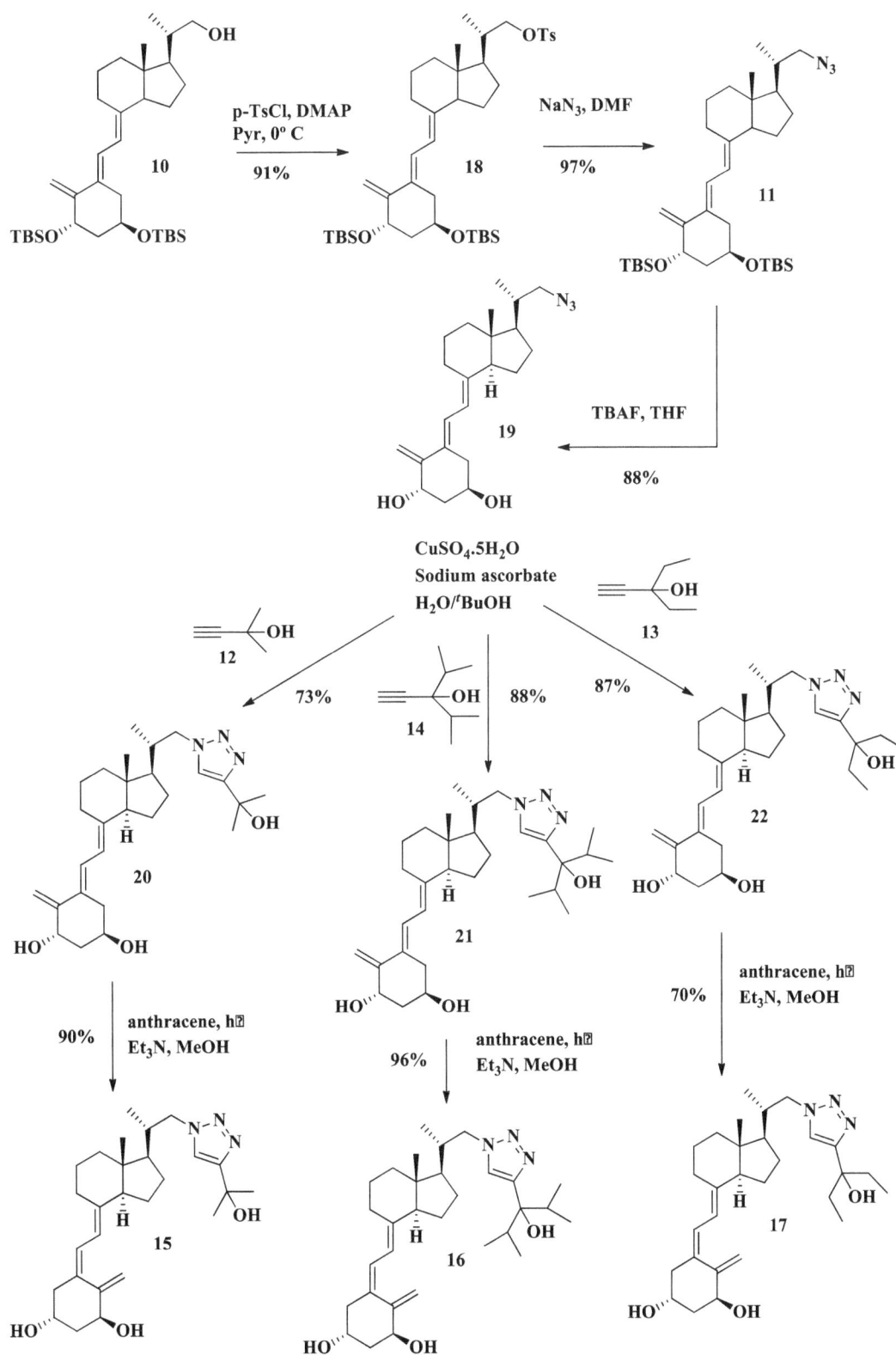

Scheme 3. Synthesis of analogues **15**, **16** and **17**

Tosylation of alcohol **10** followed by displacement of the C-22 tosylate of **18** with sodium azide, led to key azide **11** in 88% overall yield. Removal of the silyl protecting groups of **11** afforded azide **19** which underwent a [3+2]-cycloaddition[5] with alkynes **12**, **13** and **14** to afford triazoles **20**, **21** and **22** in 73, 88 and 87% yields respectively.

Computational studies carried out by Sharpless and co-workers,[6] proved that the exclusive regioselectivity of the triazole formation could be explained by a stepwise mechanism involving unprecedented matallacycle intermediates (Figure 2).

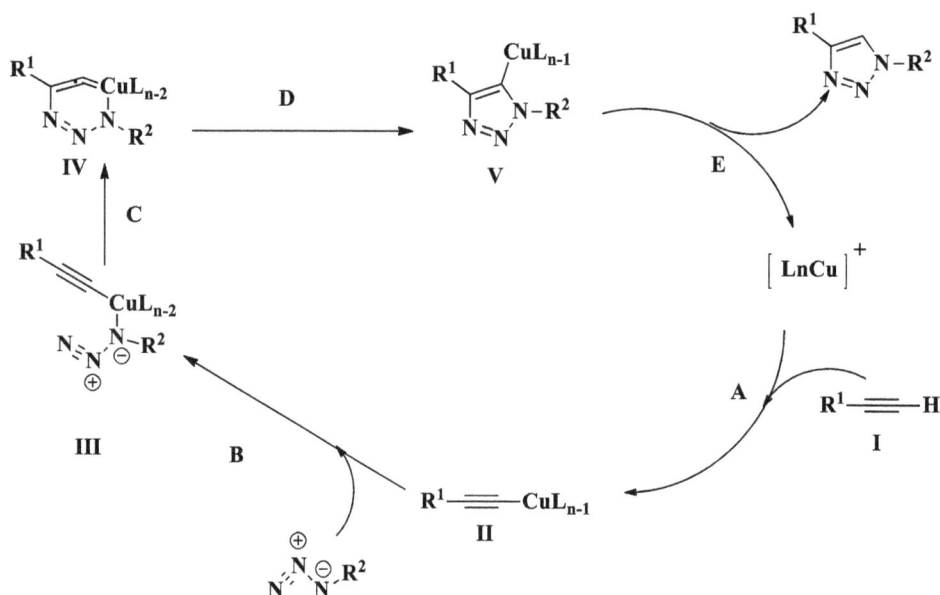

Figure 2. Sharpless proposed mechanism for the formation of 1,4-disubstituted 1,2,3-triazoles

Photosensitized isomerization of **20, 21** and **22** using anthracene as triplet sensitizer afforded target Vitamin D analogues **15, 16** and **17** in 90%, 96% and 70% yields respectively.

Conclusion

In conclusion, we have improved the method described by Calverley and Choudhry for the large scale synthesis of a late-stage intermediate which leads to a straightforward access to some calcitriol analogues with a triazole ring in their side chain. The use of intermediates mentioned above to access new calcitriol analogues is underway in our laboratory.
The preliminary results of the biological activity of some of our azavitamin D analogues showed that they had less calcemic effect than calcitriol. These results as well as the activities of the whole series of the synthesized analogues will be published in due time after patent protection.

Acknowledgements

This work was supported financially by the Xunta de Galicia (project CN 2012/184).The work of the NMR, SC-XRD and MS divisions of the research support services of the University of Vigo (CACTI) is also gratefully acknowledged. A.F; M.S.; M.G. and O.D. thank the University Cheikh Anta Diop (Dakar, Sénégal) for financial support for a research stay at the University of Vigo.

Experimental Section

General Procedures

Solvents were purified and dried by standard procedures. Flash chromatography was performed on silicagel (Merck 60, 230–400 mesh). Analytical TLC was performed on plates precoated with silica gel (Merck 60 F254, 0.25 mm). Melting points were obtained using a Gallenkamp apparatus and are uncorrected. Optical rotations were obtained using a Jasco P-2000 polarimeter. IR spectra were obtained using a Jasco FT/IR-6100 Type A spectrometer. ^{1}H NMR (400 MHz) and ^{13}C NMR (100 MHz) spectra were recorded on a Bruker ARX-400 spectrometer using TMS as the internal standard; chemical shifts (δ) are quoted in ppm and coupling constants (J) in Hz. Mass spectrometry (MS and HRMS) was carried out using a Hewlett-Packard 5988A spectrometer. The reactions were carried out protecting the glassware from light using aluminum foil.

(6S)-6-((tert-butyldimethylsilyl)oxy)-1-((E)-((3aS,7aR)-1-((2R,5R,E)-5,6-dimethylhept-3-en-2-yl)-7a-methylhexahydro-1H-inden-4(2H)-ylidene)methyl)-1,3,4,5,6,7-hexahydrobenzo[c]thiophene 2,2-dioxide (3)

In a three neckround-bottom flask at -25 °C was condensed SO_2 (100 mL, 2.00 mol) and a solution of vitamin D_2 (100 g, 0.25 mol) in CH_2Cl_2 (250 mL) was added via cannula. At the end of the addition the orange mixture was stirred at -10 °C for 90 min and allowed to reach room temperature, thus removing excess SO_2. The solvent was evaporated and the residue was dissolved in CH_2Cl_2 and cooled to 5 °C. Imidazole (22.5 g, 0.33 mol), TBSCl (50 g, 0.33 mol) and a catalytic amount of DMAP were added to the mixture which was stirred overnight, reaching room temperature. H_2O (100 mL) was added and the aqueous phase was extracted with CH_2Cl_2 (3 x 25 mL). The combined organic phases were washed with brine (25 mL) and dried. Solvent evaporation afforded 140 g (96%) of known compound **3**[4b].

(2S)-2-((3aS,7aR, E)-4-((E)-2-((S)-5-((tert-butyldimethylsilyl)oxy)-2-methylenecyclohexylidene)ethylidene)-7a-methyloctahydro-1H-inden-1-yl)propan-1-ol (6)

A solution of **3** (5 g, 8.7 mmol) in MeOH (66 mL) and CH_2Cl_2 (170 mL) was subjected to ozonolysis using the best conditions described in Table 1 and after 5 h at -78° C the mixture was allowed to reach -10° C. PPh_3 (3 g, 11.5 mmol) was added and stirring was continued for 30 min. The mixture was allowed to reach 0 °C and an aqueous saturated solution of $NaHCO_3$ (40 mL) was added. After extraction with CH_2Cl_2 (3 x 100 mL), the combined organic phases were washed with brine (3 x 50 mL), dried (Na_2SO_4), filtered and concentrated to give 4.43 g of a residue (aldehyde **4**) which was used for the next reaction without further purification. The residue was dissolved in ethanol (45 mL) and $NaHCO_3$ (4.43 g) was added and the mixture was refluxed for 5 h. The solvent was rotatory evaporated affording a residue which was dissolved in CH_2Cl_2 (25 mL) and filtered in order to remove excess $NaHCO_3$. To the filtrate was added brine (50 mL). After extraction with CH_2Cl_2 (3 x 25 mL), the combined organic phases were dried (Na_2SO_4), filtered and concentrated to afford a residue (compound **5**) which was used for the next reaction without further purification. The residue was dissolved in MeOH (45 mL) and cooled to 0 °C. $NaBH_4$ (400 mg, 10.44 mmol) was added portionwise to the mixture and stirring was continued for 15 min. H_2O (50 mL) was added and the product extracted with CH_2Cl_2 (3 x 50 mL). The combined organic phases were dried (Na_2SO_4), filtered and concentrated to afford a residue which was chromatographed on silica gel using 10% EtOAc/Hexane as eluent, affording 3.3 g (85% , 3 steps) of alcohol **6**, as a white solid, M.p.: 53-55 °C, Rf = 0.59 (30% EtOAc/Hexane);

¹H-NMR (CDCl₃, δ): 4.74 (1H; d; J=10,08 Hz; H-7); 4.58 (1H; d; J=9,6 Hz; H-6); 3.92 (1H; s; H-3); 3.50 (2H; m; H-22);3.32 (2H; s; CH₂-19); 2.85 (1H, m); 2.43 (2H, m), 2.36 (2H, m); 2.29 (1H, m); 2.06 (1H, m); 1.98 (1H, m); 1.86 (2H, m), 1.65 (7H, m); 1.36 (3H, m),0.99 (3H; d; J=6,48 Hz; CH₃-21); 0.82 (9H; s; tert-BuSi); 0.53(3H; s; CH₃-18); 0.01(3H; s; CH₃-TBS); 0.00 (3H; s; CH₃-TBS);

¹³C-NMR (CDCl₃, δ): 150.3 (C-8); 131.1 (C-5); 127.1 (C-10); 116.1 (C-7); 110.2 (CH₂-19); 68.0 (CH₂-22); 67.2 (CH-3 y CH-6); 58.5 (CH₂); 56.2 (C-14); 53.1 (C-17); 46.0 (C-13); 40.3 (CH₂); 39.4 (C-20); 34.5 (CH₂); 31.4 (CH₂); 29.8 (CH₂); 27.4 (CH₂); 26.2 (CH₃-terc-BuSi); 25.0 (CH₂); 23.9 (CH₂); 22.6 (CH₂); 18.5 (C-tert-Bu); 17.3 (C-21); 12.1 (C-18); -4.6 (CH₃-TBS); -4.7 (CH₃-TBS);

MS (m/z (%)): 445.22 (M⁺+1, 20); 311.13 (32); 281.09 (20); 267.08 (21); 209.10 (41); 193.15 (100);

HRMS: Calcdfor $C_{28}H_{48}O_2Si$: 445.3502, found: 445.3508.

(2S)-2-((3aS,7aR, E)-4-((E)-2-((S)-5-((tert-butyldimethylsilyl)oxy)-2-methylenecyclohexylidene)ethylidene)-7a-methyloctahydro-1H-inden-1-yl)propyl acetate (7)

To a solution of alcohol **6** (3.0 g, 6.75 mmol) in CH_2Cl_2 (30 mL) were added pyr (1.2 mL, 14.8 mmol), Ac_2O (0.7 mL, 7.43 mmol) and a catalytic amount of DMAP. The mixture was stirred for 2 h at room temperature and cooled to 0 °C before adding an aqueous saturated solution of NH_4Cl (30 mL). After extraction with CH_2Cl_2 (3 x 40 mL), the combined organic phases were washed with an aqueous saturated solution of $CuSO_4$ (3 x 40 mL), dried (Na_2SO_4), filtered and concentrated to afford a residue which was chromatographed on silica gel using 5% EtOAc/Hexane affording 2.9 g (94%) of acetate **7**, as a yellowish oil; Rf = 0.66(10% EtOAc/Hexane);

¹H-NMR(CDCl₃, δ):6.50 (1H; d; J=11,45 Hz; H-6); 5.88 (1H; d; J=11,50; H-7); 4.94 (1H; s; H-19); 4.65 (1H; s; H-19); 4.10 (1H, dd, J=3.8 y 7.4 Hz, H-22); 3.86 (1H, t, J=3.8 Hz, H-3); 3.81 (1H, dd, J=3.8 y 7.4 Hz, H-22); 2.85 (1H, m); 2.68 (1H, m); 2.52 (1H, m); 2.22 (2H, m); 2.15 (1H, m); 2.07 (3H, s, CH₃-Ac); 2.00 (1H, m); 1.86 (2H, m); 1.65 (7H, m); 1.35 (3H, m); 1.05 (3H; d; J=6.54 Hz; CH₃-21); 0.89 (9H; s; tert-BuSi); 0.59 (3H; s; CH₃-18); 0.08(3H, s, CH₃-TBS); 0.07(3H, s, CH₃-TBS);

¹³C-NMR (CDCl₃, δ): 171.4 (C=O); 150.0 (C-10); 143.2 (C-8); 136.5 (C-5); 119.9 (CH-6); 116.2 (CH-7); 107.6 (CH₂-19); 69.6 (CH-3); 69.5 (CH₂-22); 56.1 (CH-14); 53.1 (CH-17); 45.9 (C-13); 40.3 (CH₂); 37.5 (CH₂); 36.2 (CH-20); 35.2 (CH₂); 31.2 (CH₂); 28.9 (CH₂); 27.2 (CH₂); 26.0 (CH₃-tert-BuSi); 23.5 (CH₂); 22.3 (CH₂); 23.0 (CH₃-Ac); 18.2 (C-tert-BuSi); 17.3 (CH₃-21); 12.1 (CH₃-18); -4,6 (CH₃-TBS);

MS (m/z (%)): 487.33 (M⁺+1, 12); 486.33 (M⁺, 23); 295.20 (21); 193.15 (100); 171.21 (32);

HRMS: Calcd for $C_{30}H_{50}O_3Si$: 486.3529, found: 486.3518.

(2S)-2-((3aS,7aR, E)-4-((E)-2-((3S,5R)-5-((tert-butyldimethylsilyl)oxy)-3-hydroxy-2-methylenecyclohexylidene)ethylidene)-7a-methyloctahydro-1H-inden-1-yl)propyl acetate (8)

A solution of SeO_2 (0.7 g, 6.32 mmol) in MeOH (50 mL) was refluxed for 45 min. A solution of acetate **7** (2.18 g, 4.47 mmol) in CH_2Cl_2 (52 mL) was also refluxed for 15 min before adding it via cannula to the previous solution of SeO_2. After the addition, the mixture was refluxed for 2 h and allowed to reach room temperature. H_2O (10 mL) was added and the product extracted with CH_2Cl_2 (3 x 25 mL). The combined organic phases were dried (Na_2SO_4), filtered and concentrated to afford a residue which was chromatographed on silica gel using 10% EtOAc/Hexane affording 1.37 g (61%) of alcohol **8**, as a white solid, M.p.: 44° C, Rf = 0.42 (20% EtOAc/Hexane);

^{1}H-NMR (CDCl$_3$, δ): 6.45 (1H, d, J=11.4 Hz; H-6); 5.80 (1H, d, J=11.4 Hz, H-7); 4.42(1H, s, H-19); 4,12 (1H, s, H-19); 4,01 (1H, dd, J=7,43 y 3,24 Hz, H-22); 3,74 (1H, dd, J=7,5 y 3,20Hz, H-22); 2.75 (2H, m); 2.43 (1H, m); 2.33 (1H, m); 1,98(3H, s, CH$_3$-Ac); 1.86 (4H, m); 1.62 (4H, m); 1.46 (3H, m); 1.22 (2H, m); 0,96 (3H, d, J=6,6 Hz;CH$_3$-21); 0,79 (9H, s, *tert*-BuSi); 0,50(3H, s, CH$_3$-18); 0,00(6H, s, CH$_3$-TBS);

^{13}C-NMR (CDCl$_3$, δ): 171.8 (C=O); 153.5 (C-10); 144.0 (C-8); 135.0 (C-5); 122.6 (C-6); 116.8 (C-7); 108.1 (CH$_2$-19); 70.9 (C-3); 69.8 (CH$_2$-22); 67.2 (C-1); 56.5 (C-14); 53.5 (C-17); 46.4 (C-13); 43.3(CH$_2$); 40.7 (CH$_2$); 37.3 (CH$_2$); 36.5 (C-20); 29.4 (CH$_2$); 27.5 (CH$_2$); 26.2 (CH$_3$-*terc*-BuSi); 23.9 (CH$_2$); 22.7 (CH$_2$); 21.4 (CH$_3$-Ac); 18.5 (C-*tert*-BuSi); 17.7 (C-21); 12.5 (C-18); -4.3 (CH$_3$-TBS);

MS (m/z (%)): 503.25 (M$^+$+1, 18); 502.34 (M$^+$, 23); 307.16 (20); 171.21 (32);

HRMS: Calcdfor C$_{30}$H$_{50}$O$_4$Si: 502.3478, found: 502.3495.

(2S)-2-((3aS,7aR,E)-4-((E)-2-((3S,5R)-3,5-bis((*tert*-butyldimethylsilyl)oxy)-2-methylenecyclohexylidene)ethylidene)-7a-methyloctahydro-1H-inden-1-yl)propyl acetate (9)

To a solution of alcohol **8** (1.90 g, 3.76 mmol) in CH$_2$Cl$_2$ (10 mL), at 0 °C was added imidazole (343 mg, 5.0 mmol), TBSCl (760 mg, 5.0 mmol) and a catalytic amount of DMAP and the mixture was left stirring at room temperature for 5 h. H$_2$O (10 mL) was added and the product extracted with CH$_2$Cl$_2$ (3 x 20 mL). The combined organic phases were washed with brine (20 mL), dried (Na$_2$SO$_4$), filtered and concentrated to afford a residue which was chromatographed on silica gel using 3% EtOAc/Hexane as solvent, affording 2.1 g (90%) of acetate **9**, as a white solid, M.p.: 77-80 °C, Rf = 0.51 (20% EtOAc/Hexane);

^{1}H-NMR (CDCl$_3$, δ):6.40 (1H, d, J=11.32 Hz; CH-6); 5.78 (1H, d, J=11.16 Hz; CH-7); 4.92(1H, s; CH$_2$-19); 4.88(1H, s; CH$_2$-19); 4.48 (1H, d; J=4.7Hz; CH-3); 4.16 (1H, s; CH-1); 4.05(1H, dd; J=3.25 Hz; J=7.34 Hz; H-22); 3.74(1H, dd; J$_1$=3.22 Hz; J$_2$=7.39 Hz; H-22); 2.86 (1H, m); 2.55 (1H, m); 2.22 (1H, m); 1.98(3H, s; CH$_3$-Ac); 1.95 (1H, m); 1.85 (3H, m); 1.65 (4H, m); 1.55 (3H, m); 1.33 (3H, m); 1.01 (3H, s, CH$_3$-21); 0.84(9H, s; CH$_3$-*tert*-BuSi); 0.80(9H, s; CH$_3$-*tert*-BuSi); 0.51(3H, s; CH$_3$-18); 0.00 (12H, s; CH$_3$-TBS);

^{13}C-NMR (CDCl$_3$, δ): 171.6 (C=O); 154.0 (C-10); 143.1 (C-8); 136.0 (C-5); 122.0 (CH-6); 117.0 (CH-7); 107.0 (CH$_2$-19); 70.6 (CH-3); 69.8 (CH$_2$-22); 67.6 (CH-1); 56.5 (C-14); 53.5 (C-17); 46.4 (C-13); 44.3 (CH$_2$); 40.8 (CH$_2$); 36.9 (CH$_2$); 36.5 (CH-20); 29.3 (CH$_2$); 27.5 (CH$_2$); 26.3 (CH$_3$-*terc*-BuSi); 23.8 (CH$_2$); 22.7 (CH$_2$); 21.3 (CH$_3$-Ac); 18.6

(C-*tert*-BuSi); 18.4 (C-*tert*-BuSi); 17.7(CH$_3$-21); 12.4 (CH$_3$-18); -4.4 (CH$_3$-TBS); -4.5 (CH$_3$-TBS);

MS (m/z (%)): 617.44 (M$^+$+1,38); 616.42 (23); 485.33 (52); 284.16 (100); 171.21 (32); 163.22 (20);

HRMS: Calcdfor C$_{36}$H$_{64}$O$_4$Si$_2$: 616.4298, found: 616.4262.

(2S)-2-((3aS,7aR, E)-4-((E)-2-((3S,5R)-3,5-bis((*tert*-butyldimethylsilyl)oxy)-2-methylenecyclohexylidene)ethylidene)-7a-methyloctahydro-1H-inden-1-yl)propan-1-ol (10)

To a solution of acetate **9** (1.1 g, 1.83 mmol) in MeOH (30 mL) at room temperature was added K$_2$CO$_3$ (500 mg, 3.67 mmol) and the mixture was stirred for 10 h. Excess K$_2$CO$_3$ was eliminated by firtration and H$_2$O (100 mL) was added to the filtrate and the product extracted with CH$_2$Cl$_2$ (3 x 50 mL). The combined organic phases were washed with brine (3 x 20 mL), dried (Na$_2$SO$_4$), filtered and concentrated to afford a residue which was chromatographed on silica gel using 3% EtOAc/Hexane as solvent, affording 990 mg (94%) of alcohol **10**, as a white solid, M.p.: 110-113° C, Rf = 0.63 (30% EtOAc/Hexane);

^{1}H-NMR (CDCl$_3$, δ): 6.41 (1H, d, J=11,3 Hz; H-6); 5.78 (1H, d, J=11.3;CH-7); 4.92(1H, s, H-19); 4.87(1H, s, H-19); 4.45 (2H, m, H-22); 4.15 (1H, m, H-1); 3.61 (1H, m, H-3); 2.75 (1H, m); 2.45 (1H, m); 2.27 (1H, m); 1.95 (2H, m); 1.63 (4H, m); 1.51 (3H, m); 1.22 (6H, m);1.00 (3H, d, J=7,6 Hz; CH$_3$-21); 0.83(9H, s, *tert*-BuSi); 0.80(9H, s,*tert*-BuSi); 0.50(3H, s, CH$_3$-18); 0.00(12H, s, CH$_3$-TBS);

^{13}C-NMR (CDCl$_3$, δ): 154.0 (C-10); 143.5 (C-8); 135.9 (C-5); 122.1 (C-6); 116.9 (C-7); 107.1(CH$_2$-19); 70.7 (C-3); 68.3 (CH$_2$-22); 67.6 (C-1); 56.6 (C-14); 53.3 (C-17); 46.3 (C-13); 44.3 (CH$_2$); 40.8 (CH$_2$) 39.5 (C-20); 37.0(CH$_2$); 29.3 (CH$_2$); 27.6(CH$_2$); 26.3(CH$_3$-*tert*-BuSi); 26.2 (CH$_3$-*tert*-BuSi); 23.9 (CH$_2$); 22.7 (CH$_2$); 18.6 (C-*tert*-BuSi); 18.5(C-*tert*-BuSi); 17.3 (C-21); 12.5 (C-18); -4,4 (CH$_3$-TBS);

MS (m/z (%)): 575.36 (M$^+$+1,51); 442.25 (77); 249.11 (45); 247.10 (27);

HRMS: Calcd for C$_{34}$H$_{62}$O$_3$Si$_2$: 575.4316, found: 575.4317.

(S)-2-((1R,3aS,7aR, E)-4-((E)-2-((3S,5R)-3,5-bis((tert-butyldimethylsilyl)oxy)-2-methylenecyclohexylidene)ethylidene)-7a-methyloctahydro-1H-inden-1-yl)propyl 4-methylbenzenesulfonate (18).

To a solution of **10** (990 mg, 1.72 mmol) in Py (9 mL) at 0 °C was added p-TsCl (660 mg, 3.44 mmol) and DMAP (c.c.). The mixture was stirred at this temperature for 9 h, quenched with NH$_4$Cl (10 mL), then allowed to reach room temperature. The product was extracted with EtOAc (3×15 mL). The organic phase was washed with CuSO$_4$ (3×20 ml).

After drying (Na$_2$SO$_4$) and solvent evaporation, the residue was chromatographed on silicagel using 3% EtOAc-hexane as eluent, to afford 1.1 g of tosylate **18** [91%, white solid; Mp= 50-53 °C; Rf: 0.87 (30% EtOAc-hexane)].

^{1}H-NMR (CDCl$_3$, δ): 7.72 (2H, d, *J*=8.2 Hz, H-Ts), 7.27 (2H, d, *J*=8.0 Hz, H-Ts), 6.37 (1H, d, *J*=11.3 Hz, H-6), 5.74 (1H, d, *J*=11.3 Hz, H-7), 4.91 (1H, s, H-19), 4.88 (1H, s, H-19), 4.48 (1H, m, H-1), 4.45 (1H, m, H-3), 3.92 (1H, m, H-22), 3.90 (1H, m, H-22), 2.75 (1H, m), 2.43 (1H, m), 2.37 (3H, s, CH$_3$-Ts), 2.17 (1H, m), 1.85 (3H, m), 1.55 (5H, m), 1.37 (3H, m), 1.15 (3H, m), 0.93 (3H, d, *J*=6.5 Hz, H-21), 0.84 (9H, s, CH$_3$-*terc*-BuSi), 0.80 (9H, s, CH$_3$-*terc*-BuSi), 0.44 (3H, s, H-18), 0.00 (12H, s, CH$_3$-SiMe);

^{13}C-NMR (CDCl$_3$, δ): 154.0 (C-10), 145.0 (C-8), 142.9 (C-Ts), 136.1 (C-5), 133.6 (C-Ts), 130.2 (CH-Ts), 128.3 (CH-6), 122.0 (CH-7), 107.1 (CH$_2$-19), 75.9 (CH$_2$-22), 70.6 (CH-1), 67.2 (CH-3), 56.4 (CH-14), 52.6 (CH-17), 46.2 (C-13), 44.3 (CH$_2$), 40.6 (CH$_2$), 37.0 (CH-20), 36.9 (CH$_2$), 29.2 (CH$_2$), 27.3 (CH$_2$), 26.3 (CH$_3$-*terc*-BuSi), 26.3 (CH$_3$-*terc*-BuSi), 26.2 (CH$_3$-*terc*-BuSi), 23.8 (CH$_2$), 22.6 (CH$_2$), 22.0 (CH$_3$-Ts), 18.6 (C-*terc*-BuSi), 18.4 (C-*terc*-BuSi), 17.4 (CH$_3$-18), 12.4 (CH$_3$-21), -4.4 (CH$_3$-SiMe), -4.5 (CH$_3$-SiMe), -4.5 (CH$_3$-SiMe);

LRMS: : [m/z %]:729.35 [(M+1)$^+$, (23)], 728.35 [M$^+$, (22)], 727.34 [M$^+$-1, (14)], 597.28 (32), 596.28 (46), 425.27(35), 379.19 (16), 249.13 (47), 248.13(100), 247.11 (33).

HRMS: m/z calcd C$_{41}$O$_5$Si$_2$SH$_{68}$ for: 729.4404; found: 729.4418.

(((1R,3S,*E*)-5-((*E*)-2-((3aS,7aR)-1-((S)-1-azidopropan-2-yl)-7a-methylhexahydro-*1H*-inden-4(*2H*)-ylidene)ethylidene)-4-methylenecyclohexane-1,3-diyl)bis(oxy))bis(*tert*-butyldimethylsilane) (11)

To a solution of tosylate **18** (933 mg, 1.28 mmol) in DMF (15 mL) was added NaN$_3$ (832 mg, 12.8 mol) and the mixture was stirred at room temperature for 42 h. CH$_2$Cl$_2$ (25 mL) was added and the organic phase was washed with H$_2$O (3 x 15 mL), dried (Na$_2$SO$_4$), filtered and concentrated in vacuo to afford a residue which was chromatographed on silica gel using 5% EtOAc/Hexane as solvent, affording 750 mg (97%) of azide **11**, as a white solid, M.p.: 104 °C, Rf = 0.75 (10% EtOAc/Hexane);

^{1}H-NMR (CDCl$_3$, δ): 6.43 (1H, d, J=11.4 Hz, H-6), 5.82 (1H, d, J=11.3 Hz, H-7), 4.95 (2H, d, J=16.2 Hz, H-19), 4.63 (1H, m, H-22), 4.43 (1H, m, H-22), 3.46 (1H, dd, J=,11.9 y 3.1 Hz, H-1), 3.05 (1H, m, H-3), 2.81 (1H, m); 2.77 (1H, m); 2.56 (1H, m); 2.12 (1H, m); 1.92 (4H, m); 1.85 (4H, m); 1.63 (2H, m); 1.53 (3 H, s, CH$_3$-18), 1.21 (3H, m); 1.05 (3 H, d, J= 6.5 Hz, CH$_3$-21), 0.89 (9H, s, *tert*-BuSi), 0.85 (9H, s, *tert*-BuSi), 0.05 (12 H, s, CH$_3$-TBS);

^{13}C-RMN (CDCl$_3$, δ):153.6 (C-10), 142.7 (C-8), 135,7 (C-5), 121.6 (C-6), 116.2 (C-7), 106.7 (CH$_2$-19), 70.2 (C-3), 67.2 (C-22), 58.0 (CH$_2$), 56.2

(C-14), 53.6 (C-17), 43.9 (CH$_2$), 40.3 (CH$_2$), 37.2 (C-20), 36.6 (CH$_2$), 28.9 (CH$_2$), 27.4 (CH$_2$), 25.9 (CH$_3$-*tert*-BuSi), 23.4 (CH$_2$), 22.3 (CH$_2$), 18.2 (C-*tert*-BuSi), 17.9 (C-21), 12.1 (C-18), -4.8 y -4.9 (CH$_3$-TBS);

MS (m/z (%)): 600,43 (M+1, 40); 599.44 (M+, 46); 570.43 (21); 542.37 (27); 467.34 (73); 440.32 (20); 248.15 (100);

HRMS: Calcd for C$_{34}$H$_{61}$N$_3$O$_2$ Si$_2$: 599.4302, found: 599.4302.

(1R,3S, *E*)-5-((*E*)-2-((3aS,7aR)-1-((S)-1-azidopropan-2-yl)-7a-methylhexahydro-*1H*-inden-4(*2H*)-ylidene)ethylidene)-4-methylenecyclohexane-1,3-diol (19)

To a solution of azide **11** (116 mg, 0.19 mmol) in THF (2 mL) was added TBAF (1.16 mL, 1.16 mmol, 1M sln in THF) and the mixture was stirred for 16 h. Aqueous saturated solution of NH$_4$Cl (10 mL) was added and the product was extracted with EtOAc (3 x 10 mL). The combined organic phases were dried (Na$_2$SO$_4$), filtered and concentrated to afford a residue which was chromatographed on silica gel using 50% EtOAc/Hexane as solvent, affording 72 mg (99%) of azide **19**, as a colourless oil, Rf = 0.11 (20% EtOAc/Hexane);

^{1}H-NMR (CDCl$_3$, δ): 6.55 (1H, d, J= 11.5 Hz, H-6); 5.86 (1H, d, J= 11.5 Hz, H-7); 5.10 (1H, s; H-19); 4.95 (1H, s, H-19); 4.47 (1H, m, CH$_2$-22); 4.22 (1H, m, CH$_2$-22); 3.37 (1H, dd, J=11.9 y 3.1 Hz, H-1); 3.06 (1H, m, H-3); 2.86 (1H, m); 2.75 (1H, m), 2.66 (1H, m); 2.43 (1H, m); 1.85 (5H, m); 1.66 (5H, m); 1.55 (2H, m); 1.32 (3H, m), 1.05, (3H, d, J= 6.6 Hz, CH$_3$-21); 0.56 (3 H, s, CH$_3$-18);

^{13}C-NMR (CDCl$_3$, δ):151.7 (C-10); 144.5 (C-8); 133,0 (C-5); 123.2 (C-6); 116.1 (C-7); 106.7 (CH$_2$-19); 71.1 (C-3); 67.5 (C-22); 57.9 (CH$_2$); 56.2 (C-14); 53.6 (C-17); 42.0 (CH$_2$); 40.2 (CH$_2$); 37.2 (C-20); 36.4 (CH$_2$); 29.0 (CH$_2$); 27.3 (CH$_2$); 23.4 (CH$_2$); 22.3 (CH$_2$); 17.9 (C-21); 12.1 (C-18);

MS (m/z (%)): 371.25 (M$^+$,56); 354.24 (35); 322.22 (23); 307.07 (100), 289.07 (44), 273.08 (20);

HRMS: Calcd for C$_{22}$H$_{33}$N$_3$O$_2$:371.2573, found: 371.2565.

General procedure for the click chemistry reaction of azide **19** with alkynes **12**, **13** and **14** to afford compounds **20**, **21** and **22**.

To a solution of azide **19** (70 mg, 0.19 mmol) in *tert*-BuOH (2 mL) and H$_2$O (1 mL) was added a catalytic amount of CuSO$_4$.5H$_2$O, sodium ascorbate (13υL of 1M aqueous sln), and the chosen alkyne (0.20 mmol). The mixture was stirred at room temperature for 7 h. H$_2$O (10 mL) was added and the product was extracted with EtOAc (3 x 15 mL). The combined organic phases were dried (Na$_2$SO$_4$), filtered and concentrated to afford a residue which was chromatographed on silica gel using 50% EtOAc/Hexane as solvent to afford the corresponding triazoles **20**, **21** or **22**.

(1R,3S,E)-5-((E)-2-((3aS,7aR)-1-((S)-1-(4-(2-hydroxypropan-2-yl)-*1H*-1,2,3-triazol-1-yl)propan-2-yl)-7a-methylhexahydro-*1H*-inden-4(*2H*)-ylidene)ethylidene)-4-methylenecyclohexane-1,3-diol (20)

Yield 73%, Brownish solid, M.p.: 106 ℃, Rf = 0.31 (EtOAc);

¹H-NMR (CDCl₃, δ) : 7.38 (1H, s, H-23); 6.55 (1H, d, J=11.4 Hz, H-6); 5.89 (1H, d, J=11.4 Hz, H-7); 5.19 (1H, s, H-19); 4.96 (1H, s, H-19); 4.57-4.50 (2H, m, H-22); 4.21 (1H, m, H-1); 4.04 (1H, m, H-3); 2.99 (2H, m); 2.76 (1H, m); 2.66 (1H, m); 2.43 (1H, m); 1.89 (5H, m); 1.75 (4H, m); 1.63 (6H, s, CH₃-25); 1.43 (2H, m); 1.21 (2H, m); 0.87 (3 H, d, J= 6.5 Hz, CH₃-21); 0.60 (3H, s, CH₃-18);
¹³C-NMR (CDCl₃,δ): 155.4 (C-10); 151.6 (C-24); 144.2 (C-8); 133.2 (C-5);123.6 (C-23); 123.0 (C-6); 116.3 (C-7);109.7 (CH₂-19); 76.7 (C-25); 71.1 (C-3); 67.5 (C-1);56.1 (C-13); 55.9 (CH₂-22); 54.1 (C-17);42.0 (CH₂); 40.2 (CH₂); 38.2 (C-26 o C-27); 36.7 (CH₂), 30.5 (C-26 o C-27); 28.9 (CH₂); 27.6 (CH₂); 25.8 (C-25); 23.3 (CH₂); 22.4 (CH₂); 17.2 (C-21); 12.2 (C-18);
MS (m/z (%)): 456.34 (M+1,40); 307.11 (20); 289.10 (23); 155.27 (38);
HRMS: Calcd for C₂₇H₄₂N₃O₃: 456.3226, found: 456.3224.

(1R,3S,E)-5-((E)-2-((3aS,7aR)-1-((S)-1-(4-(3-hydroxy-2,4-dimethylpentan-3-yl)-*1H*-1,2,3-triazol-1-yl)propan-2-yl)-7a-methylhexahydro-*1H*-inden-4(*2H*)-ylidene)ethylidene)-4-methylenecyclohexane-1,3-diol (21)

Yield 86%, White solid, M.p.: 123 ℃, Rf = 0.41 (EtOAc);

¹H-NMR (CDCl₃, δ): 7.30 (1H, s, H-23); 6.53(1H, d, J= 11.3 Hz, H-6); 5.87 (1H, d, J= 11.3 Hz, H-7); 5.07 (1H, s, H-19); 4.93 (1H, s, H-19); 4.47 (1H. m. H-1); 4.36 (1H, m, H-3); 4.33 (2H, m, H-22); 2.82 (2H, m); 2, 76 (2H, m); 2.66 (2H, m); 2.44 (2H, m);1.66 (5H, m); 1.43 (4H, m); 1.22 (2H, m); 1.11 (3H, m), 1.08(3H, d, J=12.8 Hz, CH₃-21); 0.80 (12H, m, CH₃-*iso*propyl); 0.60 (3H, s, CH₃-18);
¹³C-NMR (CDCl₃,δ): 171.7 (C-10); 151.7 (C-8); 141.0 (C-24); 133.3 (C-5); 123.4 (C-23); 122.8 (C-6); 117.2 (C-7); 109.6 (CH₂-19); 72.0 (C-25); 70.0 (C-3);65.7 (C-1); 56.1 (C-13); 55.8 (CH₂-22); 54.2 (C-17); 41.9 (CH₂); 40.1 (CH₂); 38.1; 36.6 (CH₂); 34.1 (CH-*iso*propyl); 30.5; 28.9 (CH₂); 27.5 (CH₂); 23.3 (CH₂); 22.4 (CH₂); 17.1 (C-21); 14.1 (CH₃-*iso*propyl); 12.2 (C-18);
MS (m/z (%)): 512.53 (M⁺+1,100); 511.46 (M⁺, 30) 394.43 (20); 322.31 (34); 307.16 (20); 154.24 (96);
HRMS: Calcd for C₃₁H₄₉N₃O₃: 511.3852, found: 511.3868.

(1R,3S,E)-5-((E)-2-((3aS,7aR)-1-((S)-1-(4-(3-hydroxypentan-3-yl)-*1H*-1,2,3-triazol-1-yl)propan-2-yl)-7a-methylhexahydro-*1H*-inden-4(*2H*)-ylidene)ethylidene)-4-methylenecyclohexane-1,3-diol (22)

Yield 87%, White solid, M.p.: 110 ℃, Rf = 0.41 (EtOAc);

¹H-NMR (CDCl₃, δ): 7.37 (1H, s, H-23); 6.55 (1H, d, J= 11.4 Hz, H-6); 5.89 (1H, d, J= 11.4 Hz, H-7); 5.09 (1H, s, H-19); 4.96 (1H, s, H-19); 4.47 (1H, m, H-1), 4.36 (1H, m, H-3); 4.33 (2H, m, H-22); 2.82 (2H, m); 2.76 (2H, m), 2.66 (1H, m); 2.43 (1H, m); 2.05 (4H,m, CH₂-Et); 1.88 (4H, m); 1.66 (5H, m); 1.43 (3H, m); 1.22 (3H, m); 0.85(9 H, m,CH₃-Et y CH₃-21); 0.60 (3H, s, CH₃-18);
¹³C-NMR (CDCl₃,δ): 151.7 (C-10); 144.6 (C-24); 144.6 (C-8); 133.3 (C-5); 123.0 (C-23); 119.6 (C-6); 116.3 (C-7);109.7 (CH₂-19); 76.7(C-25); 71.1 (C-3);65.8 (C-1); 56.2 (C-13); 55.9 (CH₂-22); 54.2 (C-17); 42.0 (CH₂); 40.2 (CH₂); 38.2 ; 36.7 (CH₂); 33.9 (CH₂-Et); 30.5; 28.9 (CH₂); 27.5 (CH₂); 23.3 (CH₂); 22.4 (CH₂); 17.1 (C-21); 12.2 (C-18); 7.8 (CH₃-Et);
MS (m/z (%)): 484.51 (M⁺+1, 37); 307.17 (28); 235.25 (20); 155.26 (33);
HRMS: Calcd for C₂₉H₄₆N₃O₃: 484.3539, found: 484.3546.

Photosensitized isomerization of **20,21** and **22** to afford **15**, **16** and **17**, using anthracene as triplet sensitizer was carried out following the general procedure described for compound **20**.

To a solution of **20** (10 mg, 0.02 mmol) in MeOH (10 mL) was added anthracene (3 mg, 0.01 mmol) and a catalytic amount of Et₃N. The mixture was irradiated with a 200 W lamp for 6 h, diluted with CH₂Cl₂ (15 mL) and washed with brine (10 mL). After solvent evaporation the resulting residue was chromatographed on silica gel using CH₂Cl₂ and 10% MeOH/CH₂Cl₂ as solvent, affording 9 mg (90%) of analogue **15**.

(1R,3S,Z)-5-((E)-2-((3aS,7aR)-1-((S)-1-(4-(2-hydroxypropan-2-yl)-*1H*-1,2,3-triazol-1-yl)propan-2-yl)-7a-methylhexahydro-*1H*-inden-4(*2H*)-ylidene)ethylidene)-4-methylenecyclohexane-1,3-diol (15)

Yield 90%, White solid, M.p.: 110 ℃, Rf = 0.31 (10% MeOH/CH₂Cl₂);

¹H-NMR (CDCl₃, δ): 7.34 (1H, s, H-23); 6.35 (1H, d, J=11.1 Hz, H-6), 6.02 (1H, d, J=11.2 Hz, H-7), 5.32 (1H, s, H-19); 4.99 (1H, s, H-19), 4.57-4.50 (2H, m, H-22), 4.21 (1H, m, H-1), 4.04 (1H, m, H-3); 2.99 (2H, m); 2.76 (1H, m); 2.66 (1H, m);2.43 (1H, m); 1.89 (5H, m); 1.75 (4H, m); 1.63 (6H, s, CH₃-25); 1.43 (2H, m); 1.21 (3H, m); 0.86 (3 H, d, J= 6.5 Hz, CH₃-21); 0.58 (3H, s, CH₃-18);

^{13}C-NMR (CDCl$_3$,δ): 155.4 (C-10), 147.6 (C-24), 142.2 (C-8); 133.4 (C-5); 124.6 (C-23),119.6 (C-6); 117.4 (C-7); 111.8 (CH$_2$-19); 76.7(C-25); 70.8 (C-3); 67.5 (C-1);56.3 (C-13); 55.7(CH$_2$-22); 54.1 (C-17);42.3 (CH$_2$); 40.2 (CH$_2$); 38.2 (C-26 o C-27); 36.7 (CH$_2$); 30.5 (C-26 o C-27); 28.9 (CH$_2$); 27.6 (CH$_2$); 25.8 (C-25); 23.3 (CH$_2$); 22.4 (CH$_2$); 17.2 (C-21); 12.2 (C-18);

MS (m/z (%)): 456.34 (M+1, 40), 307.11 (20), 289.10 (23); 155.27 (38);

HRMS: Calcd for C$_{27}$H$_{42}$N$_3$O$_3$: 456.3226, found: 456.3224.

(1R,3S,Z)-5-((E)-2-((3aS,7aR)-1-((S)-1-(4-(3-hydroxy-2,4-dimethylpentan-3-yl)-1H-1,2,3-triazol-1-yl)propan-2-yl)-7a-methylhexahydro-1H-inden-4(2H)-ylidene)ethylidene)-4-methylenecyclohexane-1,3-diol (16)

Yield 96%, colourless oil, Rf = 0.41 (10% MeOH/CH$_2$Cl$_2$);

^{1}H-NMR (CDCl$_3$, δ): 7.28 (1H, s, H-23); 6.36 (1H, d, J=11.4 Hz, H-6); 6.03 (1H, d, J=11.4 Hz, H-7); 5.33 (1H, s, H-19); 4.99 (1H, s, H-19); 4.39 (1H, m, H-1); 4.36 (1H, m, H-3); 4.33 (2H, m, H-22); 2.82 (2H, m), 2, 76 (2H, m); 2.66 (2H, m); 2.44 (2H, m);1.66 (5H, m); 1.43 (4H, m); 1.22 (2H, m); 1.11 (3H, m), 1.08 (3H, d, J=12.8 Hz, CH$_3$-21); 0.80 (12H, m, CH$_3$-isopropyl); 0.59 (3H, s, CH$_3$-18);

^{13}C-NMR (CDCl$_3$,δ): 150.7 (C-10); 147.7 (C-24); 142.3 (C-8); 133.5 (C-5); 124.3 (C-23); 121.4 (C-6); 117.4 (C-7); 111.6 (CH$_2$-19); 72.0 (C-25); 70.7 (C-3); 65.7 (C-1); 56.1 (C-13); 55.8 (CH$_2$-22); 54.2 (C-17); 41.9 (CH$_2$); 40.1 (CH$_2$); 38.2; 36.6 (CH$_2$); 34.1 (CH-isopropyl), 30.48; 28.9 (CH$_2$); 27.5 (CH$_2$); 23.4 (CH$_2$); 22.4 (CH$_2$); 17.1 (C-21); 14.1(CH$_3$-isopropyl); 12.2 (C-18);

MS (m/z (%)): 512.53 (100); 394.43 (20); 322.31 (34); 307.16 (20); 154.24 (96);

HRMS: Calcd for C$_{31}$H$_{50}$N$_3$O$_2$: 512.3852, found: 512.3868.

(1R,3S,Z)-5-((E)-2-((3aS,7aR)-1-((S)-1-(4-(3-hydroxypentan-3-yl)-1H-1,2,3-triazol-1-yl)propan-2-yl)-7a-methylhexahydro-1H-inden-4(2H)-ylidene)ethylidene)-4-methylenecyclohexane-1,3-diol (17)

Yield 90%, colourless oil, Rf = 0.41 (10% MeOH/CH$_2$Cl$_2$);

^{1}H-NMR (CDCl$_3$, δ): 7.33 (1H, s, H-23); 6.36 (1H, d, J= 11.4 Hz, H-6), 6.02 (1H, d, J= 11.4 Hz, H-7); 5.32 (1H, s, H-19); 4.99 (1H, s, H-19); 4.47 (1H. m. H-1), 4.36 (1H, m, H-3); 4.33 (2H, m, H-22);2.82 (2H, m); 2.76 (2H, m), 2.66 (1H, m); 2.43 (1H, m); 2.05 (4H,m, CH$_2$-Et);1.88 (4H, m); 1.66 (5H, m), 1.43 (3H, m); 1.22 (3H, m); 0.85(9 H, m,CH$_3$-Et y CH$_3$-21); 0.59 (3H, s, CH$_3$-18);

^{13}C-NMR (CDCl$_3$,δ): 151.7 (C-10); 147.6 (C-24); 142.6 (C-8); 133.4 (C-5); 124.0 (C-23); 120.6 (C-6); 117.3 (C-7);111.7 (CH$_2$-19); 76.7(C-25); 71.1 (C-3);65.8 (C-1); 56.3 (C-13); 56.2 (CH$_2$-22); 54.2

(C-17); 42.0 (CH$_2$); 40.2 (CH$_2$); 38.2; 36.7 (CH$_2$); 33.9 (CH$_2$-Et); 31.5; 28.9 (CH$_2$); 27.6 (CH$_2$); 23.2 (CH$_2$); 22.3 (CH$_2$); 17.2 (C-21); 12.2 (C-18); 7.8 (CH$_3$-Et);

MS (m/z (%)): 484.51 (M$^+$+1, 37); 307.17 (28); 235.25 (20); 155.26 (33);

HRMS: Calcd for C$_{29}$H$_{46}$N$_3$O$_3$: 484.3539, found: 484.3546.

References

1 - A. W. Norman, *Vitamin D the Calcium Homeostatic Steroid Hormone*; Academic Press: New York, 1979.

2 - (a) A. W. Norman, R. Bouillon, M. Thomasset, *Vitamin D: Chemistry, Biology and clinical Application of the Steroid Hormone;* Eds., Vitamin D Workshop: Riverside, CA, 1997; (b) D. Feldman, F. H. Glorieux, J. W. Pike, Vitamin D; Academic Press: San Diego, CA, 1997; (c) R. Pardo, M. Santelli, Synthesis of vitamin D metabolites. *Bull. Soc. Chim. Fr.* **1985**, 98-114; H. Dai, G. H. Posner, Synthetic approaches to vitamin D. *Synthesis* **1994**, *12*, 1383-1398; (d) G.-D. Zhu, W. H. Okamura, Synthesis of Vitamin D (Calciferol). *Chem. Rev.* **1995**, *95*, 1877-1952; (e) G. H. Posner, M. Kahraman, Organic chemistry of vitamin D analogues (deltanoids). *Eur. J. Org. Chem.* **2003**, *20*, 3889-3895.

3 - (a) Y. Fall, M. Torneiro, L. Castedo, A. Mouriño, *Tetrahedron Lett.***1992**, *33*, 6683-6686; (b) Y. Fall, M. Torneiro, L. Castedo, A. Mouriño, *Tetrahedron,* **1997**, *53*, 4703-4714; (c) Y. Fall, *Tetrahedron Lett.* **1997**, *38*, 4909-4912; (d) M. Torneiro, Y. Fall, L. Castedo, A. Mouriño, *J. Org. Chem.* **1997**, *62*, 6344-6352; (e) M. De los Angeles Rey, J. A. Martínez-Pérez, A. Fernández-Gacio, K. Halkes, Y. Fall, J. Granja, A. Mouriño, *J. Org. Chem.* **1999**, *64*, 3196-3206; (f) Y. Fall, C. Fernández, C. Vitale, A. Mouriño, *Tetrahedron Lett.* **2000**, *41*, 7323-7326; (g) Y. Fall, C. Fernández, V. González, A. Mouriño, *Tetrahedron Lett.* **2001**, *42*, 7815-7817; (h) Y. Fall, C. Fernández, V. González, A. Mouriño, *Synlett.* **2001**, *10*, 1567-1568; (i) Y. Fall, V. González, B. Vidal, A. Mouriño, *Tetrahedron Lett.* **2002**, *43*, 427-429; (j) Y. Fall, C. Barreiro, C. Fernández, A. Mouriño, *Tetrahedron Lett.* **2002**, *43*, 1433-1436; (k) Y. Fall, O. Diouf, G. Gómez, T. Bolaño, *Tetrahedron Lett.* **2003**, *44*, 6069-6072; (l) Suárez, P. L.; Z. Gándara, G. Gómez, Y. Fall, *Tetrahedron Lett.* **2004**, *45*, 4619-4621; (m) C. Fernández, O. Diouf, E. Momán, G. Gómez, Y. Fall, *Synthesis* **2005**, *10*, 1701-1705; (n) C. Fernández, G. Gómez, C. Lago, E. Momán, Y. Fall, *Synlett.* **2005**, 14, 2163-2166; (o) C. Fernández, Z. Gándara, G. Gómez, B. Covelo, Y. Fall, *TetrahedronLett.* **2007**, *48*, 2939-2942; (p) Z. Gándara, M. Pérez, X. Pérez-García, G.

Gómez, Y. Fall, *TetrahedronLett.* **2009**, *50*, 4874-4877; (q) D. G. Salomón, S. M. Grioli, M. Buschiazzo, E. Mascaró, C. Vitale, G. Radivoy, M. Pérez, Y. Fall, E. A. Mesri, A. C. Curino, M. M. Facchinetti, *ACS Med. Chem. Lett.,***2011**, *2*, 503-508; (r) M. L. Rivadulla, X. Pérez-García, M. Pérez, G. Gómez, Y. Fall, *TetrahedronLett.* **2013**, *54*, 3164-3166; (s) A. Martínez, Z. Gándara, M. González, G. Gómez, Y. Fall, *Tetrahedron Lett.* **2013**, *54*, 3514-3517; (t) M. J. Ferronato, D. G. Salomón, M. E. Fermento, N. A. Gandini, A. L. Romero, M. L. Rivadulla, X. Pérez-García, G. Gómez, M. Pérez, Y. Fall, M. M. Facchinetti, A. C. Curino, *Arch. Pharm. Chem. Life Sci.* **2015**, *348*, 315-329.

4 - (a) D. R. Andrews, D. H. R. Barton, R. H. Hesse, M. M. Pechet, *J. Org. Chem.* **1986**, *51*, 4819-4828 and references there cited; (b) M. J. Calverley, *Tetrahedron* **1987**, *43*, 4609-4619; (c) S. C. Choudhry, P. S. Belica, D. L. Coffen, A. Focella, H. Maehr, P. S. Manchand, L. Serico, R. T. Yang, *J. Org. Chem.* **1993**, *58*, 1496-1500; (d) I. Hijikuro, T. Doi, T. Takahashi, *J. Am. Chem. Soc.* **2001**, *123*, 3716-3722; (e) T. P. Sabroe, H. Pedersen, E. Binderup, *Org. Process Res. Dev.* **2004**, 8, 133-135;

(f) H. Shimizu, K. Shimizu, N. Kubodera, K. Yakushijin, D. A. Horne, *Tetrahedron Lett.* **2004**, 45, 1347-1350; (g) R. M. Moriarty, D. Albinescu, *J. Org. Chem.* **2005**, *70*, 7624-7628; (h) W-M. Xu, J. He, M-Q. Yu, G-X. Shen, *Org. Lett.* **2010**, *12*, 4431-4433; (i) C. Liu, G-D. Zhao, X. Mao, T. Suenaga, T. Fujishima, C-M. Zhang, Z-P. Liu, *Eur. J. Med. Chem.* **2014**, *85*, 569-575; (j) K. Sokolowska, R. R. Sicinski, *Steroids*, **2014**, *87*, 67-75; (k) U. Banerjee, A. M. DeBerardinis, M. K. Hadden, *Bioorg. Med. Chem.* **2015**, *23*, 548-555.

5 - (a) H. C. Kolb, M. G. Finn, K. B. Sharpless, *Angew.Chem., Int. Ed.* **2001**, *40*, 2004-2021; (b) V. V. Rostovtsev, L. G. Green, V. V. Fokin, K. B. Sharpless, *Angew.Chem., Int. Ed.* **2002**, *41*, 2596-2599; (c) C. W. Toroe, M. Christensen, M. Meldal, *J. Org. Chem.* **2002**, *67*, 3057-3064; (d) A. Krasiñski, V. V. Forkin, K. B. Sharpless, *Org. Lett.***2004**, *6*, 1237-1240; (e) T. Thirumurugan, D. Matosiuk, K. Jozwiak, *Chem. Rev.* **2013**, *113*, 4905-4979.

6 – F. Himo, T. Lowell, R. Hilgraf, V. V. Rostovtsev, L. Noodleman, K. B. Sharpless, V. V. Fokin, *J.Am.Chem.Soc.* **2005**, *127*, 210-216.

15

Reactivity of (3-Methylpentadienyl)iron(1+) Cation: Late-stage Introduction of a (3-Methyl-2Z,4-pentadien-1-yl) Side Chain

Subhabrata Chaudhury, Shukun Li, and William A. Donaldson*

Department of Chemistry, Marquette University, P. O. Box 1881, Milwaukee, WI 53201-1881 USA

Abstract: The 3-methyl-2Z,4-pentadien-1-yl sidechain is found in various sesquiterpenes and diterpenes. A route for the late stage introduction of this functionality was developed which relies on nucleophilic attack on the (3-methylpentadienyl)iron(1+) cation, followed by oxidative decomplexation. This methodology was applied to the synthesis of the proposed structure of heteroscyphic acid A methyl ester. Realization of this synthesis led to a correction of the proposed structure.

Keywords: Organoiron complexes; Alkylation; Diene ligands; Terpenoids

Introduction

The 3-methyl-2Z,4-pentadienyl sidechain is a functionality appearing in a number of naturally occurring sesquiterpenes and diterpenes. For example, (+)-striatene [1] (**1**, Figure 1), and the labdane diterpenes (+)-solidagol [2] **2** and (-)-ent-3 β-acetoxylabda-8(17),12Z,14-trien-2α-ol [3] **3** were isolated from the liverwort *Ptychanthus striatus*, from Canadian golden rod (*Solidago canadensis*) and from the ornamental plant *Plectranthus fruticosus* respectively. Similarly, the clerodane diterpene (+)-caseargrewiin E **4** [4], isolated from a Thai shrubby tree, exhibited cytotoxic activity against KB, BC1 and NCI-H187 cancer cell lines in the range 0.15-0.91 μg/mL range. In spite of these and other examples [5], only a single synthesis of a terpene containing this functionality has been reported [6]. Audran and co-workers reported the synthesis of **1** which involved enolate alkylation with 5Z-bromo-3-methylpent-3-en-1-yne, followed by hydrozirconation (Scheme 1). It should be noted that attempts at reduction of the **5** using H_2 and a poisoned catalyst were unsuccessful.

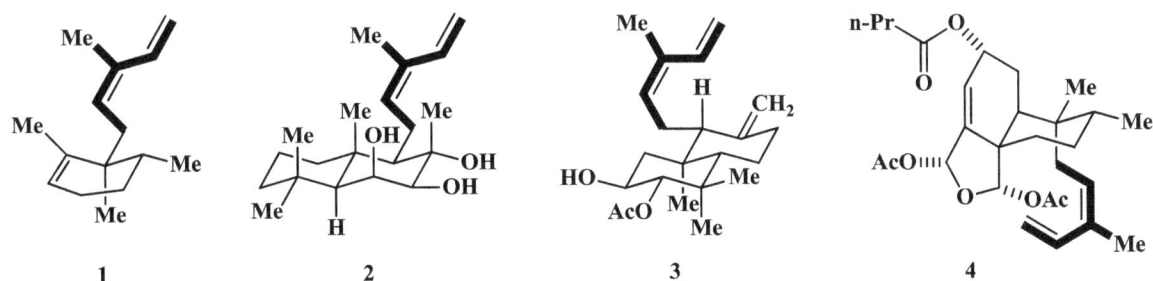

Figure 1. Sesquiterpenes and diterpenes possessing a 3-methyl-2Z,4-pentadienyl sidechain.

As part of our interest in the application of organoiron complexes to organic synthesis [7], we have examined the reactivity the (3-methylpenta-dienyl)Fe(CO)$_2$PPh$_3$$^+$ cation (**6**, Scheme 2) with nucleophiles as a means for late-stage introduction of the 3-methyl-2Z,4-pentadienyl sidechain [8].

*Corresponding author: William Donaldson
Email address: william.donaldson@marquette.edu*

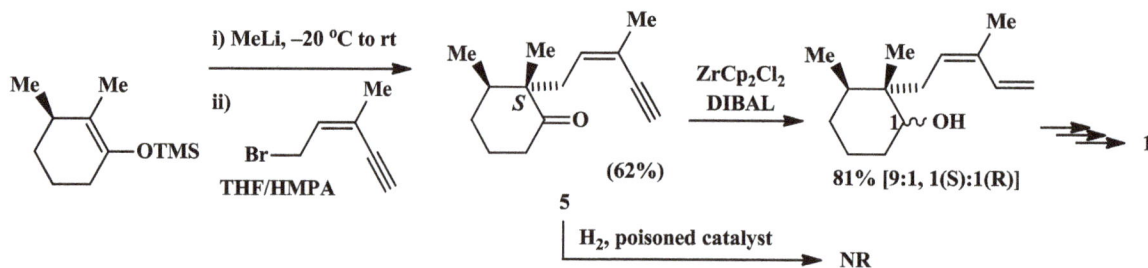

Scheme 1. Synthesis of the (3-methyl-2Z,4-pentadien-1-yl) side chain of (+)-striatene (ref. 5).

Heteroscyphic acids A, B and C, isolated from cultured cells of *Heteroscyphus planus*, were assigned the proposed structures **7a**, **7b**, and **7c** (Figure 2) containing a 3-methyl-2Z,4-pentadienyl sidechain on the basis of their spectroscopic data [9]. We have previously [8a] utilized the (3-methylpentadienyl)Fe$^+$ cation **6** to prepare the

methyl ester of the 8-desmethyl-analog (**8**) of **7a**. Comparison of the NMR spectral data for **8** with that reported for heteroscyphic acid A led to the conclusion that the structures of the heteroscyphic acids were more consistent with a 3-methyl-2E, 4-pentadienyl sidechain. We herein report the full experimental details for these studies.

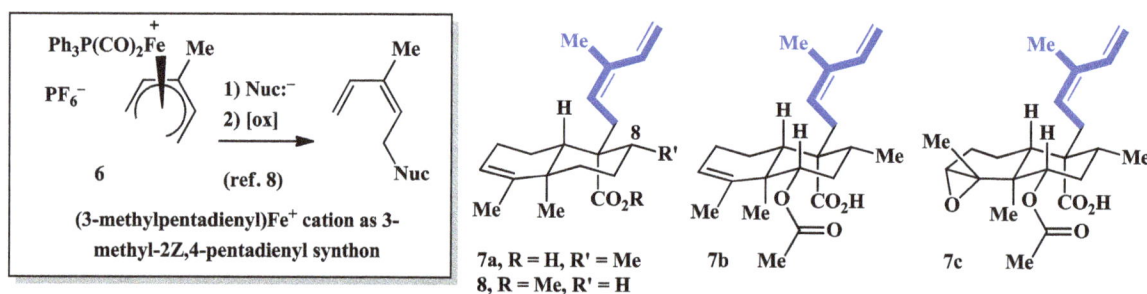

Figure 2. (3-Methylpentadienyl)Fe$^+$ cation as a synthon for 3-methyl-2Z,4-pentadien-1-yl and the "proposed" structures for heteroscyphic acids A, B and C.

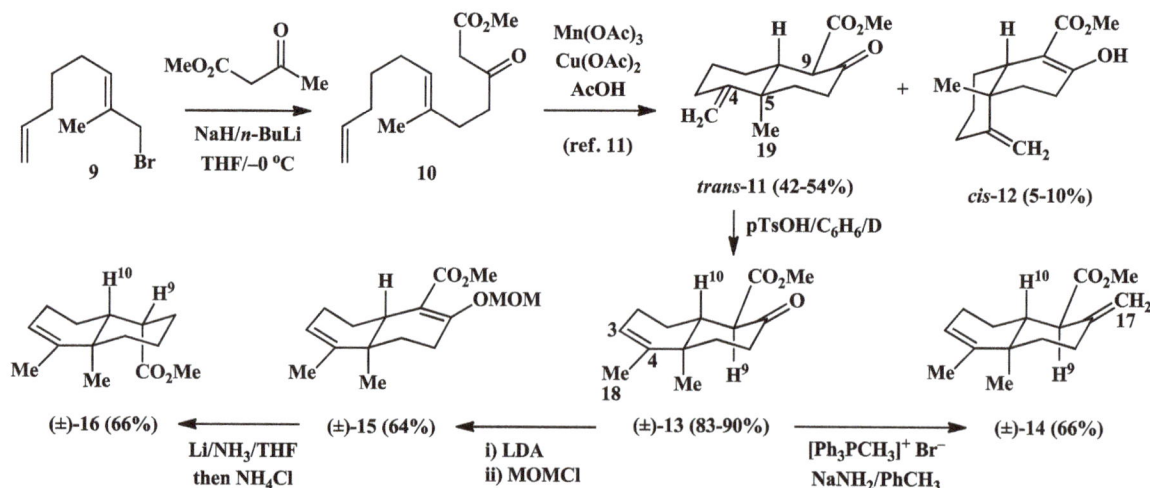

Scheme 2. Synthesis of octahydronaphthalene synthons (heteroscyphic atom numbering).

Results and Discussion

Alkylation of the dianion of methylacetoacetate with the known [10] bromide **9** gave the acyclic β-ketoester **10** (Scheme 2). Oxidative cyclization of **10**, according to the literature procedure [11] gave a

chromatographically separable mixture of *trans*-decalone (±)-**11** along with minor amounts of the *cis*-isomer (±)-**12**. Separation of these two isomers was facilitated by the fact that **12** exists almost entirely in its enol tautomer. Compounds **11** and **12** were characterized by comparison to the literature data for the corresponding ethyl esters [11]. Acid catalyzed isomerization of the exocyclic olefin of **11** gave the endocyclic isomer (±)-**13**. The structural assignment of **13** was based on its NMR spectral data. In particular signals at δ 140.1 and 122.2 ppm in the ^{13}C NMR spectrum and at δ 5.29 (1H, m) and 1.65 ppm (3H, d, J = 1.5 Hz) in the ^{1}H NMR spectrum are characteristic of the C-3 and C-4 olefinic carbons and their associated proton and methyl group respectively.

Attempted olefination of **13** with the ylide generated from the reaction of butyl lithium with methyltriphenylphosphonium bromide in THF gave recovered starting material; presumably due to deprotonation of the acidic β-ketoester in polar solvents. Alternatively, addition of the salt-free ylide generated from trimethylphosphonium bromide with sodium amide in toluene [12] to **13** gave (±)-**14** in moderate yield. The structural assignment of **14** was based on its NMR spectral data. In particular, signals at δ 146.2 and 107.9 ppm in the ^{13}C NMR spectrum and at δ 4.80 (1H, br s) and 4.49 ppm (1H, br s) in the ^{1}H NMR spectrum are characteristic of the exocyclic olefinic carbons and the attached protons.

β-Ketoester **13** was converted into its methoxymethyl vinyl ether (±)-**15** by treatment with LDA followed by reaction with MOMCl. O-alkylation (as compared to C-alkylation) was evident by the presence of two olefinic peaks in the ^{13}C NMR spectrum of **15** at δ 151.1 and 116.3 ppm. Treatment of **15** with Li/NH₃ gave (±)-**16** via reduction of the enoate of **15**, followed by elimination of $CH_3OCH_2O^-$ anion and reduction of the resultant enoate [13]. The ester substituent in **16** was assigned to occupy an axial orientation on the basis of its ^{1}H NMR spectral data. In particular, the signal for H-9 (δ 2.57-2.49 ppm) of **16** did not evidence any large couplings, and thus pointed to an equatorial orientation for H-9.

With octahydronaphthalene synthons **13**, **14**, and **16** successfully prepared, attention was turned to installation of the 3-methyl-2Z,4-pentadien-1-yl sidechain. Toward this end, the sodium salt of **13**, generated by reaction with sodium hydride, was reacted with (3-methylpentadienyl)iron(1+) cation **6** to afford a mixture of diastereomeric complexes **17/17'** (Scheme 3). While the mixture of **17/17'**

gave a satisfactory combustion analysis, interpretation of the NMR spectra was complicated due to signal overlap of the diastereomers as well as ^{31}P coupling. Nonetheless, oxidative decomplexation of this mixture gave a single product (±)-**18**. In a similar fashion, the lithium salt of **14** or **16** (generated by reaction with LDA) with **6**, gave a mixture of isomeric complexes **19/19'** or **20/20'** respectively; decomplexation of each mixture gave a single product (±)-**21** or (±)-**8**.

The structural assignments for **18**, **21** and **8** were based on their NMR spectral data. For products **18** and **21**, the pentadienyl sidechain was assigned the β-orientation, while for **8** the sidechain was assigned the α-orientation. In particular, for **18** the singlet for Me-19 appears at δ 1.03 ppm while for **8** this singlet appears at δ 0.84 ppm. The upfield chemical shift for this signal of **8** is consistent with an axial ester group at C-9 [14]. In addition, there is an nOE interaction observed between Me-19 and one of the H-12 protons of **18**, while a NOESY interaction was observed between the Me-19 and the methyl ester of **8**. For **21** the upfield chemical shifts of the H-17 olefinic methylene protons (δ 4.78 and 5.00 ppm) may be attributed to the anisotropic effect of the neighboring ester substituent in an α-orientation. Notably, these orientations are consistent with the known [15] stereoselectivity for alkylation on the α-face of other bicyclo[4.4.0]decane β-ketoesters while alkylation of the exocyclic enolate derived from a bicyclo[4.4.0]decane 2-carboxylate generally proceeds on the β-face [16]. In addition, the 3-methyl-2,4-pentadienyl side chain for **18**, **21** and **8** were all assigned the Z-configuration. In particular, the signals for H-14 appear at ca. δ 6.8-6.7 ppm while signals for the C-14, C-15 and the dienyl methyl C-16 appear at ca. δ 135, 114 and 20 ppm respectively. These chemical shifts are characteristic of a 3-methyl-2Z,4-pentadienyl group [2-5]. This was found to be in sharp contrast to the chemical shifts reported [9] for H14 (δ 6.37 ppm) C14, C15 and the dienyl methyl C16 (δ 141.7, 111.1 and 12.1 ppm) of the sidechain of heteroscyphic acid methyl ester. In fact, these chemical shifts are more consistent with those reported [17] for diterpenes which possess a 3-methyl-2E,4-pentadienyl sidechain.

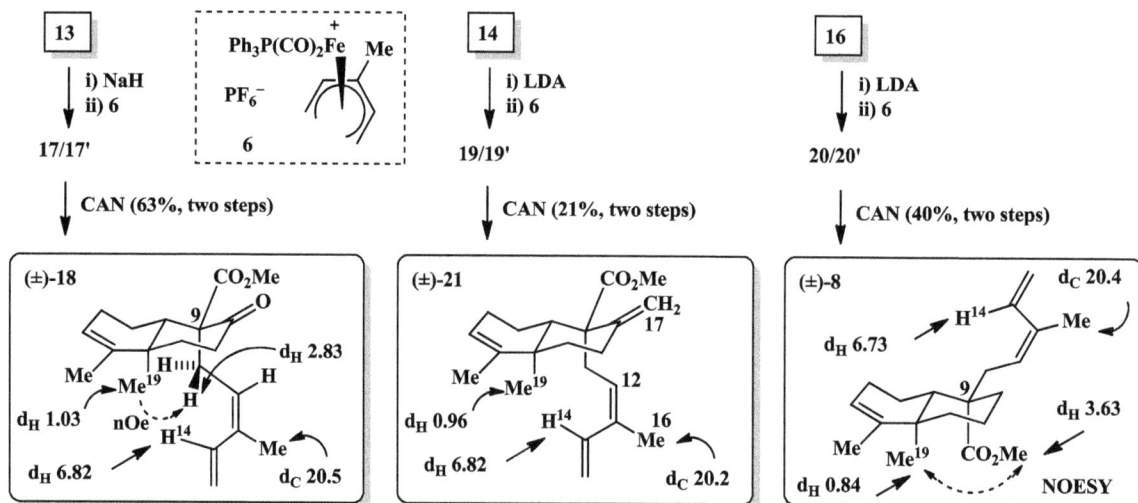

Scheme 3. Dienylation of octahydronaphthalenes **13**, **14** and **16** followed by decomplexation (heteroscyphic acid atom numbering).

Conclusion

The ability to rapidly introduce a 3-methyl-2Z, 4-pentadienyl sidechain was demonstrated by the synthesis of **8**, a nor-diterpene related to the *proposed* structure of heteroscyphic acid A, as well as **18** and **21**. While this synthetic exercise revealed that the sidechains of the heteroscyphic acids more likely possess the *E*-stereochemistry, this methodology might be applied to the synthesis of compounds such as **1-4**.

Acknowledgements

This work was supported by the National Science Foundation (CHE-0415771). Mass spectrometry was provided by the Washington University Mass Specrometry Resource, an NIH Research Resource.

Experimental Section

^{1}H and proton-decoupled ^{13}C NMR spectra were recorded at 300 MHz and 75 MHz respectively. Proton and carbon assignments refer to heteroscyphic acid skeleton numbering. High-resolution mass spectra were obtained from the Washington University Resource for Biomedical and Bioorganic mass spectrometry. Tetrahydrofuran was distilled from sodium benzophenone ketyl prior to use. Anhydrous CH_2Cl_2 and anhydrous DMF were purchased from Aldrich Chemical Company. Compounds **6**[8b] and **9**[10] were prepared by literature procedures.

Methyl 6-methyl-3-oxo-6,11-dodecadienoate (10). To a flame-dried round-bottom flask, NaH (60% dispersion in mineral oil, 2.53 g, 63.3 mmol) was suspended in dry THF (165 mL) under N_2. The suspension was cooled in an ice bath and methyl acetoacetate (6.81 g, 58.7 mmol) was added slowly

(CAUTION: hydrogen gas is evolved during the addition). The mixture was stirred for 10 min, and then a solution of *n*-butyl lithium in THF (2.5 M, 25.3 mL, 63.3 mmol) was added. During this addition, the solution became a bright orange in color. After stirring at 0 °C for 10 min, a solution of 1-bromo-2-methyl-2,7-octadiene (5.96 g, 29.4 mmol) in THF (15 mL) was added. The ice bath was removed and the solution stirred at room temperature for 30 min. A solution of 3 M HCl (50 mL) was added followed by ether (50 mL). The mixture was separated and the aqueous layer was extracted several time with ether. The combined organic layers were dried (MgSO$_4$) and concentrated under reduced pressure. The residue was purified by column chromatography (SiO$_2$, hexanes-ethyl-acetate = 8:1 → 5:1 gradient) to afford **10** (5.21 g, 75%) as a pale yellow oil;

IR (neat) 3076, 2927, 1750, 1717, 1637, 911 cm^{-1}.

^{1}**H NMR** (CDCl$_3$): δ = 5.80 (1H, dtd, *J* = 16.9, 6.8, 3.3 Hz, H-11), 5.13 (1H, br t, *J* = 7.7 Hz, H-7), 4.97 (2H, m, =CH$_2$), 3.74 (3H, s, OMe), 3.46 (2H, s, H-2), 2.64 (2H, br t, *J* = 7.5 Hz), 2.27 (2H, br t, *J* = 7.8 Hz), 2.04-1.78 (4H, m), 1.59 (3H, s, Me-6), 1.41 (2H, pent, *J* = 7.3 Hz).

^{13}**C NMR** (CDCl$_3$): δ = 202.5 (C-3), 167.7 (C-1), 138.9 (C-10), 133.3 (C-6), 125.4 (C-7), 114.6 (C-11), 52.6 (OMe), 49.2, 42.0, 33.6, 33.4, 29.2, 27.6, 16.4 (Me-6).

Anal. Calcd for C$_{14}$H$_{22}$O$_3$: C, 70.56; H, 9.30. Found: C, 70.65; H, 9.17.

Decahydro-4a-methyl-5-methylene-2-oxo-1-naphthalenecarboxylic acid methyl ester (11). To a degassed solution of **10** (3.31 g, 13.9 mmol) dissolved in glacial acetic acid (35 mL) was added solid Mn(OAc)$_3$ (1.75 g, 6.53 mmol), followed by solid Cu(OAc)$_2$ (0.590 g, 3.24 mmol). The reaction mixture was stirred under N_2 for 7 h at room temperature and then filtered through a bed of celite. The filter bed was washed several times with ether, the combined ethereal extracts were washed with saturated NaHCO$_3$, followed by water, dried (MgSO$_4$) and concentrated under reduced pressure. The residue was purified by column chromatography (SiO$_2$, hexanes-ethyl acetate = 8:1 → 5:1 gradient) to afford (±)-**11**

(1.45 g, 44%) as a colorless oil, followed by a variable but minor amount of the *cis*-isomer (**10**).

11: IR (neat) 3086, 1715, 1635 cm^{-1}.

^{1}H NMR (CDCl$_3$): δ = 4.74 (1H, br s, =CH$_2$), 4.65 (1H, br s, =CH$_2$), 3.75 (3H, s, OMe), 3.27 (1H, d, *J* = 12.9 Hz, H-9), 2.55-2.47 (2H, m), 2.40-2.32 (1H, m), 2.05-1.76 (6H, m), 1.45-1.37 (2H, m), 1.22 (3H, s, Me-19).

^{13}C NMR (CDCl$_3$): δ = 205.6 (C-8), 170.3 (CO$_2$R), 155.0 (C-4), 106.3 (=CH$_2$), 60.3 (C-9), 52.3 (OMe), 47.7, 38.1, 38.0, 36.1, 32.7, 27.6, 27.4, 17.5 (Me-19).

FAB-HRMS *m/z* 237.1485 (calcd for C$_{14}$H$_{21}$O$_3$ [M+H$^+$] *m/z* 237.1491).

1,2,3,4,4a,7,8,8a-Octahydro-4a,5-dimethyl-2-oxo-1-naphthalenecarboxylic acid methyl ester (13).

To a solution of **11** (200 mg, 0.847 mmol) in benzene (10 mL) was added *p*-toluenesulfonic acid monohydrate (30 mg, 0.16 mmol). The mixture was heated at reflux for 2 d under N$_2$. The mixture was cooled to room temperature and then a few drops of triethylamine were added to neutralize the acid. The mixture was filtered through a pad of celite, the filter bed washed with ether, and the combined organic layers were concentrated under reduced pressure. The residue was purified by column chromatography (SiO$_2$, hexanes-ethyl acetate = 8:1) to afford (±)-**13** (708 mg, 84%) as a colorless oil;

IR (neat) 1746, 1712 cm^{-1}.

^{1}H NMR (CDCl$_3$): δ = 5.29 (1H, br s, H-3), 3.74 (3H, s, OMe), 3.29 (1H, d, *J* = 13.5 Hz, H-9), 2.50-2.42 (2H, m), 2.19 (1H, td, *J* = 13.2, 2.7 Hz, H-10), 2.10-1.97 (3H, m), 1.65 (3H, d, *J* = 1.5 Hz, Me-18), 1.60-1.30 (3H, m), 1.16 (3H, s, Me-19).

^{13}C NMR (CDCl$_3$): δ = 206.0 (C-8), 170.5 (CO$_2$R), 140.1 (C-4), 122.2 (C-3), 60.0 (C-9), 52.2 (OMe), 45.2, 37.9, 36.4, 35.3, 25.7, 23.8, 18.8 (Me-18), 17.7 (Me-19).

FAB-HRMS *m/z* 237.1485 (calcd for C$_{14}$H$_{21}$O$_3$ [M+H$^+$] *m/z* 237.1491).

1,2,3,4,4a,7,8,8a-Octahydro-4a,5-dimethyl-2-methylene-1-naphthalenecarboxylic acid methyl ester (14).

To a suspension of NaNH$_2$ (637 mg, 16.3 mmol) in dry toluene (41 mL) under N$_2$, was added methyltriphenylphosphonium bromide (4.48 g, 12.5 mmol), and the mixture was heated at reflux for 3 h. During this time formation of the ylide was detected by change of the solution to a bright orange color. The warm solution was transferred by a cannula to a solution of **13** (593 mg, 2.51 mmol) in toluene (6 mL) under N$_2$. The reaction mixture was stirred at room temperature for 6 h and then filtered through a pad of celite. The filtrate was concentrated under reduced pressure and the residue was dissolved in

hexanes to induce the precipitation of triphenylphosphine oxide and then filtered again. The filtrate was washed with water, dried (MgSO$_4$) and concentrated. The residue was purified by column chromatography (SiO$_2$, hexanes-ethyl acetate = 8:1) to afford (±)-**14**

(382 mg, 66%) as a colorless oil.

IR (neat) 2937, 1740, 1645, 891 cm^{-1}.

^{1}H NMR (CDCl$_3$): δ = 5.27-5.22 (1H, narrow m, H-3), 4.80 (1H, br s, =CH$_2$), 4.49 (1H, br s, =CH$_2$), 3.75 (3H, s, OMe), 3.09 (1H, d, *J* = 12.5 Hz, H-9), 2.36-2.29 (2H, m), 2.10-1.96 (2H, m), 1.82 (1H, td, *J* = 12.3, 3.1 Hz, H-10), 1.62 (3H, br s, Me-18), 1.55-1.20 (4H, m), 1.06 (3H, s, Me-19).

^{13}C NMR (CDCl$_3$): δ = 174.1 (CO$_2$R), 146.2 (C-8), 141.7 (C-4), 121.1 (C-3), 107.9 (=CH$_2$), 52.0 (OMe), 51.6, 45.7, 37.0, 36.7, 31.4, 25.9, 23.2, 18.8, 18.2.

FAB-HRMS *m/z* 234.1615 (calcd for C$_{15}$H$_{22}$O$_3$ [M $^+$] *m/z* 234.1620).

3,4,4a,7,8,8a-Hexahydro-2-(methoxymethoxy)-4a,5-dimethyl-1-naphthalenecarboxylic acid methyl ester (15).

To a suspension of NaH (40 mg, 1.0 mmol) in HMPA (3 mL) at 0 °C under N$_2$ was added a solution of **13** (200 mg, 0.847 mmol) in HMPA (3 mL). The reaction mixture was warmed to room temperature over a 2 h period. To this solution was added chloromethyl methyl ether (82 mg, 1.0 mmol) and the reaction mixture was stirred for an additional 3 h. The resulting mixture was poured into a separatory funnel containing ice-water, saturated NaHCO$_3$ (10 mL) and ether (15 mL). The layers were separated, and the aqueous layer was extracted several times with ether. The combined organic layers were dried (MgSO$_4$) and concentrated under reduced pressure. The residue was purified by column chromatography (SiO$_2$, hexanes-ethyl acetate = 10:1) to afford (±)-**15**

(150 mg, 64%) as a colorless solid; mp 65-68 °C;

IR (neat) 2949, 2830, 1725, 1680 cm^{-1}.

^{1}H NMR (CDCl$_3$): δ = 5.23 (1H, br s, H-3), 4.91 (1H, d, *J* = 6.9 Hz, OCH$_2$OMe), 4.85 (1H, d, *J* = 6.9 Hz, OCH$_2$OMe), 3.73 (3H, s, CO$_2$Me), 3.42 (3H, s, OCH$_2$OMe), 2.57-2.49 (1H, m), 2.39-2.30 (2H, m), 2.10-2.02 (2H, m), 1.97-1.88 (1H, m), 1.64 (3H, br s, Me-18), 1.56-1.40 (3H, m), 0.98 (3H, s, Me-19).

^{13}C NMR (CDCl$_3$): δ = 169.5 (CO$_2$R), 151.1 (C-8), 141.4 (C-4), 121.2 (C-3), 116.3 (C-9), 93.2 (OCH$_2$OMe), 56.5 (OCH$_2$OMe), 51.6 (CO$_2$Me), 41.7, 35.9, 31.8, 25.7, 22.8, 21.4, 18.9, 18.2.

Anal. Calcd for C$_{16}$H$_{24}$O$_4$: C, 68.55; H, 8.63. Found: C, 68.69; H, 8.46.

1,2,3,4,4a,7,8,8a-Octahydro-4a,5-dimethyl-1-naphthalenecarboxylic acid methyl ester (16).

To a dispersion of lithium metal (80 mg, 12 mmol) in liquid NH$_3$ at -78 °C under N$_2$ was added a solution of **15** (460 mg, 1.64 mmol) in ether (8 mL). The reaction mixture was stirred at -78 °C for 15 min, and then quenched by addition of solid NH$_4$Cl (2.46 g) in one portion. The mixture was stirred for an additional 30 min at -78 °C and then slowly warmed to room temperature. Additional ether (30 mL) was added, and the mixture was filtered through a pad of filter-aid. The inorganic salts were washed several times with ether and the combined ethereal layers were concentrated under reduced pressure. The residue was purified by column chromatography (SiO$_2$, hexanes-ethyl acetate = 10:1) to afford (±)-**16** (240 mg, 66%) as a colorless oil.

IR (neat) 1728 cm^{-1}.

^{1}H NMR (CDCl$_3$): δ = 5.23 (1H, m, H-3), 3.66 (3H, s, OMe), 2.53 (1H, br t, J = 4.8 Hz, H-9), 2.21-2.12 (1H, m), 2.06-1.85 (4H, m), 1.80-1.72 (2H, m), 1.66-1.61 (1H, m), 1.57 (3H, d, J = 1.8 Hz, Me-18), 1.44-1.30 (2H, m), 1.13 (1H, dt, J = 13.2, 4.1 Hz), 0.86 (3H, s, Me-19);

^{13}C NMR (CDCl$_3$): δ = 176.2 (CO$_2$R), 143.0 (C-4), 121.2 (C-3), 51.3 (OMe), 46.1, 43.3, 38.3, 36.8, 28.6, 27.0, 25.5, 19.3, 18.2, 17.5.

EI-HRMS m/z 222.1619 (calcd for C$_{14}$H$_{23}$O$_2$ [M+H$^+$] m/z 222.1620).

1,2,3,4,4a,7,8,8a-Octahydro-4a,5-dimethyl-1-(3-methyl-2Z,4-pentadien-1-yl)-2-oxo-1-naphthalenecarboxylic acid methyl ester (18).

To a solution of NaH (25 mg, 0.64 mmol) in dry THF (10 mL) at 0 °C under N$_2$, was added a solution of **13** (150 mg, 0.635 mmol) in THF (10 mL). The mixture was stirred for 30 min, and then solid cation **6** (381 mg, 0.635 mmol) was added in one portion. The reaction mixture was stirred at 0 °C for 1 h and then 30 min at room temperature. The reaction mixture was poured into saturated NaCl solution (15 mL), and extracted several times with ethyl acetate. The combined extracts were dried (MgSO$_4$) and concentrated under reduced pressure. The residue was purified by column chromatography (SiO$_2$, hexanes-ethyl acetate = 8:1) to afford a mixture of diastereomeric diene-iron complexes **17/17'**

(325 mg, 75%) as a yellow solid. mp (decomposes) 89-100 °C.

The ^{1}H and ^{13}C NMR spectra for this product were too complex for complete interpretation due to the presence of diastereomers.

^{1}H NMR (partial, CDCl$_3$): δ = 7.56-7.30 (m, 15H, PPh$_3$), 5.32-5.26 (m, 1H), 4.24-4.07 (br m, 1H), 3.63 and 3.61 (2 x s, 3H).

Anal. calcd. for C$_{40}$H$_{43}$O$_5$PFe: C, 69.57; H, 6.27. Found: C, 69.42; H, 6.40.

To a solution of the **17/17'** (110 mg, 0.159 mmol) in methanol (10 mL) was added solid ceric ammonium nitrate [CAN] (220 mg, 0.401 mmol) in two portions. Monitoring of the reaction by TLC indicated that complete disappearance of **17/17'** in 30 min. Water (15 mL) was added and the reaction mixture was extracted several times with ether. The combined organic extracts were dried (MgSO$_4$) and concentrated under reduced pressure. The residue was purified by column chromatography (SiO$_2$, hexanes-ethyl acetate = 10:1) to afford (±)-**18**

(50 mg, 99%) as a colorless oil;

IR (neat) 2950, 1713, 1435, 1217 cm^{-1}.

^{1}H NMR (CDCl$_3$): δ = 6.82 (1H, dd, J = 17.3, 10.7 Hz, H-14), 5.30-5.18 (3H total, br m & d, J = 16.6 Hz, H-3, H12 and H-15$_Z$), 5.12 (1H, d, J = 10.9 Hz, H-15$_E$), 3.66 (3H, s, OMe), 2.83 (1H, dd, J = 14.7, 6.5 Hz, H-11), 2.62 (1H, dd, J = 14.4, 9.1 Hz, H-11'), 2.47 (1H, ddd, J = 15.5, 4.7, 2.3 Hz, H-7$_{eq}$), 2.10-1.96 (1H, m), 1.90-1.78 (2H, m), 1.72 (3H, s, Me-16), 1.65-1.63 (1H, m), 1.54 (3H, s, Me-18), 1.43 (1H, td, J = 16.5, 4.9 Hz), 1.08-1.04 (2H, m), 1.03 (3H, s, Me-19).

^{13}C NMR (CDCl$_3$): δ = 208.0 (C-8), 173.9 (CO$_2$Me), 141.2 (C-4), 135.1 (C-14), 133.3, 125.1, 122.4 (C-3), 114.6 (C-15), 61.8 (C-9), 52.3 (OMe), 50.0, 38.1, 37.7, 36.4, 31.0, 27.2, 21.1, 20.5 (Me-16), 18.6, 17.5.

FAB-HRMS m/z 323.2182 (calcd for C$_{20}$H$_{28}$O$_3$Li (M+Li$^+$) m/z 323.2199).

1,2,3,4,4a,7,8,8a-Octahydro-4a,5-dimethyl-1-(3-methyl-2Z,4-pentadien-1-yl)-2-methylene-1-naphthalenecarboxylic acid methyl ester (21).

To a solution of **14** (100 mg, 0.427 mmol) in dry THF (10 mL) at 0 °C under N$_2$, was added a solution of lithium diisopropylamine in THF (1.8 $\underline{M}$, 0.3 mL, 0.5 mmol). The mixture was stirred for 1 h at 0 °C, and then solid cation **6** (0.31 g, 0.54 mmol) was added in one portion. The reaction mixture was stirred for 3 h and then worked up in a fashion similar to that for **17/17'**. Purification of the residue by column chromatography (SiO$_2$, hexanes-ethyl acetate = 20:1) gave a diastereomeric mixture of diene-iron complexes **19/19'**

(70 mg, 30%) as a yellow oil.

The ^{1}H and ^{13}C NMR spectra for this product were too complex for complete interpretation due to the presence of diastereomers.

^{1}H NMR (partial, CDCl$_3$): δ = 7.56-7.32 (m, 15H, PPh$_3$), 5.26-5.05 (m, 3H), 4.21-4.10 (br m, 1H), 3.64 and 3.61 (2 x s, 3H).

^{13}C NMR (partial, CDCl$_3$, diastereomeric signals reported as pairs: δ = 176.6 and 175.8 ($\underline{C}$O$_2$Me), 149.7, 146.9, 142.9 and 142.6 (C3), 136.3 (d, J_{PH} = 37.5 Hz), 133.3 (d, J_{PH} = 10.2 Hz), 129.8, 128.3 (d, J_{PH} = 8.7 Hz), 121.7 and 121.2, 112.0 and 111.1, 103.1 and 102.3, 94.1 and 93.45.

Anal. calcd. for C$_{41}$H$_{45}$O$_4$PFe·H$_2$O: C, 69.70; H, 6.70. Found: C, 69.77; H, 6.95.

Decomplexation of the mixture of **19/19'** (70 mg, 0.16 mmol) in methanol (10 mL) with CAN (112 mg, 0.204 mmol) was carried out in a fashion similar to the decomplexation of **17/17'**. Purification of the residue by column chromatography (SiO$_2$, hexanes-ethyl acetate = 10:1) gave (±)-**21**

(22 mg, 70%) as a colorless oil;

IR (neat) 2963, 1718, 1436, 1265 cm^{-1}.

^{1}H NMR (CDCl$_3$): δ = 6.82 (1H, dd, J = 17.4, 10.9 Hz, H-14), 5.34 (1H, t, J = 6.5 Hz, H-12), 5.28-5.20 (2H, m, H-3 and H-15$_Z$), 5.13 (1H, d, J = 10.6 Hz, H-15$_E$), 5.00 (1H, s, H-17$_E$), 4.78 (1H, s, H-17$_Z$),

3.65 (3H, s, OMe), 2.86-2.63 (3H, m), 2.34-2.25 (1H, m), 2.02-1.78 (8H, m), 1.57 (3H, br s, Me-18), 1.30-1.20 (2H, m), 0.95 (3H, s, Me-19).

^{13}C NMR (CDCl$_3$): δ = 176.4 (CO$_2$R), 148.8 (C-8), 142.7 (C-4), 133.9, 133.7, 125.6, 121.7 (C-3), 114.5 (C-15), 110.6 (C-17), 53.7, 50.6, 38.9, 38.4, 32.4, 31.6, 29.9, 27.5, 20.8, 20.2 (Me-16), 18.4, 17.8.

FAB-HRMS m/z 314.2240 (calcd for C$_{21}$H$_{30}$O$_2$ [M+H$^+$] m/z 314.2246).

1,2,3,4,4a,7,8,8a-Octahydro-4a,5-dimethyl-1-(3-methyl-2Z,4-pentadienyl)-1-naphthalenecarboxylic acid methyl ester (8).

To a solution of 16 (100 mg, 0.451 mmol) in dry THF (4 mL) at -78 °C under N$_2$, was added a solution of lithium diisopropylamine in THF (0.5 mmol, freshly prepared from diisopropylamine and n-butyl lithium). The mixture was stirred for 30 min, and then solid cation 6 (207 mg, 0.451 mmol) was added in one portion. The reaction mixture was warmed at room temperature and stirred for an additional 3 h. The reaction mixture was quenched with 1M HCl (10 mL), and extracted several times with ether. The combined extracts were washed with water, dried (MgSO$_4$) and concentrated under reduced pressure. The residue was purified by column chromatography (SiO$_2$, hexanes-ethyl acetate = 10:1) to afford a diastereomeric mixture of diene-iron complexes 20/20'
(304 mg, 100%) as a yellow solid, which was used in the next step without further characterization.

Decomplexation of 20/20' (200 mg, 0.296 mmol) with CAN (405 mg, 0.741 mmol) was carried out in a fashion similar to that for the decomplexation 17/17' except that in DMF (15 mL) was used a solvent instead of methanol. Purification of the residue by column chromatography (SiO$_2$, hexanes-ethyl acetate = 10:1) followed by a second purification by column chromatography (SiO$_2$ impregnated with ~10 % AgNO$_3$, hexanes) gave (±)-8 (35 mg, 40%) as a colorless oil;

IR (neat) 3054, 2987, 1720, 1422, 1265 cm^{-1}.

^{1}H NMR (CDCl$_3$): δ = 6.73 (1H, dd, J = 17.3, 10.8 Hz, H-14), 5.34-5.08 (4H total, m, H-3, H-12, H15$_E$ and H-15$_Z$), 3.63 (3H, s, OMe), 2.77 (1H, dd, J = 14.1, 8.3 Hz), 2.22-2.16 (2H, m), 2.06-1.87 (3H, m), 1.83 (3H, s, Me-16), 1.80-1.70 (2H, m), 1.58 (3H, br s, Me-18), 1.50-1.40 (2H, m), 1.26-0.86 (3H, m), 0.83 (3H, s, Me-19).

^{13}C NMR (CDCl$_3$): δ = 177.3 (CO$_2$R), 143.2 (C-4), 135.2 (C-14), 133.6, 125.3, 121.5 (C-3), 114.3 (C-15), 55.9, 51.3 (OMe), 48.3, 38.8, 37.9, 37.0, 34.3, 27.5, 20.6, 20.4 (Me-16), 19.4, 18.5, 17.7.

FAB-HRMS m/z 303.2320 (calcd for C$_{20}$H$_{31}$O$_2$ [M+H$^+$] m/z 303.2324).

References

1 - R. Takeda, R. Mori and Y. Hirose, Chem. Lett. 1982, 1625-1628. R. Takeda, H. Naoki, T. Iwashita, K. Mizukawa, Y. Hirose and M. Inoue, Bull. Chem. Soc. Jpn. 1983, 1125-1132.

2 - J. Li, L. Pan, J. N. Fletcher, W. Lv, Y. Deng, M. A. Vincent, J. P. Slack, T. S. McCluskey, Z. Jia, M. Cushman and A. Douglas Kinghorn, J. Nat. Prod. 2014, 77, 1739-1743.

3 - C. Gaspar-Marques, M. F. Simoes and B. Rodriguez, J. Nat. Prod. 2004, 67, 614-621.

4 - S. Kanokmedhakul, K. Kanokmedhakul and M. Buayairaksa, J. Nat. Prod. 2007, 70, 1122-1126.

5 - For other examples of diterpenes containing a 3-methyl-2Z,4-pentadien-1-yl) side chain see: (a) T. Ohtsuki, T. Koyano, T. Kowithayakorn, N. Yamaguchi and M. Ishibashi, Planta Med. 2004, 70, 1170-1173. (b) C. Gaspar-Marques, M. F. Simoes, A. Duarte and B. Rodriguez, J. Nat. Prod. 2003, 66, 491-496. (c) J. A. Garbarino and A. Molinari, J. Nat. Prod. 1993, 56, 624-626. (d) S. Kanokmedhakul, K. Kanokmedhakul, T. Kanarsa and M. Buayairaksa, J. Nat. Prod. 2005, 68, 183-188. (e) M. S. Hunter, D. G. Corley, C. P. Carron, E. Rowold, B. F. Kilpatrick and R. C. Durley, J. Nat. Prod. 1997, 60, 894-899.

6 - P. Bremond, N. Vanthuyne and G. Audran, Tetrahedron Lett. 2009, 50, 5723-5725.

7 - (a) W. A. Donaldson and S. Chaudhury, Eur. J. Org. Chem. 2009, 3831-3843. (b) W. A. Donaldson, Curr. Org. Chem. 2000, 4, 851-882.

8 - (a) S. Chaudhury, S. Li and W. A. Donaldson, Chem. Commun. 2006, 2069-2070. (b) S. Chaudhury, S. Li, D. W. Bennett, T. Siddiquee, D. T. Haworth and W. A. Donaldson, Organometallics 2007, 26, 5295-5303.

9 - (a) K. Nabeta, T. Oohata, N. Izumi and K. Katoh, Phytochemistry 1994, 37, 1263-1268. (b) K. Nabeta, T. Ishikawa, T. Kawae and H. Okuyama, J. Chem. Soc., Chem. Commun. 1995, 651-652. (c) K. Nabeta, T. Ishikawa and H. Okuyama, J. Chem. Soc., Perkin Trans. 1 1995, 3111-3115.

10 - D. L. Boger and R. J. Mathvink, J. Org. Chem. 1992, 58, 1429-1443.

11 - P. A. Zoretic, M. Ramchandani and M. L. Caspar, Synth. Commun. 1991, 21, 915-922.

12 - M. Liapis and V. Ragoussis, J. Chem. Soc., Perkin Trans. 1 1985, 815-817.

13 - R. M. Coates and J. E. Shaw, J. Org. Chem. 1970, 35, 2601-2605.

14 - E. Wenkert, A. Afonso, P. Beak, J. W. J. Carney, P. W. Jeffs and J. D. McChesney, J. Org. Chem. 1965, 30, 713-722.

15 - M. Czarny, K. K. Maheshwari, J. A. Nelson and T. A. Spencer, J. Org. Chem. 1975, 40, 2079-2085. (b) E. Wenkert, A. Afonso, J. Bredenberg, C. Kaneko and A. Tahara, J. Am. Chem. Soc. 1964, 86, 2038-2043.

16 - (a) T. Ling, C. Chowdhury, B. Kramer, B. G. Vong, M. Palladino and E. A. Theodorakis,

J. Org. Chem. **2002**, *66*, 8843-8853. (b) S. C. Welch, C. P. Hagan, J. H. Kim and P. S. Chu, *J. Org. Chem.* **1977**, *42*, 2879-2887. (c) S. C. Welch, C. P. Hagan, W. P. Fleming and J. W. Trotter, *J. Am. Chem. Soc.* **1977**, *99*, 549-556.

17 - (a) M. Furlan, M. N. Lopes, J. B. Fernandes and J. R. Pirani, *Phytochemistry* **1996**, *41*, 1159-1161. (b) D. Herlem, F. Khuong-Huu and A. S. Kende, *Tetrahedron* **1994**, 50, 2055-2070. (c) A. F. Barrero, J. F. Sanchez, J. Altarejos, A. Perales and R. Torres, *Phytochemistry*, **1992**, *31*, 615-620. (d) J. Bastard, D. K. Duc, M. Fetizon, M. J. Francis, P. K. Grant, R. T. Weavers, C. Kaneko, G. V. Baddeley, J. M. Bernassau, R. R. Burfitt, P. M. Wuvkulich and E. Wenkert, *J. Nat. Prod.* **1984**, *47*, 592-599.

16

Selective synthesis of functionalized allyl amines and ethers from *(E)*-2,3-bisphosphonated allyl bromide

Maha Ameur and Hassen Amri [*]

Selective Organic Synthesis & Biological Activity, Faculty of Science,
University of Tunis El Manar, 2092 Tunis, Tunisia

Abstract: An efficient and straightforward approach toward functionalized allylic amines and ethers is disclosed starting from tetraethyl (3-bromoprop-1-ene-1,2-diyl)bisphosphonate *(E)*-1 and secondary and tertiary amines. The coupling reaction of allyl bromide *(E)*-1 with less bulky secondary amines provides a new family of allylamines **3**, while, its reaction with more bulky secondary and tertiary amines, in methanol at reflux, led to functional allyl ethers **4** and **5**.

Keywords: allyl bromide; secondary amine; tertiary amine; allylamine; allyl ether.

Introduction

Heteroatom-containing compounds are important synthetic targets of interest in organic and medicinal chemistry [1-8]. This is mainly because they are useful intermediates for the synthesis of biologically naturally occurring products [9], antibiotics [10], chiral auxiliaries [11] and β-amino alcohols [12]. Despite, the different methodologies discovered for the synthesis of carbon- heteroatom bonds formation reactions, much of them have many constraints such as harsh reaction conditions, high cost of reagents and long reaction times. Among these various potentialities, the addition of an amine to an unsaturated C-C bond provides an excellent route for the formation of carbon-heteroatom bond. Following this protocol, our research group has been actively working on the synthesis of new compounds containing C-N and C-O bonds [13]. In pursuit of our studies, the present paper reports an efficient protocol for the synthesis of novel functionalized allylamines **3**, allyl ethers **4** and **5** through the reaction of tetraethyl (prop-2-ene-1,2-diyl)bis-phosphonate *(E)*-1 with secondary and tertiary amines (Figure 1).

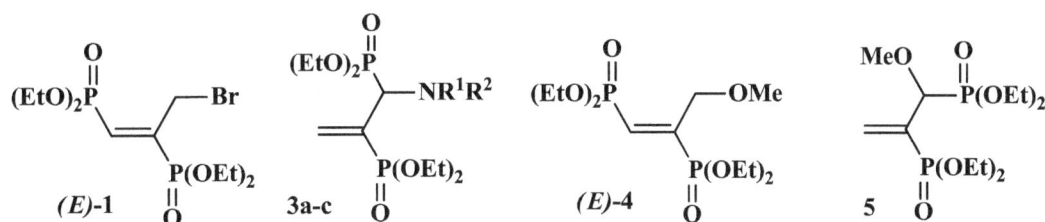

Figure 1. Structure of the new functionalized allylamines **3**, allyl ethers **4** and **5**

Results and Discussion

In our recent work, we have developed a simple and convenient synthesis of tetraethyl (3-bromoprop-1-ene-1,2-diyl)bisphosphonate *(E)*-1 [14] *via* a tandem reaction of bromination-dehydrobromination of tetraethyl (prop-2-ene-1,2-diyl)bisphosphonate in the presence of DBU in acetonitrile at room temperature (Scheme 1).

Corresponding author: Hassen Amri
Email adress : hassen.amri@fst.rnu.tn

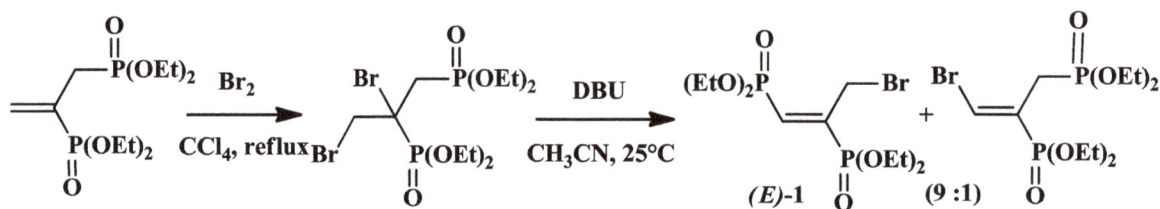

Scheme 1. Synthesis of the mixture of allyl and vinyl bromides

In a recent work [14] devoted to the study of the reactivity of allyl bromide *(E)*-1 with various primary amines in methanol at 0 °C, we have shown that the coupling reaction follows an addition-elimination or S_N2' process to provide a new family of functionalized allyl amines **3** with good yields (Scheme 2). Allylamines and their derivatives have proven to be valuable synthons in organic chemistry [15] as they are found in a range of natural products such as gabaculine [16], ocyzosymicine [17] and cytosinine [18] and serve as precursors to a large number of pharmaceutically and biologically active compounds which display a variety of pharmacological properties including anti-convulsant [19], anti-inflammatory [20] and anti-tumor agents [21].

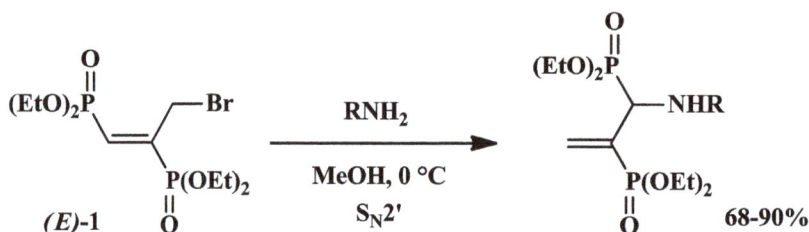

Scheme 2. Synthesis of tetraethyl (1-(alkylamino)prop-2-ene-1,2-diyl)bisphosphonates

This interesting result prompted us to study the behavior of allyl bromide *(E)*-1 towards a series of secondary and tertiary amines. We started our investigation using excess amounts of less bulky secondary amines (2 equiv.) in methanol at room temperature. Thus, their coupling reaction with allyl bromide *(E)*-1 afforded a new family of allylamines **3a-c** *via* an expected S_N2' substitution in good yields ranging from 82 to 87% (Scheme 3, Table 1).

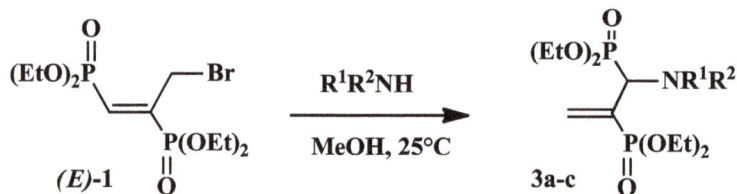

Scheme 3. Preparation of functionalized allylamines **3a-c**

Table 1. Synthesis of tetraethyl (1-(dialkylamino)prop-2-ene-1,2-diyl)bisphosphonates **3a-c**.

Entry	R^1	R^2	Product	Time (h)	Yield (%)[*]
1	Me	Me	**3a**	3	87
2	Et	Et	**3b**	5	85
3	^nPr	^nPr	**3c**	7	82

[*]Yields refer to the pure isolated products characterized by ^1H, ^{13}C NMR.

In contrast, with more bulky secondary amines such as diisopropylamine, dicyclohexylamine and diphenylamine in refluxing methanol, the reaction led to unexpected methoxy allylether *(E)*-4, resulting from an unpredictable successive S_N2'-S_N2' reactions (Scheme 4). The results are gathered in Table 2.

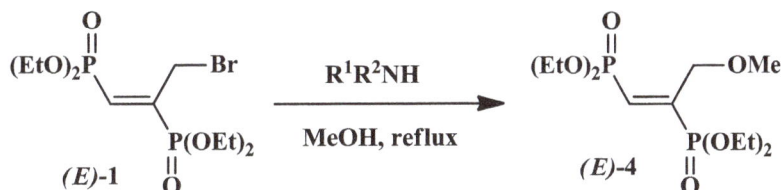

Scheme 4. Unpredictable synthesis of allyl ether (*E*)-4.

Table 2. Synthesis of functionalized primary methoxy allyl ether (*E*)-4.

Entry	R^1	R^2	Product	Time (h)	Yield (%)*
1	^iPr	$^cC_6H_{11}$	4	5	72
2	$^cC_6H_{11}$	$^cC_6H_{11}$	4	8	74
3	Ph	Ph	4	15	78

*Yields refer to the pure isolated products characterized by ^1H, ^{13}C NMR

Mechanistically, the formation of allyl ether (*E*)-4 can be achieved from the brominated derivative (*E*)-1 through successive S_N2'-type reaction with bulky secondary amines. First, secondary amine played the role of co-nucleophile in a direct nucleophilic substitution of allyl bromide (*E*)-1 to give the corresponding quaternary ammonium salt. Then, this intermediate undergoes a second S_N2' reaction with methanol which plays a dual role as reactant and as solvent (Scheme 5).

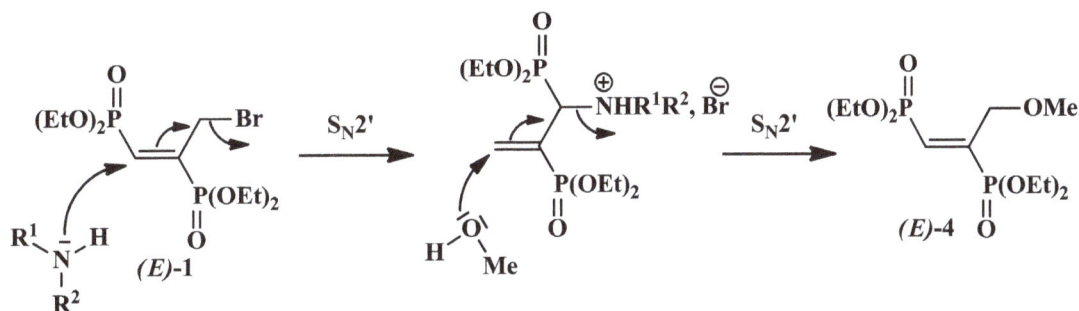

Scheme 5. Plausible mechanism for the formation of allyl ether (*E*)-4

The structure of **4** was established on the basis of its ^1H and ^{13}C NMR spectra and by heteronuclear multiple bond correlation (HMBC). The (*E*)-stereochemistry was assigned on the basis of NOESY spectra that show no correlation between the ethylenic proton (6.71 ppm) and those of group CH$_2$OMe (δ = 4.49 ppm). This result indicates that the vinylic proton and the CH$_2$ group are on opposite sides of the double bond and therefore, the alkene in **4** is (*E*) configuration. The obtained result may be explained by the steric factor of secondary amines. The results of the electrophilic behavior of allyl bromide (*E*)-1 towards different more bulky and less bulky secondary amines are summarized in Scheme 6.

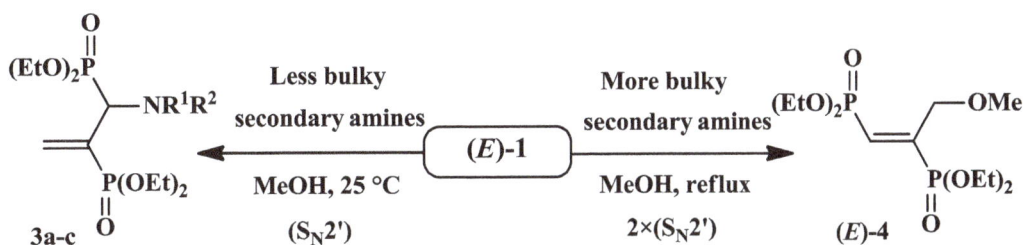

Scheme 6. Regioselective substitution of allyl bromide (*E*)-1 with some secondary amines

The diversity of previous results obtained with less or more bulky secondary amines prompted us to try other more hindered nitrogen nucleophiles such as tertiary amines (triethylamine, ethyl diisopropylamine and triethanolamine). Surprisingly, it appears that the size of these amines strongly influences their reaction course with allyl bromide (*E*)-1, while the coupling is only possible in refluxing methanol to yield a representative of secondary methoxy allyl ether: tetraethyl (1-methoxyprop-2-ene-1,2-diyl) bisphosphonate 5 in moderate to good yields (Scheme 7, Table 3).

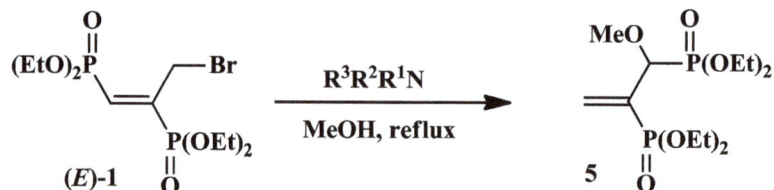

Scheme 7. Synthesis of tetraethyl (1-methoxyprop-2-ene-1,2-diyl)bisphosphonate 5

Table 3: One pot synthesis of secondary allyl ether 5.

Entry	R^1	R^2	R^3	Product	Time (d)	Yield (%)
1	Et	Et	Et	5	3	84
2	Et	^{i}Pr	^{i}Pr	5	6	78
3	$(CH_2)_2OH$	$(CH_2)_2OH$	$(CH_2)_2OH$	5	5	68

*Yields refer to the pure isolated products characterized by ^{1}H, ^{13}C NMR

The plausible mechanism which illustrates the synthesis of secondary methoxy allyl ether 5 is depicted in Scheme 8. The reaction consisted of a two-step sequence: substitution S_N2-type reaction of (*E*)-1 with tertiary amines followed an easy displacement of the resulting quaternary ammonium by methanol in a clean S_N2' reaction to provide the expected product 5. Taking into account of the limited methanol nucleophilicity, the substitution of allyl bromide (*E*)-1 with this solvent is very slow and inefficient even at reflux. The use of more hindered secondary and tertiary amines as co-nucleophilic nitrogen reagents is to provide a quaternary ammonium which can be easily displaced by the methanol.

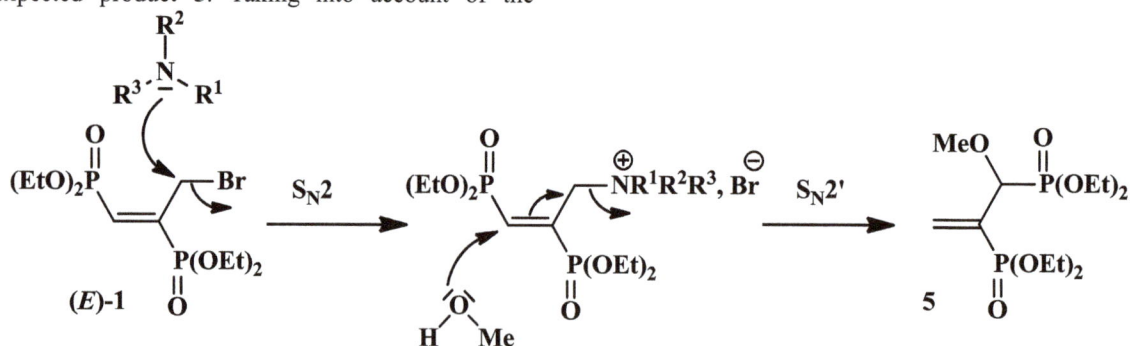

Scheme 8. Plausible mechanism for the synthesis of allyl ether 5

Conclusion

In conclusion, we have successfully developed an efficient and highly regio- and stereoselective synthesis of functionalized allyl amines 3 and allyl ethers 4 and 5 *via* an effective coupling reaction of allyl bromide (*E*)-1 with a variety of secondary and tertiary amines under mild conditions.

Experimental Section

Starting materials and solvents were used without further purification. ^{1}H-NMR, ^{31}P-NMR and ^{13}C-NMR spectra were recorded on a Bruker AMX 300 spectrometer working at 300 MHz, 121 MHz and 75 MHz respectively for ^{1}H, ^{31}P and ^{13}C with CDCl$_3$ as the solvent and TMS as the internal standard.

The chemical shifts (δ) and coupling constants (J) are, respectively, expressed in parts per million (ppm) and Hertz (Hz). All NMR spectra were acquired at room temperature. Assignments of proton (^{1}H-NMR) and carbon (^{13}C-NMR) signals were secured by DEPT 135 and HMBC experiments. Multiplicity of peaks is indicated by the following: s, singlet; d, doublet; t, triplet; dd, double doublet and m, multiplet. IR spectra were recorded on a Bruker Vertex 70 FT-IR spectrophotometer. High-Resolution Mass Spectrometry (HRMS) analyses were performed in Laberca laboratory at Oniris (*Nantes-Atlantic National* College of *Veterinary Medicine, Food Science and Engineering*) on a mass spectrometer equipped with a door coupled to a linear Orbitrap (LTQ-Orbitrap of Thermo Fisher Scientific) in positive electrospray ionization. All reactions were monitored by TLC on silica gel plates (Fluka Kieselgel 60 F$_{254}$, Merck) eluting with the solvents indicated, visualized by a 254 nm UV lamp and aqueous potassium permanganate solution. For column chromatography, Fluka Kieselgel 70-230 mesh was used.

General Procedure for the synthesis of tetraethyl1-(dialkylamino)prop-2-ene-1,2-diyl)bis-phosphonates 3a-c

To a solution of tetraethyl (3-bromoprop-1-ene-1,2-diyl)bisphosphonate *E-1* (0.2g, 0.5 mmol) diluted in 5 mL of absolute methanol was added dropwise secondary amine (2 mmol) diluted in 3 mL of methanol at 0 °C. The reaction mixture was stirred at room temperature. After stirring during the time indicated in Table 1, the excess of the solvent was evaporated under reduced pressure and the crude product was purified by chromatography on silica gel (CH$_2$Cl$_2$-MeOH, 9.5-0.5).

Tetraethyl(1-(dimethylamino)prop-2-ene-1,2-diyl)bisphosphonate 3a

Yield: 87%; Yellow liquid;

IR (neat): 1650 (C=C), 1243 (P=O).

^{1}**H-NMR (CDCl$_3$, δ):** 6.56 (d, 1H, $^3J_{HP}$ = 47.7 Hz, =CH); 6.44 (dd, 1H, $^3J_{HP}$ = 23 Hz, $^4J_{HP}$ = 3 Hz, =CH); 4.25-4.01 (m, 8H, 4OCH$_2$); 3.88 (dd, 1H, $^2J_{HP}$ = 24 Hz, $^3J_{HP}$ = 15 Hz, CHP); 2.47 (s, 3H, CH$_3$); 2.46 (s, 3H, CH$_3$); 1.33, 1.28 (2t, 12H, J = 6 Hz, J = 6 Hz, 4CH$_3$);

^{13}**C-NMR (CDCl$_3$, δ):** 135.8 (t, =CH$_2$, $^2J_{CP}$ = $^3J_{CP}$ = 7.5 Hz); 131.4 (dd, =C, $^1J_{CP}$ =172.5 Hz, $^2J_{CP}$ = 11.5 Hz,); 63.1, 62.7 (2d, 2OCH$_2$, $^2J_{CP}$ = 7.5 Hz, $^2J_{CP}$ = 7.5 Hz); 62.3, 62.0 (2d, 2OCH$_2$, $^2J_{CP}$ = 6 Hz, $^2J_{CP}$ = 6 Hz); 60.2 (dd, CHP, $^1J_{CP}$ = 156.75 Hz, $^2J_{CP}$ = 11.25 Hz); 42.9 (s, CH$_3$); 42.8 (s, CH$_3$); 16.5, 16.4 (2d, 2CH$_3$, $^3J_{CP}$ = 6.75 Hz, $^3J_{CP}$ = 6.75 Hz); 16.3, 16.2 (2d, 2CH$_3$, $^3J_{CP}$ = 6 Hz, $^3J_{CP}$ = 6 Hz);

^{31}**P-NMR (CDCl$_3$, δ):** 17.9 (d, = CP, J = 58.08 Hz); 21.9 (d, CHP, J = 58.08 Hz);

HRMS (ESI+) calculated for [M+H$^+$] 358.15429, found 358.15359.

Tetraethyl(1-(diethylamino)prop-2-ene-1,2-diyl)bis-phosphonate 3b

Yield: 85%; Yellow liquid;

IR (neat): 1648 (C=C), 1233 (P=O).

^{1}**H-NMR (CDCl$_3$, δ):** 6.58 (d, 1H, $^3J_{HP}$ = 48 Hz, =CH); 6.35 (dd, 1H, $^3J_{HP}$ = 24 Hz, $^4J_{HP}$ = 3 Hz, =CH); 4.20- 3.92 (m, 9H, CHP + 4OCH$_2$); 2.97- 2.74 (m, 2H, CH$_2$); 2.58- 2.47 (m, 2H, CH$_2$); 1.27 (3t, 9H, J = 6 Hz, 3CH$_3$); 1.19 (t, 3H, J = 6 Hz, CH$_3$); 1.00 (t, 6H, J = 6 Hz, 2CH$_3$);

^{13}**C-NMR (CDCl$_3$, δ):** 136.4 (t, =CH$_2$, $^2J_{CP}$ = $^3J_{CP}$ = 6.75 Hz); 132.4 (dd, =C, $^1J_{CP}$ =170.25 Hz, $^2J_{CP}$ = 10.5 Hz,); 63.1, 62.8 (2d, 2OCH$_2$, $^2J_{CP}$ = 7.5 Hz, $^2J_{CP}$ = 7.5 Hz); 62.4, 62.0 (2d, 2OCH$_2$, $^2J_{CP}$ = 7.5 Hz, $^2J_{CP}$ = 7.5 Hz); 55.5 (dd, CHP, $^1J_{CP}$ = 159 Hz, $^2J_{CP}$ = 11.25 Hz); 45.1 (s, CH$_2$); 45.1 (s, CH$_2$); 16.6, 16.4 (2d, 2CH$_3$, $^3J_{CP}$ = 6.5 Hz, $^3J_{CP}$ = 6.5 Hz); 16.3, 16.2 (2d, 2CH$_3$, $^3J_{CP}$ = 6 Hz, $^3J_{CP}$ = 6 Hz); 14.0 (s, CH$_3$); 14.0 (s, CH$_3$);

^{31}**P-NMR (CDCl$_3$, δ):** 18.5 (d, =CP, J = 62.92 Hz); 22.6 (d, CHP, J = 62.92 Hz);

HRMS (ES+) calculated for [M+Na$^+$] 408.1681, found 408.1684.

(1-(dipropylamino)prop-2-ene-1,2-diyl)bis-phosphonate 3c

Yield: 82%; Yellow liquid;

IR (neat): 1642 (C=C), 1244 (P=O);

^{1}**H-NMR (CDCl$_3$, δ):** 6.60 (d, 1H, $^3J_{HP}$ = 48 Hz, =CH); 6.50 (dd, 1H, $^3J_{HP}$ = 21 Hz, $^4J_{HP}$ = 3 Hz, =CH); 4.22- 3.94 (m, 9H, CHP + 4OCH$_2$); 2.76 (t, 4H, J = 6 Hz, 2CH$_2$); 1.67- 1.56 (m, 4H, 2CH$_2$); 1.30, 1.22 (2t, 12H, J = 6.5 Hz, J = 6.5 Hz, 4CH$_3$); 0.90 (t, 6H, J = 6 Hz, 2CH$_3$);

^{13}**C-NMR (CDCl$_3$, δ):** 137.4 (t, =CH$_2$, $^2J_{CP}$ = $^3J_{CP}$ = 7.5 Hz); 133.5 (dd, =C, $^1J_{CP}$ =171.5 Hz, $^2J_{CP}$ = 11.5 Hz,); 63.4, 62.9 (2d, 2OCH$_2$, $^2J_{CP}$ = 7.5 Hz, $^2J_{CP}$ = 7.5 Hz); 62.5, 62.1 (2d, 2OCH$_2$, $^2J_{CP}$ = 7.5 Hz, $^2J_{CP}$ = 7.5 Hz); 55.4 (dd, CHP, $^1J_{CP}$ = 161 Hz, $^2J_{CP}$ = 11.5 Hz); 45.2 (s, CH$_2$); 45.1 (s, CH$_2$); 16.6, 16.5 (2d, 2CH$_3$, $^3J_{CP}$ = 7.5 Hz, $^3J_{CP}$ = 7.5 Hz); 16.3, 16.3 (2d, 2CH$_3$, $^3J_{CP}$ = 6 Hz, $^3J_{CP}$ = 6 Hz); 11.0 (s, CH$_3$); 11.0 (s, CH$_3$);

^{31}**P-NMR (CDCl$_3$, δ):** 18.7 (d, =CP, J = 61.42 Hz); 22.9 (d, CHP, J = 61.42 Hz);

HRMS (ES+) calculated for [M+ Na $^+$] 436.1994, found 436.1997.

General Procedure for the synthesis of tetraethyl(3-methoxyprop-1-ene-1,2-diyl)bis-phosphonate 4

Compound **4** was similarly prepared as **3a-c** from *(E)*-1. The reaction mixture was stirred in refluxed methanol for 5 to 15 hours (Table 2). The crude product **4** was purified by chromatography on silica gel (CH$_2$Cl$_2$-MeOH, 9.4-0.6).

Yellow liquid;

^{1}**H-NMR (CDCl$_3$, δ):** 6.71 (dd, 1H, $^2J_{HP}$ = 27 Hz, $^3J_{HP}$ = 15 Hz, =CH); 4.49 (d, 2H, $^3J_{HP}$ = 15 Hz, CH$_2$OCH$_3$); 4.20- 4.08 (m, 8H, 4OCH$_2$); 3.39 (s, OCH$_3$); 1.34, 1.28 (2t, 12H, J = 6 Hz, J = 6 Hz, 4CH$_3$);

^{13}C-NMR (CDCl$_3$, δ): 148.2 (dd, =C, $^1J_{CP}$ = 164.25 Hz, $^2J_{CP}$ = 7.5 Hz); 133.1 (dd, =CH, $^1J_{CP}$ = 172.5 Hz, $^2J_{CP}$ = 9.75 Hz); 68.0 (t, CH$_2$OCH$_3$, $^2J_{CP}$ = $^3J_{CP}$ = 9 Hz); 62.7 (d, 2OCH$_2$, $^2J_{CP}$ = 6 Hz); 62.4 (d, 2OCH$_2$, $^2J_{CP}$ =6 Hz); 58.6 (s, CH$_3$); 16.3, 16.2 (2d, 4CH$_3$, $^3J_{CP}$ = 6 Hz, $^3J_{CP}$ =6 Hz);
^{31}P-NMR (CDCl$_3$, δ): 12.0 (d, =CP, J = 95.59 Hz); 15.0 (d, CHP, J = 95.59 Hz);
HRMS (ES+) calculated for [M+ Na $^+$] 367.1051, found 367.1053.

General Procedure for the synthesis of tetraethyl(1-methoxyprop-2-ene-1,2-diyl)bis-phosphonate 5

To a solution of tetraethyl (3-bromoprop-1-ene-1,2-diyl)bisphosphonate (E)-1 (0.2g, 0.5 mmol) diluted in 5 mL of absolute methanol, was added dropwise tertiary amine (2 mmol) diluted in 3 mL of methanol at room temperature. The reaction mixture was stirred at reflux for 3 to 6 days (Table 3). After completion, the excess of the solvent was evaporated under reduced pressure and the crude product 5 was purified by column chromatography on silica gel (CH$_2$Cl$_2$-MeOH, 9.5-0.5).

Yellow liquid;
^{1}H-NMR (CDCl$_3$, δ): 6.49 (dd, 1H, $^3J_{HP}$ = 48 Hz, $^4J_{HP}$ = 6 Hz, =CH); 6.45 (dd, 1H, $^3J_{HP}$ = 24 Hz, $^4J_{HP}$ = 3 Hz, =CH); 4.32 (dd, 1H, $^2J_{HP}$ = 15 Hz, $^3J_{HP}$ = 9 Hz, CHP); 4.23- 4.09 (m, 8H, 4OCH$_2$); 3.43 (s, OCH$_3$); 1.35, 1.33, 1.28 (3t, 12H, J = 6 Hz, J = 6 Hz, J = 6 Hz, 4CH$_3$);
^{13}C-NMR (CDCl$_3$, δ): 135.2 (t, =CH$_2$, $^2J_{CP}$ = $^3J_{CP}$ = 7.5 Hz); 134.3 (d, =C, $^1J_{CP}$ 177.75 Hz); 74.7 (dd, CHP, $^1J_{CP}$ =168 Hz, $^2J_{CP}$ = 15.75 Hz); 63.2, 63.0 (2d, 2OCH$_2$, $^2J_{CP}$ = 7.5 Hz, $^2J_{CP}$ = 7.5 Hz); 62.5, 62.2 (2d, 2OCH$_2$, $^2J_{CP}$ =6 Hz, $^2J_{CP}$ = 6 Hz); 58.8 (d, OCH$_3$, $^3J_{CP}$ =12.75 Hz); 16.4, 16.2 (2d, 4CH$_3$, $^3J_{CP}$ = 6 Hz, $^3J_{CP}$ =6 Hz);
^{31}P-NMR (CDCl$_3$, δ): 16.0 (d, =CP, J = 26.62 Hz); 18.2 (d, CHP, J = 26.62 Hz);
HRMS (ES+) calculated for [M+H$^+$] 345.12265, found 345.12183.

References

1- L. Aiwen, L. Xiyan, Org. Lett., 2000, 2, 2357-2360.

2- (a) M. Hagithara, N. J. Anthony, T. J. Stout, J. Clardy, S. L. Schreiber, J. Am. Chem. Soc., 1992, 114, 6568- 6570; (b) S. Devadder, P. Verheyden, H. C. M. Jaspers, G. Van Binst,D. Tourué, Tetrahedron Lett., 1996, 37, 703-706.

3- Y. Yamamoto, M. Schmid, J. Chem. Soc., Chem. Commun., 1989, 1310-1312.

4- R. Yamaguchi, M. Moriyasu, M. Kawanisi, J. Org. Chem., 1985, 50, 287-288.

5- R. M. Borzilleri, X. Zheng, R.J.Schmidt, J. A. Johnson, S. H. Kim, J. D. Marco, C. R. Fairchild, J. Z. Gougoutas, F. Y. F. Lee, B. H. Long, G. D. Vite, J. Am. Chem. Soc., 2000, 122, 8890-8897.

6- S. Laschat, H. Kunz, J. Org. Chem., 1991, 56, 5883-588.

7- (a) D. L. Wright, J. P. Schulte II; M. A. Page, Org. Lett., 2000, 2, 1847-1850; (b) F. X. Felpin, S. Girard, G. Vo-Thanh, R. J. Robins, J. Villiéras, J. Lebreton, J. Org. Chem., 2001, 66, 6305-6312; (c) R. P. Jain, R. M. Williams, J. Org. Chem., 2002, 67, 6361-6365; (d) R. Di Fabio, G. Alvaro, B. Bertani, D. Donati, S. Giacobbe, C. Marchioro, C. Palma, S. M. Lynn, J. Org. Chem., 2002, 67, 7319-7328; (e) P. V. Ramachandran, T. E. Burghardt, L. Bland-Berry, J. Org. Chem., 2005, 70, 7911-7918; (g) P. Besada, L. Mamedova, C. J. Thomas, S. Costanzi, K. A. Jacobson, Org. Biomol. Chem., 2005, 3, 2016-2025.

8- (a) K. C. Nicolaou, M. O. Frederick, R. J. Aversa, Angew. Chem. Int. Ed., 2008, 47, 7182-7225; (b) S. P. Roche, J. A. Porco Jr., Angew. Chem. Int. Ed., 2011, 50, 4068-4093.

9- (a) G. Bartoli, C. Cimarelli, E. Marcantoni, G. Palmieri, M. Petrini, J. Org. Chem. 1994, 59, 5328-5335; (b) M. Liu, M. P. Sibi, Tetrahedron, 2002, 58, 7991-8035; (c) S. M. Hecht, Acc. Chem. Res., 1986, 19, 383-391.

10- (a) Y. -F. Wang, T. Izawa, S. Kobayashi, M. Ohno, J. Am. Chem. Soc., 1982, 104, 6465-6466; (b) S. Hashiguchi, A. Kawada, H. Natsugari, Perkin Trans I, 1991, J. Chem. Soc., 1991, 2435-2444; (c) G. Cardillo, C. Tomasini, Chem. Soc. Rev., 1996, 25, 117-128.

11- (a) E. L. Eliel, X. -C. He, J. Org. Chem., 1990, 55, 2114-2119; (b) Y. Hayashi,; J. J. Rode, E. J. Corey, J. Am. Chem. Soc., 1996, 118, 5502-5503; (c) M. Genov, V. Dimitrov, V. Ivanova, Tetrahedron: Asymmetry, 1997, 8, 3703-3706.

12- (a) E. F. Kleinmann, Comprehensive Organic Synthesis, B. M.,Trost, (Ed.), Pergamon: New York, NY, 1991;, pp 893-951; (b) E. J. Corey, G. A. Reichard, Tetrahedron Lett., 1989, 30, 5207-5210.

13- (a) S. Ben Gharbia, R. Besbes,J. Villiéras, H. Amri, Synth. Commun., 1996, 26, 1685-1692; (b) S. Hbaïeb, Z. Latiri, H. Amri, Synth. Commun., 1999, 6, 981-988; (c) H. Kraiem, I. M. Abdullah, H. Amri, Tetrahedron Lett., 2003, 44, 553-555; (d) A. Arfaoui, F. Béji, T. Ben Ayed, H. Amri, Synth. Commun., 2008, 38, 3717-3725; (e) A. Fray, J. Ben Kraïem, H. Amri, Heteroatom Chem., 2013, 24, 460-465; (f) A. Arfaoui, H. Amri, Synth. Commun., 2015, 45, 2627-2635; (g) K. Jebali, A. Planchat, H. Amri, M. Mathé-Allainmat, J. Lebreton, Synthesis, 2016, 48, 1502-1517.

14- M. Ameur, A.Arfaoui, H. Amri, Arkivoc, **2014**, (v), 199-209.

15- D. Banerjee, R. V. Jagadeesh, K. Junge, H. Junge, M. Beller, ChemSusChem, **2012**, 5, 2039-2044.

16- (a) K. Kobayashi, K. A. Miyazama, H. Terrahara, H. Mishime, H. Kurihare, Tetrahedron Lett., **1976**, 17, 537-540; (b) R. R. Rsndo,; F. *J*. Bangerter, Am. Chem. Soc., **1976**, 98, 6762-6764; (c) R. R. Rando, F. Bangaeter, J. Am. Chem. Soc., **1977**, 99, 5141-5145; (d) R. D. Allan, G. A. R. Johnstone, B. Twitchin, Neurosci. Lett., **1977**, 4, 51-54, CA 1977, 86, p 15039o.

17- (a) T. Hashimoto, S. Kondo, H. Naganawa, T. Takita, K. Maede, H. Umezawa, J. Antibiot., **1974**, 27, 86-87; (b) T. Hashimoto, S. Takahashi, H. Naganawa, T. Kakita, H. Umezawa, J. Antibiot., **1972**, 25, 350-355; (c) C. Y. P. Teng, B. Garmen, Tetrahedron Lett., **1982**, 23, 313-316.

18- T. Kondo, H. Nakai, Tetrahedron, **1973**, 29, 1801-1806.

19- D.; E. Natalie, S. C. Donna, M. Khurana, N. S. Noha, P. S. James, J. H. Sylvia, N. Abraham, S. T. Robert, A. M. Jacqueline, Eur. J. Med. Chem., **2003**, 38, 49-64.

20- G. Dannhardt, A. Bauer, Nowe, U. J. Prakt. Chem., **1998**, 340, 256-263.

21- D. L. Boger, T. Ishizaki, J. R. Wysoki, S. A. Munk, O. Suntornwat, J. Am. Chem. Soc., **1989**, 111, 6461-6463.

17

Green and efficient method for the synthesis of 1,5-benzodiazepines using phosphate fertilizers as catalysts under solvent-free conditions

Sarra Sibous, Touriya Ghailane, Serrar Houda, Rachida Ghailane, Said Boukhris and Abdelaziz Souizi [*]

Laboratory of Organic, Organometallic and Theoretical Chemistry, University of Ibn Tofail,
B.P. 133, 14000 Kenitra, Morocco

Abstract: Three-component reaction in one pot transformation of aldehydes, ethylacetoacetate and o-phenylenediamine was employed for the synthesis, under solvent-free, of 1,5-benzodiazepine derivatives using fertilizers mono-ammonium phosphate (MAP), di-ammonium Phosphate (DAP) and triple super phosphate (TSP) as safe, clean, and recyclable catalysts. The synthesis method seems to be operationally simple and provides access to a variety of 1,5-benzodiazepines with excellent yields in a short reaction time.

Keywords: 1,5-benzodiazepine derivatives, recyclable catalysts, phosphate fertilizers, MAP, DAP, TSP.

Introduction

Benzodiazepines form an important class of heterocyclic compounds containing nitrogen atoms. They have attracted special attention due to their important biological or pharmacological properties [1-8] and their wide field of application in different areas of medicine [9-12] and agriculture [13]. It was reported that the 1,5-benzodiazepines and its derivatives present interesting pharmacological and therapeutic properties. Their use has been extended to treat various diseases such as cancer, viral infection (non-nucleoside inhibitors of HIV-1 reverse transcriptase), and cardiovascular disorders. For this reason, their synthesis has received great attention, especially in the field of medicinal chemistry [10,14]. Therefore, the research on synthetic methods for 1,5-benzodiazepines has become one of the hot issues.

It is known that the most classical synthesis method for 1,5-benzodiazepines is the condensation of o-phenylenediamines with α,β-unsaturated carbonyl compounds in the presence of a wide variety of Lewis-acid catalysts [15-26].

In order to develop environmentally friendly organic reactions, the synthesis of benzodiazepines using different solid acid catalysts [27-29] and heteropolyacid catalysts using THF or acetonitrile as reaction solvent [30-31] has also been investigated. Recently, a crystalline iron-based metal-organic framework MOF-235 was synthesized, characterized by different techniques and used as a heterogeneous catalyst for the synthesis of 1,5-benzodiazepines by the cyclo-condensation of 1,2-diamines with ketones [32].

However, developing an efficient and recyclable heterogeneous catalyst system for the synthesis of 1,5-benzodiazepines still needs to be explored.

On the other hand, the solvents used in organic synthesis are obviously harmful to environment; the solvent-free organic reactions have attracted the attention of organic chemists in order to reduce the pollution and to bring down the experimental cost. Indeed, it has been established that multicomponent reactions (MCRs) are a powerful tool for the synthesis of a wide variety of compounds in a single pot assuring good yields and low costs, short reaction times, minimizing energy, and avoiding expensive purification procedures. Moreover, the MCRs are generally much more environmentally friendly than the multi-step reactions and allow a rapid access to various molecules [33-44].

According to our interest in the green protocols in organic synthesis, the aim of this work is to use the fertilizers mono-ammonium phosphate (MAP), di-ammonium phosphate (DAP) and triple super phosphate (TSP) [45] as solid heterogeneous catalysts for the synthesis of 1,5-benzodiazepines (Scheme 1).

These fertilizers are produced in industrial quantities and are used mainly in agriculture as nutrient sources for nitrogen and phosphorus.

*Corresponding author: Abdelaziz Souizi
Email address: souizi@yahoo.com*

Morocco has the most important phosphate worldwide reserves and is thus a major country for the production of phosphate fertilizers whose physicochemical characteristics are shown on the website of the OCP (Office Cherifien des Phosphates) [46], the first Moroccan company.

Furthermore, these fertilizers are widely available and their use as catalysts is undoubtedly an added value for Moroccan phosphates.

Results and Discussion

Herein we focus our interest on the synthesis of 1,5-benzodiazepine derivatives using phosphate fertilizers as catalysts which have been employed as promoters of the reaction to achieve higher yields. This synthesis was accomplished by three-component reaction in one pot cyclocondensation of aldehydes, ethylacetoacetate and o-phenylenediamine in the presence of catalytic amount (1mol%) of phosphate fertilizers (Scheme 1).

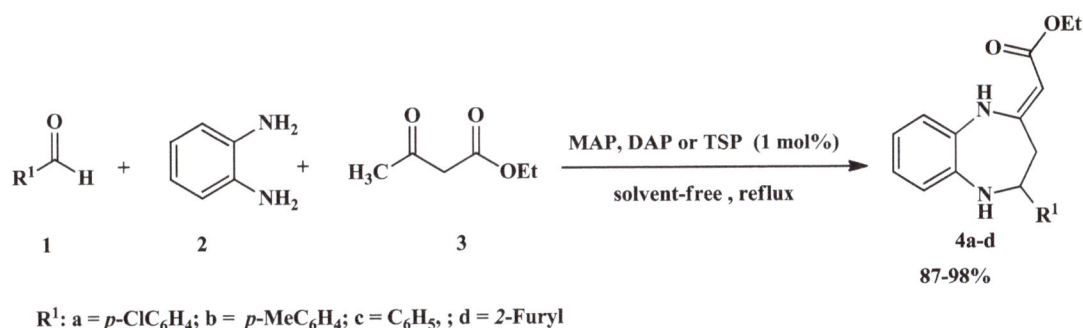

R^1: a = p-ClC$_6$H$_4$; b = p-MeC$_6$H$_4$; c = C$_6$H$_5$, ; d = 2-Furyl

Scheme 1. Synthesis of 1,5-Benzodiazepines 1 a-d.

We envisage to evaluate the catalytic activity of the mentioned fertilizers [45] as solid heterogeneous catalysts in the synthesis of 1,5-benzodiazepine derivatives. Initially, in order to optimize the reaction conditions, we chose o-phenylenediamine, ethylacetoacetate and benzaldehyde as the model reaction which occurs in the presence of the three catalysts (MAP, DAP and TSP). Considering that the

reaction medium is one of the most important factors influencing any process, several solvents are used to accomplish further insight at the solvent effect on the synthesis of 1,5- benzodiazepines. Thus, the synthesis of 1,5-benzodiazepine derivative **4b**, in various organic solvents and under solvent-free, is performed. The yields and the reaction times are summarized in Table 1.

Table 1. The solvent effect on the synthesis of 1,5-benzodiazepines [a].

Entry	Solvent	Time (min)			Yield %[b]		
		MAP	**DAP**	**TSP**	**MAP**	**DAP**	**TSP**
1	Ethanol	180	120	90	89	86	80
2	Methanol	110	90	120	90	91	94
3	Isopropanol	210	65	110	50	61	72
4	Butanol	90	60	70	51	42	49
5	DMF	60	55	90	47	61	57
6	Acetonitrile	120	80	180	77	67	71
7	THF	180	120	190	50	73	62
8	AcOEt	250	233	240	57	58	79
9	Dioxane	130	150	210	52	61	37
10	CH$_2$Cl$_2$	180	184	240	41	44	56
11	CH$_3$Cl	270	230	250	67	45	57
12	Cyclohexane	310	240	350	39	55	54
13	Solvent-free	60	55	90	92	91	96

[a] Reaction of o-phenylenediamines (**1 mmol**) b-keto esters (**1 mmol**) and benzaldehyde (**1 mmol**) catalyzed MAP or DAP or TSP under reflux conditions.
[b] Isolated yield.

The results indicate that the best yields are obtained in the solvent-free (92, 91 and 96%) or in the presence of methanol (90, 91 and 94%) using

respectively MAP, DAP and TSP as catalysts. Although methanol is the most appropriate solvent for this transformation which is accomplished with high yields compared to the other solvents. The reaction under solvent-free conditions is carried out in the shortest times which are 60, 55 and 90 min relative to those obtained with methanol (110, 90 and 120 min). This result is interesting as it is taking into account the requirements of green chemistry and minimizing the risk accompanied by the use of organic solvents such as toxicity and flammability.

The efficiency of the reaction is mainly affected by the amount of catalyst; therefore we propose to optimize this amount for the model reaction under solvent free conditions. For this purpose, we vary the amount of catalyst from 1 to 10 mol% by a step of 1mol%. The obtained yields and reaction times are reported in Table 2.

It is noticeable that the great yields of 94, 96 and 97% of the reaction are obtained for the amount of 1mol% in presence of three catalysts MAP, DAP and TSP respectively. At this amount of catalyst, the reaction time is estimated by 60, 45 and 65 min with the MAP, DAP and TSP catalysts respectively, thus theses values are also the shortest ones. In conclusion 1mol% of catalyst is sufficient to afford the best results with high yields in short reaction time.

Furthermore, it is interesting to notice that the yields of the reaction diminish when the amount of the catalyst increase, that is probably due to the dispersion effect of the reagents on the catalyst's surface.

Table 2. Effect of catalytic amount on synthesis of 1,5-benzodiazepines[a].

Entry	Amount of catalyst Mol%	Time (min)			Yield %[b]		
		MAP	**DAP**	**TSP**	**MAP**	**DAP**	**TSP**
1	1	60	45	65	94	97	96
2	2	60	45	65	94	97	96
3	3	60	50	80	94	94	96
4	4	60	55	85	92	92	96
5	5	60	55	90	92	91	96
6	6	90	60	90	90	91	94
7	7	120	70	95	87	89	90
8	8	120	90	110	82	77	88
9	9	130	120	120	79	72	80
10	10	130	180	120	79	65	80

[a] Reaction of o-phenylenediamines (**1 mmol**) b-keto esters (**1 mmol**) and benzaldehyde (**1 mmol**) catalyzed MAP or DAP or TSP under solvent-free conditions at reflux .[b] Isolated yield.

In addition, the principal disadvantage of most of the existing methods is that the catalysts are destroyed in work- up procedure and could not be recovered or reused whereas their reusability is central to their utility. It is well known that the reusability of the heterogeneous catalyst is one of the major significant parameters and is of great

importance in industrial uses. In order to study the reusability of the three catalysts used in this work, regeneration experiments were carried out. Thereby, the recovery and reusability of the supported catalyst using ethyl acetoacetate, o-phenylenediamine and benzaldehyde as model substrates were studied.

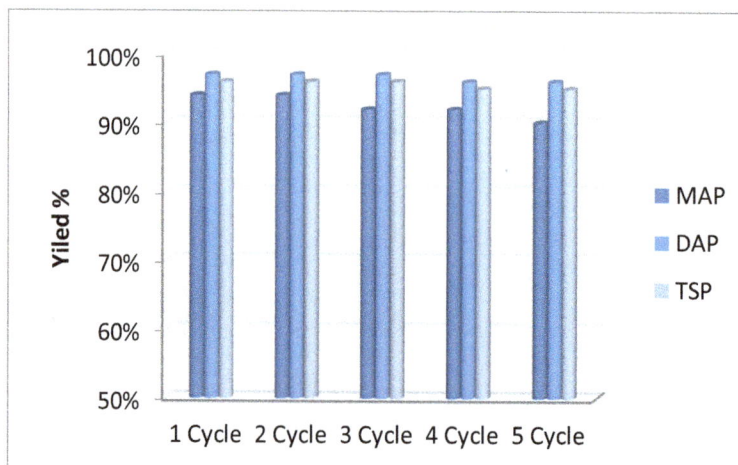

Figure 1. Reusability of MAP, DAP and TSP catalysts in synthesis of 1,5-benzodiazepine **4b**

To point out the advantages of DAP, MAP and TSP catalysts used in the synthesis of 1,5-benzodiazepine derivatives, we opt to compare their catalytic activity with that of other catalysts given in the literature. The reaction times and the yields of the isolated product are collected in Table 3. These previous studies have shown that organic and inorganic catalysts [26,46,47] were used for 1, 5-benzodiazepine derivatives synthesis by a one-pot three-component reaction in solvents leading to low yields and long reaction times compared to the ones obtained with DAP, MAP and TSP catalysts used in the present work. These catalysts were found to be prominent catalysts and provided the highest yield and short reaction times for the model reaction. Thus, they are proved to be effective catalysts for the synthesis of 1, 5-benzodiazepine derivatives.

Table 3. Optimum conditions for 1, 5- benzodiazepine **4b** synthesis catalyzed by MAP; DAP, TSP and for other derivatives synthesis using various catalysts given in previous studies.

Catalyst	Catalyst amount	Solvant	Time (min)	Yield %	Ref.
MAP	1 mol%	Solvant-free	60	94	This work
DAP	1 mol%)	Solvant-free	45	97	This work
TSP	1 mol%	Solvant-free	65	96	This work
BDMS	10 mol%	DCE	270	72	46
Benzoic acid	20 mol%	DCE	480	21	26
2-Nitrobenzoic acid	20mol%)	DCE	480	26	26
2,6-Pyridinedicarboxylic acid	20 mmol%	DCE	300	70	26
p-TsOH	10 mmol%	EtOH	480	85	47
HOAc (acetic acid)	10 mmol%	EtOH	390	79	47
STA (silicotungstic acid)	10 mmol%	EtOH	720	70	47
PMA(phosphomolybdic acid)	10 mmol%	EtOH	420	89	47
$CeCl_3 \cdot 7H_2O$	10 mmol%	EtOH	600	85	47
$NiCl_2 \cdot 6H_2O$	10 mmol%	EtOH	600	86	47
I_2	10 mmol%	EtOH	420	75	47

[a] Reaction conditions: o-phenylenediamine(**1 mmol**), b-keto esters (**1 mmol**) and benzaldehyde (**1 mmol**).
[b] Isolated yields.

Next, under the optimized conditions, we examined the generality and scope of this reaction using ethyl acetoacetate, *o*-phenylenediamine, and various substituted aromatic aldehydes, which smoothly underwent the one-pot reaction to afford the corresponding products **4a-4d** in good to excellent yields (Table 4).

Table 4. One pot three-component synthesis of 1,5-benzodiazepines **4a-4d** catalyzed by MAP, DAP and TPS

Entry	R^1	Time (min)			Yield %[b]			Mp°C	
		MAP	DAP	TSP	MAP	DAP	TSP	Found	Reported
4a	$4\text{-}ClC_6H_4$	60	45	65	94	97	96	176-178	176-180[46]
4b	$4\text{-}MeC_6H_4$	150	120	90	87	90	94	95-97	95-98[46]
4c	C_6H_5	240	180	90	89	95	98	73-75	73-77[46]
4d	2-Furyl	300	240	180	92	96	97	84-86	84-88[46]

[b] Isolated yield.

Conclusion

To sum up, we described the three-component one-pot transformation of ethyl acetoacetate, o-phenylenediamine, and various substituted aromatic aldehydes in solvent-free conditions using three phosphate fertilizers as catalysts to synthesize four 1,5-benzodiazepine derivatives. It is found that the strategy developed in this work is an efficient procedure for the synthesis of these derivatives. The advantages of this method are easy work-up procedure, good yields, short reaction times, contribution to the green chemistry development, gain on the cost of the experiment since phosphate fertilizers are very cheap and are available everywhere. Further applications of this method for various syntheses using these catalysts are in progress in our laboratory.

Experimental Section

Materials and methods

All reagents were obtained commercially from Merck or Fluka Chemical Companies. The known products were identified by comparison of their melting points and spectral data with those reported in the literature [46]. The purity determination of the substrates and reactions monitoring were accomplished by TLC on silica gel polygram SILG/UV 254 plates. Melting points were detected on a Kofler hot stage apparatus and are uncorrected.

Typical experimental procedure

To a mixture of 1,2-phenylenediamine (1 mmol) and ß-keto esters (1 mmol), 1 mol% of catalyst (MAP, DAP or TSP) was added. The mixture was stirred for 15 min under solvent-free conditions at room temperature. Then, the aromatic aldehyde (1 mmol) was added and the mixture was stirred under reflux for the specific times indicated in Table 4. After completion of the reaction (monitored by TLC), the reaction mixture was cooled to room temperature, the catalyst was separated from the mixture by filtration and washed several times with ethyl acetate and dried for a later use. The filtrate was evaporated and the obtained 1,5-benzodiazepines were recrystallized from ethanol. The structures of compounds 4a-d were confirmed by the comparison of their melting points and spectral data with those reported in the literature [26,46].

References

1- H. Schutz, Benzodiazepines , Publisher: *Springer*-Verlag, Berlin, *Heidelberg*, New York, **1982**, PP. 439.

2- J.K. Landquist, A.R. Katritzky, C.W. Rees, Comprehensive heterocyclic chemistry (Eds), Pergamon, Oxford, 1984, 1, 166-170.

3- H. Nakano, T. Inoue, N. Kawasaki, H. Miyataka, H. Matsumoto, T. Taguchi, N. Inagaki, H. Nagai, T. Satoh, Synthesis and biological activities of novel antiallergic agents with 5-lipoxygenase inhibiting action, Bioorg. Med. Chem. **2000**, 8, 373-380.

4- L.O. Rundall, B. Kiappel, in: S. Garattini, E. Mussini, L.O. Randall (Eds.), Benzodiazepines, Raven Press, New York, **1973**, 27.

5- A.H. Jadhav, H. Kim,Solvent free synthesis of 1, 5-benzodiazepine derivatives over the heterogeneous silver salt of silicotungstic acid under ambient conditions. RSC Adv, **2013**, 3, 5131-5140.

6- X. Pan, Z. Zou, W. Zhang, Ga(OTf)3-promoted condensation reactions for 1,5-benzodiazepines and 1,5-benzothiazepines. Tetrahedron Lett.**2008**, 49, 5302-5308.

7- M.C. Aversa, A. Ferlazzo, P. Giannetto, F.H. Kohnke, A convenient synthesis of novel [1, 2, 4] triazolo [4, 3-a][1, 5] benzodiazepine derivatives, Synthesis, **1986**, 1986, 230-231.

8- W.K. Huang, C.W. Cheng, S.M. Chang, Y.P. Lee, E.W.G. Diau, Synthesis and electron-transfer properties of benzimidazole-functionalized ruthenium complexes for highly efficient dye-sensitized solar cells, Chem. Commun. **2010**, 46, 8992-8994.

9- K. S. Atwal, J. L. Bergey, A. Hedberg, S. Moreland, Synthesis and biological activity of novel calcium channel blockers: 2, 5-dihydro-4-methyl-2-phenyl-1, 5-benzo-thiazepine-3-carboxylic acid esters and 2,5-dihydro-4-methyl-2-phenyl-1, 5-benzo-diazepine-3-carboxylic acid esters, J. Med. Chem. **1987**, 30, 635-640.

10- M. D. Braccio, G. Grossi, G. Roma, L. Vargiu, M. Mura, M. E. Marongiu, 1, 5-Benzo-diazepines. Part XII. Synthesis and biological evaluation of tricyclic and tetracyclic 1, 5-benzodiazepine derivatives as nevirapine analogues, Eur. J.Med. Chem. **2001**, 36, 935-949.

11- M. Abdollahi-Alibeik, I. Mohammadpoor-Baltork, Z. Zaghaghi, B. Yousefi, Efficient synthesis of 1, 5-benzodiazepines catalyzed by silica supported 12-tungstophosphoric acid, Catal. Commun. **2008**, 9, 2496-2502.

12- Ç. Radatz, R. Silva, G. Perin, E. Lenardao, R. Jacob, D. Alves, Catalyst-free synthesis of benzodiazepines and benzimidazoles using glycerol as recyclable solvent. Tetrahedron Lett. **2011**, 52, 4132-4136.

13- D. P. Clifford, D. Jackson, R. V. Edwards, P. Jeffrey, Herbicidal and pesticidal properties of some 1, 5-benzodiazepines, 1, 3, 5-benzotriazepines and 3, 1, 5-benzothia-diazepines, Pestic. Sci. **1976**, 7, 453-458.

14- V. Merluzzi, K. D. Hargrave, M. Labadia, K. Grozinger, M. Skoog, J. C. Wu, C-K.

Shih, K. Eckner, S. Hattox, J. Adams, A. S. Rosenthal, R. Faanes, R. J. Eckner, R. A. Koup, J. L. Sullivan, Inhibition of HIV-1 replication by a nonnucleoside reverse transcriptase inhibitor, Science. **1990**, *250*, 1411-1413.

15- M. Curini, F. Epifano, Ytterbium triflate promoted synthesis of 1,5-benzodiazepine derivatives. Tetrahedron Lett. **2001**, *42*, 3193-3195.

16- M. Kodomari, T. Noguchi, Solvent-free synthesis of 1, 5-benzothiazepines and benzodiazepines on inorganic supports, Synth. Commun. **2004**, *34*, 1783-1790.

17- C.A. Cortes, A. L. Valencia, New derivatives of dibenzo [b, e][1, 4] diazepin-1-ones by an efficient synthesis and spectroscopy, J. Heterocycl. Chem. **2007**, *44*, 183-184.

18- D. McGowan, O. Nyanguile, M. D. Cummings, S. Vendeville, K. Vandyck, W. Van den Broeck, J. F. Bonfanti, 1, 5-Benzodiazepine inhibitors of HCV NS5B polymerase, Bioorg. Med. Chem. Lett. **2009**, *19*, 2492-2496.

19- A. V. Vijayasankar, S. Deepa, B. R. Venugopal, N. Nagaraju, Amorphous mesoporous iron alumino-phosphate catalyst for the synthesis of 1, 5-benzodiazepines, Chin.J.Catal. **2010**, *31*, 1321-1327.

20- S. L. Wang, C. Cheng, F.Y. Wu, B. Jiang, F. Shi, S. J. Tu, G. Li, Microwave-assisted multi-component reaction in water leading to highly regioselective formation of benzo [f] azulen-1-ones, Tetrahedron. **2011**, *67*, 4485-4493.

21- L. D. Fader, R. Bethell, P. Bonneau, M. Bös, Y. Bousquet, M. G. Cordingley, N. Goudreau, Discovery of a 1, 5-dihydrobenzo [b][1, 4] diazepine-2, 4-dione series of inhibitors of HIV-1 capsid assembly, Bioorg. Med. Chem. Lett. **2011**, *22*, 398-404.

22- A. Maleki, Fe_3O_4/SiO_2 nanoparticles: an efficient and magnetically recoverable nanocatalyst for the one-pot multicomponent synthesis of diazepines. Tetrahedron. **2012**, *68*, 7827-7833.

23- I. E. Tolpygin, N. V. Mikhailenko, A. A. Bumber, E. N. Shepelenko, U. V. Revinsky, A. D. Dubonosov, V. I. Minkin, 11-R-dibenzo [b, e][1, 4] diazepin-1-ones, the chemosensors for transition metal cations. Russ. J. Gen. Chem. **2012**, *82*, 1141-1147.

24- B. Jiang, Q. Y. Li, H. Zhang, S. J. Tu, S. Pindi, G. Li, Efficient domino approaches to multifunctionalized fused pyrroles and dibenzo [b, e][1, 4] diazepin-1-ones, Org. Lett. **2012**, *14*, 700-703.

25- J. Schimer, P. Cígler, J. Veselý, K. GrantzŠašková, M. Lepsik, J. Brynda, H. G. Kraeusslich, Structure-Aided design of novel inhibitors of HIV protease based on a benzodiazepine scaffold, J. Med. Chem. **2012**, *55*, 10130-10135.

26- M. Lal, R. S. Basha, S. Sarkar, A. T. Khan, 2, 6-Pyridinedicarboxylic acid as organocatalyst for the synthesis of 1, 5-benzodiazepines through one-pot reaction. Tetrahedron Lett. **2013**, *54*, 4264-4272.

27- M. A. Chari, K. Syamasundar, Polymer (PVP) supported ferric chloride: an efficient and recyclable heterogeneous catalyst for high yield synthesis of 1, 5-benzodiazepine derivatives under solvent free conditions and microwave irradiation, Catal. Commun. **2005**, 6, 67-70.

28- A. Hegedüs, Z. Hell, A. Potor, A simple environmentally-friendly method for the selective synthesis of 1, 5-benzodiazepine derivatives using zeolite catalyst, Catal. Lett. **2005**, *105*, 229-232.

29- M. Muñoz, G. Sathicq, G. Romanelli, S. Hernández, C. I. Cabello, I. L. Botto, M. Capron, Porous modified bentonite as efficient and selective catalyst in the synthesis of 1, 5-benzodiazepines, J. Porous Mater. **2013**, 20, 65-73.

30- J. S. Yadav, B. S. Reddy, S. Praveenkumar, K. Nagaiah, N. Lingaiah, P. S. Saiprasad, $Ag_3PW_{12}O_{40}$: a novel and recyclable heteropoly acid for the synthesis of 1, 5-benzo-diazepines under solvent-free conditions, Synthesis. **2004**, *2004*, 901-904.

31- M. M. Heravi, S. Sadjadi, H. A. Oskooie, R. Hekmatshoar, F. F. Bamoharram, An Efficient Synthesis of 3H-1, 5-Benzodiazepine Derivatives Catalyzed by Heteropolyacids as a Heterogeneous Recyclable Catalyst, J. Chin. Chem. Soc. **2008**, *55*, 842-845.

32- T. D. Le, K. D. Nguyen, V. T. Nguyen, T. Truong, N. T. Phan, 1, 5-Benzodiazepine synthesis via cyclocondensation of 1, 2-diamines with ketones using iron-based metal-organic framework MOF-235 as an efficient heterogeneous catalyst. J. Catal., **2016**, 333, 94-101.

33- N. K. Terrett, Combinatorial Chemistry, Oxford University Press, New York, NY, **1998**.

34- J. Zhu, H. Bienayme, Multicomponent Reactions (Ed.), Wiley-VCH, Weinheim, Germany, **2005**.

35- L. F. Tietze, G. Brasche, K. Gericke, Domino Reactions inOrganic Synthesis, Wiley-VCH, Weinheim, **2006**.

36- K. Murai, R. Nakatani, Y. Kita, H. Fujioka, One-pot three-component reaction providing 1,5-benzodiazepine derivatives, Tetrahedron, **2008**, *64*, 11034-11040.

37- L. Banfi, A. Basso, L. Giardini, R. Riva, V. Rocca, G. Guanti, Tandem Ugi MCR/Mitsunobu Cyclization as a Short, Protecting-Group-Free Route to Benzoxazinones with Four Diversity Points, Eur. J.Org. Chem., **2011**, *2011*, 100-109.

38- G. K. Verma, K. Raghuvanshi, R. K. Verma, P. Dwivedi, M. S. Singh, An efficient one-pot solvent-free synthesis and photophysical properties of 9-aryl/alkyl-octahydroxanthene-1, 8-diones. Tetrahedron, **2011**, 67, 3698-3704.

39- M. Ghandi, T. Momeni, M. T. Nazeri, N. Zarezadeh, M. Kubicki, A one-pot three-component reaction providing tricyclic 1, 4-benzoxazepine derivatives, Tetrahedron Lett., **2013**, *54*, 2983-2985.

40- Z. Karimi-Jaberi, M. Barekat M, One-pot synthesis of tri-and tetra-substituted imidazoles using sodium dihydrogen phosphate under solvent-free conditions. Chin Chem Lett., **2010**, *21*, 1183-1186.

41- Y. Han, Y. Sun, J. Sun, CG. Yan, Efficient synthesis of pentasubstituted pyrroles via one-pot reactions of arylamines, acetylenedicarboxylates, and 3-phenacylidene-oxindoles. Tetrahedron, **2012**, *68*, 8256-8260.

42- M. Li, H. Cao, Y. Wang, X-L. Lv, LR.Wen, One-pot multicomponent cascade reaction of N, S-ketene acetal: Solvent-free synthesis of imidazo [1,2-a] thiochromeno [3,2-e] pyridines. Org. Lett., **2012**, 14, 3470-3473.

43- F. Shirini, SS. Beigbaghlou, SV. Atghia, SAR. Mousazadeh, Multi-component one-pot synthesis of unsymmetrical dihydro-5H-indeno [1, 2-b] quinolines as new pH indicators. Dyes Pigm., **2013**, 97, 19-25.

44- K. Murai, R.Nakatani, Y. Kita, H. Fujioka, One-pot three-component reaction providing 1, 5-benzodiazepine derivatives. Tetrahedron, **2008**, 45, 11034-11040.

45- I. Bahammou, A. Esaady, S. Boukhris, R. Ghailane, N. Habbadi, A. Hassikou, A. Souizi, Direct use of mineral fertilizers MAP, DAP, and TSP as heterogeneous catalysts in organic reactions, Mediterr. J. Chem., **2016**, 5, 615-623.

46- S. Sarkar, J. K. R. Deka, J. P. Hazra, A. T. Khan, Bromodimethylsulfonium Bromide (BDMS)-Catalyzed Synthesis of 1, 5-Benzo-diazepines Using a Multi-Component Reaction Strategy. Synlett., **2013**, *24*, 2601-2605.

47- Xiao-Qing Li, Lan-Zhi Wang, Highly efficient one-pot, three-component synthesis of 1, 5-benzodiazepine derivatives. Chin. Chem. Lett. **2014**, 25, 327-332.

18

Zinc-Chromium oxide catalyst for gas-phase ketonisation of pentanoic acid

Mohammed Saad Mutlaq Al-ghamdi [*] and **Hossein Bayahia**

Department of Chemistry, Faculty of Science, Albaha University, Albaha, PO Box 1988, 65431, Kingdom of Saudi Arabia

Abstract: Oxides of Zinc and Chromium mixed catalyst with different atomic ratios of Zinc and Chromium were tested in the ketonisation of pentanoic acid in the gas phase. These catalysts were active to form 5-nonanone, at 300 – 400 °C and ambient pressure. It was found that Zn-Cr with an atomic ratio (10:1) gave the best catalytic performance in comparison with other oxides with higher or lower atomic ratio of Zn and Cr mixed oxides, ZnO and Cr_2O_3. In this test, Zn-Cr (10:1) gave 82% of selectivity for 5-nonanone as the main product at 86% of conversion of the acid at 350°C. The catalyst showed stable performance at the best selected conditions with a small decrease in acid conversion. For catalyst characterization, BET surface area and porosity technique, X-ray diffraction and DRIFTS of pyridine adsorption were used.

Keywords: Ketonisation; Pentanoic acid; Zinc-Chromium oxide catalyst; 5-Nonanone, gas-phase.

Introduction

The conversion of carboxylic acid to ketones, CO_2 and H_2O is known as the kenotic decarboxylation or ketonisation of carboxylic acid [1-6] as shown in the following equation:

$$2RCOOH \longrightarrow R_2CO + CO_2 + H_2O \quad (1)$$

Additionally, the ketonisation reaction is an environmentally benign clean reaction since it gives rise to ketones without and by-product pollutants [7,8].

The first ketonisation reaction was carried out in 1858 for the production of acetone from calcium acetate [8]. Ketonisation of acetic to produce acetone first took place in 1895 as the first industrial application [9].

The Ketonisation reaction is a significant upgrading reaction by converting the biomass and achieving partial deoxygenation. Bio-oil compounds can be formed from fast pyrolysis of biomass and that is an example of upgrading of biomass-derived oxygenates. It is a promising reaction due to its importance in industrial processes of upgrading of biomass-derived oxygenates. It is a promising reaction due to its importance in industrial processes [10].

Stonkus et al.[11] used Cr-Zn-Mn mixed oxide catalyst in the ketonisation reaction of some aliphatic acids in the vapor phase in temperature ranging between 300-400 °C. In this reaction and, under

optimum condition of 325 °C, the maximum yield of ketones was obtained.

Several bulk and supported mixed oxide catalysts were used in the ketonisation of carboxylic acid in both liquid and gas phase reactions. Strong proton sites catalyst such as heteropoly acid catalyst were active in the ketonisation of propionic acid. Ru/TiO_2 catalyst has been tested in the liquid phase ketonisation of carboxylic acids and showed good catalytic activity [8].

Bulk Co-Mo and Co-Mo/ Al_2O_3 catalysts were used in the deoxygenation of propionic acid in the gas phase in the presence of N_2 and H_2. Co-Mo/Al_2O_3 and the catalysts showed a very good performance in the presence of N_2 at 200-400 °C to form 3-pentanone. This catalyst gave around 44% of 3-pentanone and only 16% of 3-pentanone which were obtained in the presence of H_2 at the same conditions [12].

Hexanoic acid ketonisation was carried out using the fermentation of sugars to produce high carbon numbers as diesel bendable as 6-undecanone over magnesium and manganese oxides and zirconia catalysts. It was found that the zirconia catalyst was more active than the other catalysts used. It gave 92% of ketone selectivity at 72% of acid conversion[13]. For pentanoic acid ketonisation to form 5-nonanone over Ceria-zirconia at 350°C, it was seen that the ketonisation reaction was thermodynamically favorable at 250°C. This reaction was carried out using the same catalyst at 350°C and,

Corresponding author: Mohammed Saad Mutlaq Al-ghamdi
Email address: i-73amodi@hotmail.com, hos1397@hotmail.com

it was reported that the catalyst was active in the ketonisation of pentanoic acid to form 5-nonanone with 60% of yield [14].

It has been reported that 90% of 5-nonanone with lower ketones such as 2-hexanone and 3-heptanone were produced from an aqueous γ-valerolactone (GVL) over Ce-Zr catalysts using single reactor. The same reaction was carried out over 0.1% Pd/Nb₂O₅ and this catalyst was selective for producing pentanoic acid [15].

Carboxylic acids such as butanoic, pentanoic and hexanoic acids were tested in the ketonisation reaction over Ceria-Zirconia catalyst in the range of temperature between 548 and 623 K. The results showed that the reactivity of ketonisation reaction decreased when increasing the length of the acids carbon chains [16]. These catalysts have also been reported as being active in ketonisation reactions, producing fuels by the upgrading of biomass-derived intermediates [17].

Crystalline silicalite zeolite[18] and Zn-Cr [19] and its support have been used in the ketonisation of acetic and propionic acids in the gas phase at 300-450 °C [18,19]. Strong proton site catalyst such as heteropoly acid was an active catalyst in the ketonisation of propionic acid in the gas phase [20].

In the current work, ketonisation of pentanoic acid was carried out using 0.2 g of zinc-chromium mixed oxide with different atomic ratio of Zn/Cr which was prepared by co-precipitation method in the range of 300-400 °C, 20 ml min⁻¹ of N₂ flow and 2 vol% of acid.

Experimental Section

Chemicals and materials
The chemical and solvent used in this research work were purchased from sigma Aldrich with no further purification. For the reaction and calibration, pentanoic acid (99.0%), 5-nonanone (>99.0%), $Zn(NO_3)_2.6H_2O$, $Cr(NO_3)_3.9H_2O$, $Cu(NO_3)_2.3H_2O$ were used.

Catalyst synthesis
Pure oxides of chromium, zinc and mixed oxides of zinc and chromium oxides were prepared by co-precipitation method as previously discussed in literature [19,21-23]. In this method, chromium and zinc nitrates were dissolved in distilled water. After that, 10 wt% aqueous ammonia was added to 0.2 M of mixture metal nitrates solution. The mixture was stirred at 70 °C until pH = 7 was achieved [18,21-23]. At 70 °C, after 2-3 hours the result was filtered and washed with distilled water until it became ammonia-free. The presence of ammonia will contaminate the system at the expense of the desired product. Next, the obtained material was air-dried at 120 °C overnight. Finally, the precipitated material was calcined at 300 °C for 3h in the presence of N₂ flow [24].

Catalyst characterisation
For catalyst characterization, various techniques were used such as nitrogen adsorption-desorption to determine the surface area, X-ray Diffraction and DIFTR- Pyridine adsorption. The procedure for adsorbing pyridine was carried out by grounding the sample of catalyst with KBr as following; firstly, the mixtures were ground and pretreated up to a temperature of 150°C and kept constant for 1 h. For the removal of pyridine, the sample was exposed to pyridine vapor at room temperature for 30 minutes and then heated at 150 °C. Finally, DFTIR spectra were recorded for all samples used in this work [18,24].

Catalyst testing
The reaction of ketonisation of pentanoic acid in the gas phase was carried out using 0.2 g of catalyst, 20 mL/min of N₂ flow, ambient pressure, and 2% vol of acid in the quartz fixed-bed reactor. Prior to the reaction, the amount of catalyst used has been pre-heated to the reaction condition for 1 h. The product was analyzed by using online GC to calculate acid conversion and ketone selectivity and yield. The amount produced of CO₂ was not quantified and not included in the product selectivity calculation.

Result and Discussion

Catalyst characterization
Catalyst surface area and porosity are presented in Table 1. BET results show that pure chromium with the highest surface is in comparison to zinc oxide and zinc-chromium mixed oxides.

Fig.1. shows nitrogen adsorption-desorption isotherm for the best catalyst which is bulk Zn-Cr (10:1) oxide catalyst. This catalyst is mesoporous (between 2 to 50 nm pore diameter) with IV isotherm type. Zn-Cr (10:1) mixed oxide catalyst showed an adsorption isotherm with an H3 hysteresis loop and a mononodal pore size distribution with pore diameter at 94 Å [19].

Table 1. Catalyst characterisation.

Catalyst[a]	S_{BET}[b] [m² g⁻¹]	Pore vol.[c] [cm³g⁻¹]	Pore size[d] [Å]
Cr₂O₃	241	0.28	45
Zn-Cr(1:30)	230	0.35	56
Zn-Cr(1:20)	227	0.36	54
Zn-Cr(1:10)	216	0.33	57
Zn-Cr(1:6)	196	0.35	56
Zn-Cr(1:1)	140	0.13	71
Zn-Cr(10:1)	46	0.11	89
Zn-Cr(20:1)	15	0.02	85
Zn-Cr(30:1)	12	0.02	88
ZnO	14	0.03	97

[a] Zinc chromium fresh catalyst, calcination condition (300°C and N₂ flow for 3 h) ; [b] BET surface area; [c] single point total pore volume; [d] average BET pore diameter.

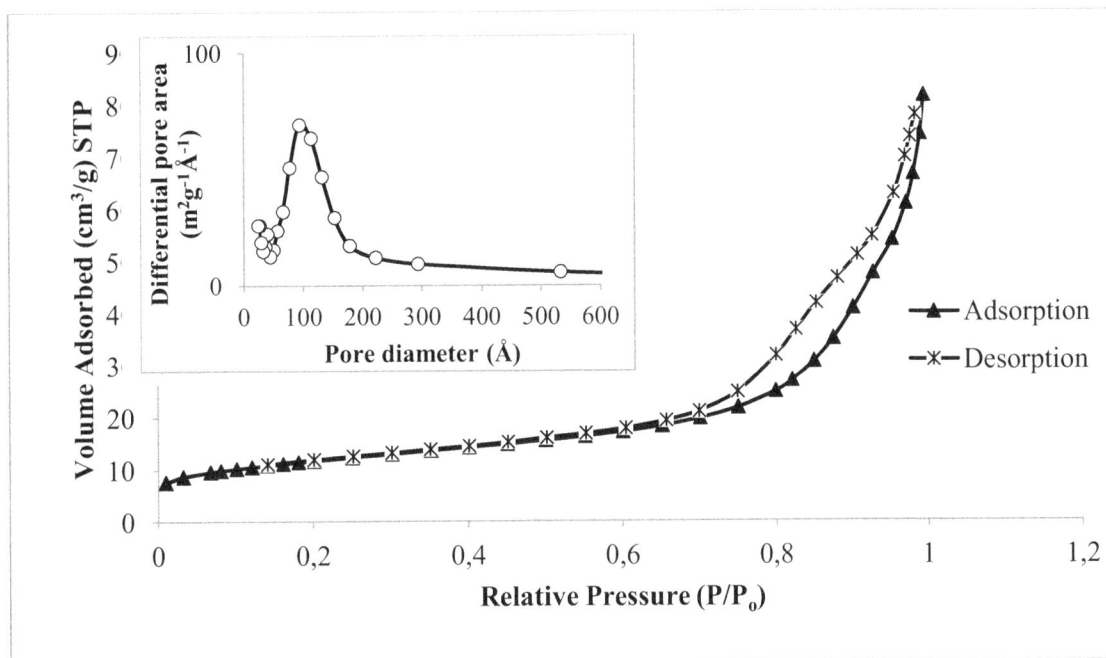

Figure 1. Fresh Zn-Cr (10:1) nitrogen adsorption and desorption isotherm.

X-ray diffraction of zinc-chromium catalyst

X-ray diffraction for pure zinc and chromium and zinc-chromium mixed oxide catalysts calcined at 300 °C in the presence of N_2 shows that rich zinc catalyst is crystalline. However, rich chromium catalyst is amorphous as shown in Fig. 2.

It has been reported by Hossein Bayahia et. al.[19] that Zn-Cr mixed oxides calcined at > 350 °C exhibited Cr_2O_3 and $ZnCr_2O_4$ spinel crystalline phase.

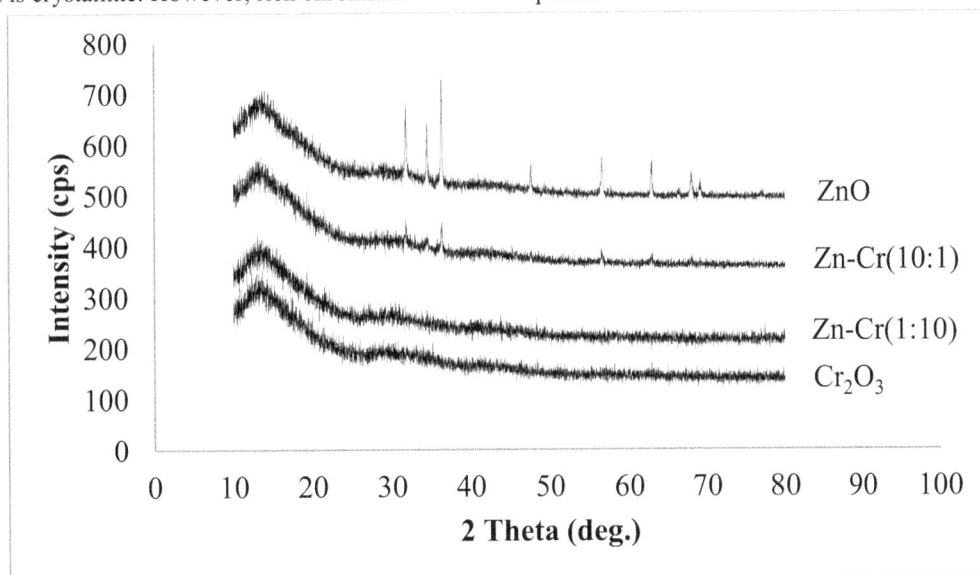

Figure 2. X-ray diffraction for the zinc-chromium catalyst.

Catalyst acidity

The acidity of zinc-chromium catalyst has been measured by DFTIR study of pyridine adsorption as shown in Fig. 3. The bands at 1540 and 1450 cm^{-1} are attributed to BrØnsted and Lewis acid sites respectively [19,24]. Moreover, the band at 1490 cm-1 is attributed to the adsorption of pyridine on BrØnsted and Lewis acid sites together as well as Hydrogen-bonded pyridine. All catalysts of Cr_2O_3, Zn-Cr (10:1) and ZnO with strong bands at 1450 cm^{-1} have Lewis acid sites. However, only pure chromium and Zn-Cr (10:1) mixed oxides have a small quantity of BrØnsted acid sites in their spectra, while, ZnO has no BrØnsted acid sites (Fig. 3). Both Lewis and BrØnsted acid sites are responsible for the ketonisation of pentanoic acid to for 5-nonanone as IR patterns for pyridine adsorption show (Fig. 3) and this in agreement with the literature [25]. This catalytic property can affect the catalytic activity.

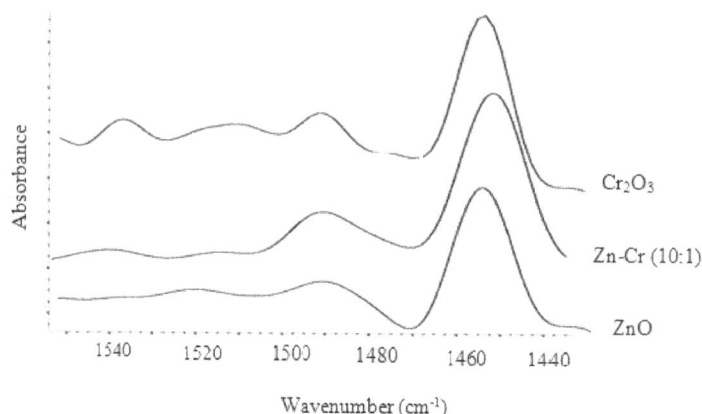

Figure 3 DRIFT spectra of pyridine adsorption for zinc-chromium catalysts.

Catalyst performance

Table 2 illustrates catalytic ketonisation of pentanoic acid over zinc-chromium catalysts. The reaction was carried out at 350 °C, with 20 mLmin⁻¹ flow of N₂, 0.2 of catalyst and 2 % vol of acid used. All catalysts were active, but zinc-chromium (10:1) catalyst was the most active one. The catalyst gave 82% of selectivity for 5-nonanone at 86% conversion (70.52% yield) with only 4% of hydrocarbons produced. These hydrocarbons might be C1-C4 or higher alkanes and alkenes [15] obtained by acid-catalysted craking of 5-nonanoe [19]. CO and CO₂ were not monitored and the unknown remaining might be some deoxygenates products [19]. It can be noted that zinc and chrome ratio play a role in ketonisation of pentanoic acid. The acid conversion and ketone selectivity are slightly decreased by increasing the ration of Zn to Cr and this is in agreement with previous report [19].

Table 2. Ketonisation of pentanoic acid over 0.2 g Zinc-Chromium mixed oxide catalysts at 350°C, 1 bar, 2 vol%20 mLmin⁻¹ N₂, 4.0 h g mol⁻¹ space-time, and 4 h TOS.

Catalyst	Conversion (%)	Selectivity (%)		
		5-Nonanone	Hydrocarbons	Unknown
Cr₂O₃	82	45	9	46
Zn-Cr(1:30)	76	60	11	29
Zn-Cr(1:20)	75	64	13	23
Zn-Cr(1:10)	71	58	9	38
Zn-Cr(1:6)	68	51	7	42
Zn-Cr(1:3)	82	42	11	47
Zn-Cr(1:1)	78	49	9	42
Zn-Cr(10:1)	86	82	4	14
Zn-Cr(20:1)	84	52	9	39
Zn-Cr(30:1)	79	61	7	32
ZnO	84	56	11	33

Table 3 Zinc-Chromium (10:1) oxide catalyst performance in pentanoic acid ketonisation, 1 bar, 20 mLmin⁻¹ N₂, 0.2 g of catalyst, 2 vol% of acid, 4.0 h g mol⁻¹, and 4 h TOS.

Temperature, °C	Conversion (%)	Selectivity (%)		
		5-nonanone	Hydrocarbons	Unknown
300	3	94	2	4
320	42	87	6	7
350	86	82	4	14
380	88	59	9	32
400	95	15	55	30

Effect of temperature

Table 3 shows the ketonisation of pentanoic acid in the range of temperature (300-400 °C). At 300 °C, only 3% of pentanoic acid was converted to 5-nonanone. However, the conversion of acid increased with increasing the temperature from 300 to 400 °C. At 350 °C the catalyst showed the best performance. Increasing the temperature to 400°C, acid conversion reached 95%, but 5-nonanot selectivity decreased to only 15% and the highest amount of hydrocarbons formed by cracking. Generally, ketonisation reaction was affected by temperature. Increasing the temperature increased the activity, and it can be seen that the selectivity of ketone was reduced.

Activation Energy for ketonisation of pentanoic acid over Zinc-Chromium oxide (10:1)

Activation energy for the pentanoic acid over 0.2 g zinc-chromium catalyst (10:1) at 300-400 °C under 20 mL/min of N_2 flow and 2 vol % of acid was measured as shown in fig. 4. The activation energy is 100 kJ mol^{-1}

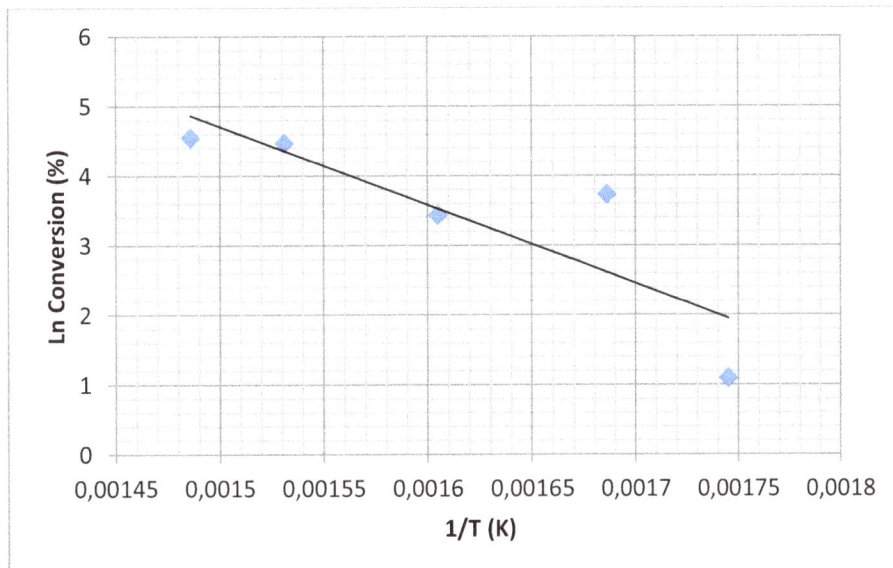

Figure 4. Arrhenius plots of ketonisation of pentanoic acid over Zinc-Chromium (10:1) oxide (300-400°C, 1 bar pressure, 0.2 g catalyst, 20 mL min^{-1} N_2 flow rate and 2 vol % of pentanoic acid).

Stability of Zinc-Chromium catalyst for ketonisation of pentanoic acid

Zinc-Chromium (10:1) oxide showed the best performance at 350 °C, 20 mL of N_2 flow and 20 vol% of pentanoic acid in the ketonisation of pentanoic acid. In this case, the catalyst was tested for at least 20 hours' time on stream. The catalyst reached the steady state from first hour TOS. The

catalyst was stable with a small decrease in acid conversion. After 18 hours, pentanoic acid conversion decreased from 86 to 73% as shown in Fig. 5. The decrease in catalytic performance with time could be due to the deposition of coke. The catalyst was regenerated by heating under oxidizing conditions. The regenerated catalyst showed similar activity as the fresh catalyst.

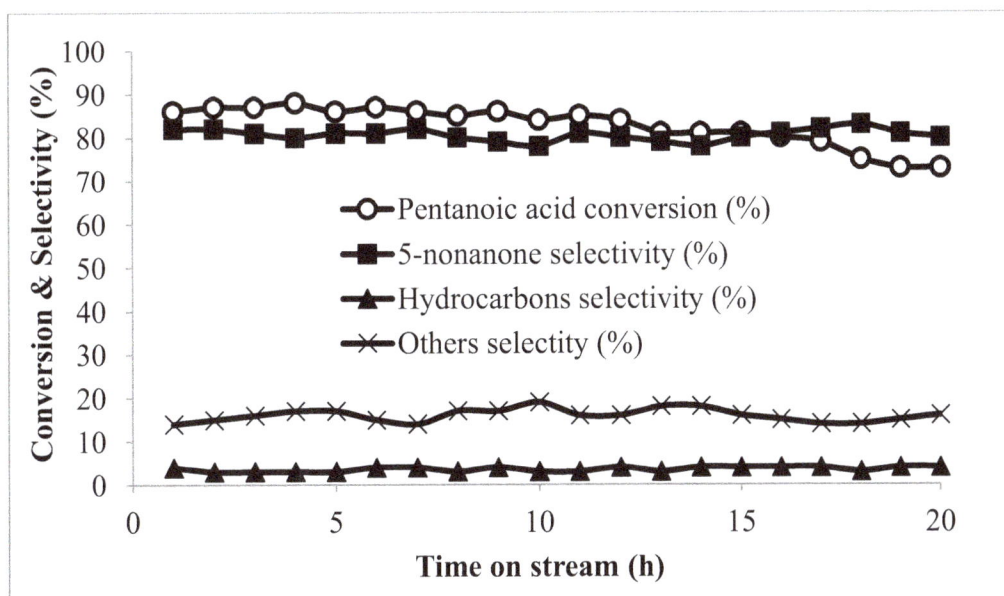

Figure 5. Ketonisation of pentanoic acid on Zinc-Chromium (10:1) oxide (350°C, 1 bar pressure, 0.2 g catalyst, 2 vol % of acid, 20 mL min^{-1} N_2 flow rate and 4.0 h g mol^{-1} space-time).

This work has demonstrated that zinc-chromium oxide catalysts are active and durable for the

ketonisation of pentanoic acid in the gas phase at 300-400 °C in the flow of N_2 and at ambient pressure. Zinc-chromium (10:1) oxide catalysts showed the best catalytic activity with around 71% of 5-nonanone yield. This catalyst was stable for 20 h TOS with small deactivation due to the coke deposition on the catalyst surface. BET, XRD, DRIFTS pyridine adsorption techniques were used for the characterization of the catalysts used in this work.

Acknowledgment

The authors would like to thank Albaha University, Albaha, Saudi Arabia for providing facilities at Physical Chemistry Laboratory in the Faculty of Science at the University City in Alaqeeq. Thanks are due to The University of Liverpool for providing GC equipment. Finally, sincere thanks to Mr. Joel Mcglone, PhD student at the University of Liverpool for proofreading this article.

References

1- A. Corma, S. Iborra and A. Velty, Chem. Rev., **2007**, 107, 2411-2502.

2- E. L. Kunkes, D. A. Simonetti, R. M. West, J. C. Serrano-Ruiz, C. A. Gärtner and J. A. Dumesic, Science, **2008**, 322, 417-421.

3- M. Snåre, I. Kubičková, P. Mäki-Arvela, K. Eränen and D. Y. Murzin, Ind. Eng. Chem. Res., **2006**, 45, 5708-5715.

4- H. Bernas, K. Eränen, I. Simakova, A. R. Leino, K. Kordás, J. Myllyoja, P. Mäki-Arvela, T. Salmi and D. Y. Murzin, Fuel, **2010**, 89, 2033-2039.

5- P. T. Do, M. Chiappero, L. L. Lobban and D. E. Resasco, Catal. Lett., **2009**, 130, 9-18.

6- J. G. Immer, M. J. Kelly and H. H. Lamb, Appl. Catal. A, **2010**, 375, 134-139.

7- M. Renz, Eur. J. Org. Chem., 2005, 979-988.

8- T. N. Pham, T. Sooknoi, S. P. Crossley and D. E. Resasco, ACS Catal., **2013**, 3, 2456-2473.

9- A. D. Murkute, J. E. Jackson and D. J. Miller, J. Catal., **2011**, 278, 189-199.

10- T. Pham, D. Shi and D. Resasco, Top. Catal., **2014**, 57, 706-714.

11- W. Stonkus, J. Yuskovets, L. Leite, M. Fleisher, K. Edolfa, I. Liepina, A. Mishnev and A. Shmidlers, Russ. J. Gen. Chem., **2011**, 81, 1523-1528.

12- H. Bayahia, Asian J. Chem., **2016**, 28, 2744-2748.

13- Yunsu Lee, Jae-Wook Choi, Dong Jin Suh, Jeong-Myeong Ha, Chang-Ha Lee, Appl. Catal. A, **2015**, 506, 288-293.

14- Paolo Fornasiero, Paolo Fornasiero, Mauro Graziani, Paolo Fornasiero, Mauro Graziani, Mauro Graziani, Renewable Resources and Renewable Energy: A Global Challenge, Second Edition, **2011**, p. 64.

15- Kai Yan, Yiyi Yang, Jiajue Chai, Yiran Lu, Appl. catal. B, 2015, **179**, 292-304.

16- Christian A. Gaertner, Juan Carlos Serrano-Ruiz, Drew J. Braden, and James A. Dumesic, Ind. Eng. Chem. Res. **2010**, 49, 6027–6033.

17- Ryan W. Snell and Brent H. Shanks, ASC Catal., **2014**, 4, 512−518.

18- H. Bayahia, E. Kozhevnikova and I. Kozhevnikov, Chem. Commun, **2013**, 49, 3842-3844.

19- H. Bayahia, E. F. Kozhevnikova and I. V. Kozhevnikov, Appl. Catal. B, **2015**, 165, 253-259.

20- M. A. Alotaibi, E. F. Kozhevnikova and I. V. Kozhevnikov, Appl. Catal. A, **2012**, 447-448, 32-40.

21- F. Al-Wadaani, E. F. Kozhevnikova and I. V. Kozhevnikov, J. Catal., **2008**, 257, 199-205.

22- F. Al-Wadaani, E. F. Kozhevnikova and I. V. Kozhevnikov, Appl. Catal. A, **2009**, 363, 153-156.

23- Hossein Bayahia, PhD thesis, the University of Liverpool (**2015**).

24- A.M. Alsalme, P.V. Wiper, Y.Z. Khimyak, E.F. Kozhevnikova, I.V. Kozhevnikov, J. Catal., **2010**, 276, 181–189.

25- H. Knözinger, in: G. Ertl, H. Knözinger, F. Schüth, J. Weitkamp (Eds.), Handbook of Heterogeneous Catalysis, vol. 2, Wiley-VCH, **2008**, p. 1154.

19

Synthesis, structural characterization and DNA studies of trivalent cobalt complexes of (2*E*)-⁴*N*-substituted-2-[4-(propan-2-yl)benzylidene] hydrazinecarbothioamide

P. Murali Krishna [1,*], **K. Hussain Reddy** [2,*]

[1] Department of Chemistry, Ramaiah Institute of Technology, Bengaluru-560 054, Karnataka, India

[2] Department of Chemistry, Sri Krishnadevaraya University, Anantapur-515 003, AP, India

Abstract: This paper describes trivalent cobalt complexes of hydrazinecarbothioamides derived from 4-(propan-2-yl) benzaldehyde and substituted thiosemicarbazides $NH_2NHC(S)NHR$, where R = H (**1**), Me (**2**), Et (**3**) or Ph (**4**) have been synthesized and characterized. The prepared ligands and complexes were characterized using various physicochemical techniques viz. elemental analysis, molar conductance, magnetic susceptibility measurements, IR, electronic absorption spectral studies and cyclic voltammetry. The electronic spectra in DMSO solvent and magnetic susceptibility data of complexes reveal that the complexes are diamagnetic with low spin octahedral cobalt(III) complexes. The absorption titration studies revealed that each of these complexes is an avid binder to calf thymus-DNA. The apparent binding constants are in the order of 10^7–10^8 M^{-1}. The nucleolytic cleavage activities of the ligands and their complexes were assayed on pUC18 plasmid DNA using gel electrophoresis in the presence and absence of H_2O_2. The ligands showed increased nuclease activity when administered as cobalt complexes. All the complexes behave as efficient chemical nucleases with hydrogen peroxide activation. These studies revealed that the complexes exhibit both oxidative and hydrolytic chemistry in DNA cleavage.

Keywords: 4-(Propan-2-yl)benzaldehyde; thiosemicarbazones; Co(III) complexes; DNA studies.

Introduction

Thiosemicarbazones (TSCs) are a class of compounds obtained by condensation of thiosemicarbazide with suitable aldehydes or ketones. Designing of novel thiosemicarbazone ligands have been growing interest due to their diverse chelating properties and pharmacological activities viz. antibacterial [1], antifungal [2] antihypertensive, antineoplastic, antiproliferative activity [3-6], anticancer activity [7,8] etc. The biological activity of the ligands is due to the ability to form chelates with transition metal ions bonding with azomethine nitrogen and sulphur. The presence of additional groups makes them potential polydentate ligands. In most complexes thiosemicarbazones behave as bidentate ligands because they can bond to metals through sulphur and the hydrazinic nitrogen atoms [9], although in a few cases they behave as unidentate ligands and bond through only sulphur atom [10]. In some cases, thiosemicarbazones act as a C, N, S donor, forming cyclometallated complexes[11, 12]. The metal complexes of thiosemicarbazones are not only the bioinorganic relevance but also the chemistry of transition metal complexes of the thiosemicarbazones is receiving significant current attention as potent Analytical agents [13-15], Photocatalysts [16,17], intermediates for the synthesis of pharmaceutical, dyes, photographic films, plastic and in textile industry.

It is well established that the transition metal complexes of TSCs are more biologically active than the free ligands, probably due to the increased lipophilicity (which controls the rate of entry into the cell) of the complexes. The presence of metal ions does not only improve upon their biological activities, selectivity, chemical stability, and their usually low water solubility, but also mitigates their side effects [18].

Recently, Pd(II), Pt(II) [19,20], Zn(II), Cd(II) [21] and Cu(II) [9], Cu(I)[10], Ni(II) [22] complexes of (2*E*)-⁴*N*-substituted-2-[4-(propan-2-yl)benzylidene]hydrazinecarbothio-amide have been synthesized, characterized, and found to exhibit strong to moderate biological activities. In the view of these finding and continuation of our work on thiosemicarbazones [9,10,16,17,24-28], we report the

Corresponding authors: P. Murali Krishna, K. Hussain Reddy
Email address: muralikp21@gmail.com, khussainreddy@yahoo.co.in

use of ^{4}N-substituted 2-[4-(propan-2-yl)benzylidene] hydrazinecarbothioamides.

Experimental

Materials and methods

Thiosemicarbazide, 4-methyl-3-thiosemicarba-zide, 4-ethyl-3-thiosemicarbazide, 4-phenyl-3-thio-semi-carbazide and cuminaldehyde (p-isopropyl benzaldehyde) were of reagent grade purchased from Sigma-Aldrich. All other chemicals were of AR grade and used as supplied. The solvents were distilled before use. Calf thymus DNA was purchased from Genie Bio labs, Bangalore, India. The plasmid pUC18 DNA was isolated from E. coli DH5a strains in Lusbria Broth (LB) medium supplemented by ampicillin cells from 5 ml culture by Qiagen column following the manufacturer's protocol.

Physical measurements

Elemental analysis was carried out on a Perkin-Elmer 2400 CHNS elemental analyzer. Magnetic susceptibility measurements were carried out on a magnetic susceptibility balance (Sherwood Scientific,

Cambridge, England), high purity $CuSO_4.5H_2O$ was used as a standard. Molar conductance (10^{-3}M) in DMF at 30±2°C was measured with a CC180 model (ELICO) direct reading conductivity bridge. The electronic spectra were recorded in DMSO with a Shimadzu UV-160A spectrophotometer. FT-IR spectra were recorded in the range 4,000–270 cm^{-1} in KBr discs on a Nicolet protégé 460 IR Spectrometer. The cyclic voltammetric measurements were performed on a Bio Analytical System (BAS) CV-27 assembly equipped with an X-Y recorder. Measurements were made on degassed (N_2 bubbling for 5 min) ligand/complex solutions (10^{-3} M) in DMF and ethanol containing tetrabutylammonium perchlorate (0.1 M) as a supporting electrolyte. The three-electrode system consisted of a glassy carbon (working), platinum wire (auxiliary) and Ag/AgCl (reference). The ^{1}H- and ^{13}C{1H}-NMR spectra were recorded on a Bruker Spectrospin DPX-300 NMR spectrometer at 300.13 and 75.47 MHz, respectively.

Preparation of the thiosemicarbazones

The ligands were prepared (**Scheme 1**) according to published procedure [9].

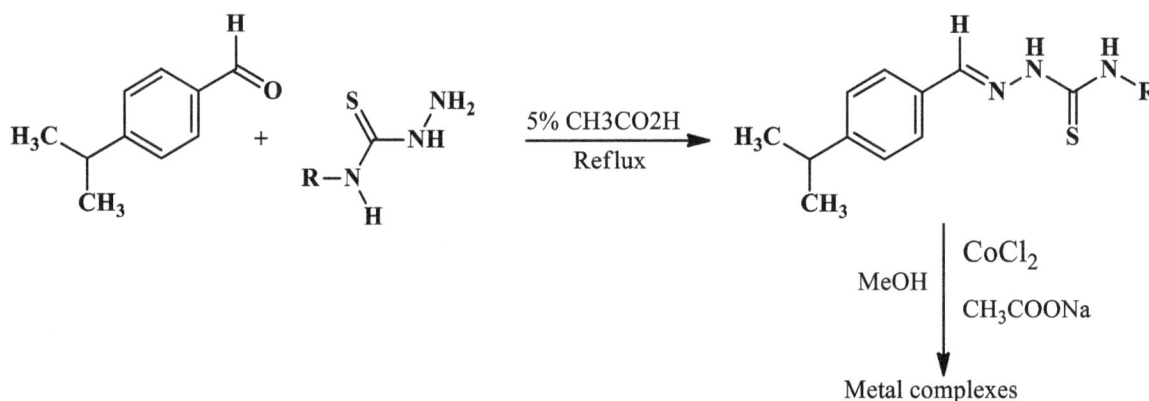

Where R= H (1), CH$_3$ (2), C$_2$H$_5$ (3), C$_6$H$_5$ (4)

Scheme 1. Preparation of the ligands and their metal complexes

Preparation of the complexes

To a methanolic solution of appropriate ligand (2mol) added 1 gm of sodium acetate to maintain pH (8-9) of the solution. Then added a solution of cobalt(II) chloride (1mol) in methanol. The reaction mixture was refluxed for about 1 hr., during which time a solid complex formed was cooled to room temperature, the resulting product was washed with hot water and finally with diethyl ether and dried in vacuum desiccators over anhydrous CaCl$_2$.

DNA binding experiments

A solution of CT-DNA in 0.5mM NaCl/5mM Tris–HCl (pH 7.0) gave a ratio of UV absorbance at 260 and 280 nm (A260/A280) of 1.8–1.9, indicating that the DNA was sufficiently free of proteins [29]. A concentrated stock solution of DNA was prepared in 5 mM Tris–HCl/50 mM NaCl in water at pH 7.0 and the concentration of

CT-DNA was determined per nucleotide by taking the absorption coefficient (6,600 dm^{3}mol^{-1}cm^{-1}) at 260 nm [30]. Stock solutions were stored at 4°C and were used after no more than 4 days. Doubly distilled water was used to prepare buffer solutions. Solutions were prepared by mixing the complex and CT-DNA in DMF medium. After equilibrium was reached (ca. 5 min) the spectra were recorded against an analogous blank solution containing the same concentration of DNA.

The data were then fitted into Eq. 1 to obtain the intrinsic binding constant (Kb) [31]:

$$\frac{[DNA]}{\varepsilon A-\varepsilon B}=\frac{[DNA]}{\varepsilon B-\varepsilon F}+\frac{1}{Kb(\varepsilon B-\varepsilon F)} ----(1)$$

Where [DNA] is the concentration of DNA in base pair, ε_A, ε_B, ε_F corresponds to the molar extinction coefficients of apparent, bound and free metal complexes respectively. A plot of

[DNA]/(ε_A-ε_F) Vs [DNA], gave a slope 1/(ε_B-ε_F) and a Y-intercept equal to 1 / K_b (ε_B-ε_F); K_b is the ratio of slope to the intercept.

Assay of nuclease activity

DMF solutions of the complexes were placed in clean Eppendorf tubes and 1 µg of pUC18 DNA was added. The contents were incubated for 30 min at 37⁰C and loaded on 0.8% Agarose gel after mixing 5 µl of loading buffer (0.25% bromophenol blue + 0.25% Xylene cyanol + 30% glycerol sterilized distilled water). Electrophoresis was performed at constant voltage (100 V) until the bromophenol blue reached to the 3/4th of the gel. The gel was stained for 10 min by immersing in an ethidium bromide solution. The gel was then destained for 10 min by keeping in sterilized distilled water and the plasmid bands visualized by photographing the gel under a UV Transilluminator. The efficiency of DNA cleavage was measured by determining the ability of the complex to form open circular (OC) or nicked circular (NC) DNA from its super coiled (SC) form. The reactions were carried out under oxidative and/or hydrolytic conditions. Control experiments were done in the presence of hydroxyl scavenger, DMSO.

Results and discussion

Characterization of the free thiosemicarbazones and their metal complexes

The detailed characterization of the ligands was reported from the same group [9, 10, 22]. The analytical data of the thiosemicarbazones are given in Table 1.

A Brown colored cobalt(III) complexes of thiosemicarbazones (1-4) are stable at room temperature, non-hygroscopic, sparingly soluble in methanol, ethanol, and readily soluble in chloroform, pyridine, dimethylformamide (DMF) and dimethylsulphoxide (DMSO). The analytical data (Table 1) suggest 1: 3 (M: L) composition for the complexes.

Conductivity and Magnetic susceptibility measurements:

All cobalt complexes are highly soluble in DMF. Therefore, the metal complexes were dissolved in DMF to perform conductivity measurements. The molar conductivities of cobalt complexes in DMF at room temperature are found in the range of 26-46 $Ohm^{-1}cm^{-2}mol^{-1}$ suggest the non-electrolytic nature [32] of complexes. The Magnetic susceptibility of cobalt complexes are presented in Table 1. The data reveal that the complexes are diamagnetic in favour of formation of low spin octahedral cobalt(III) complexes.

Electronic spectra

Typical electronic spectra of complexes 1a and 2a are given in Fig. 1. The electronic spectra of complexes were recorded in DMSO solvent. Spectral data and their assignments are given in Table 2. Two d-d bands are observed in the electronic spectra present cobalt(III) complexes. High energy bands overlap considerably with CT band.

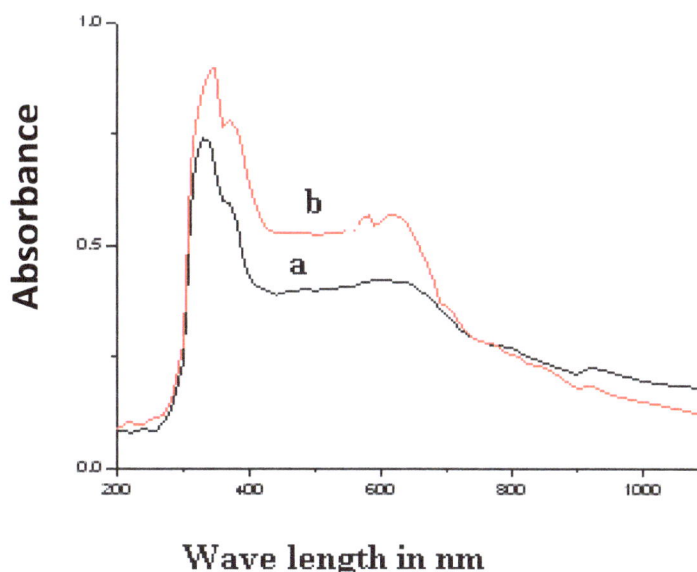

Figure 1. Electronic spectra of (a) complex 1a (b) complex 2a in DMSO

Table 1. The physical and analytical data of the ligand and its metal complexes.

Compound	Yield %	M.P (°C)	Elemental analysis Cal (Found)%				Magnetic susceptibility(χ_g)# $\times 10^{-6}$ #	Λ_M*
			C	H	N	S		
CTH (1)	87	144–147	59.50 (59.70)	6.20 (6.80)	18.9 (18.9)	14.40 (14.50)	-	-
Co(CT)₃ (1a)	82	195-198	55.29 (55.06)	5.58 (5.88)	16.94 (17.51)	12.55 (13.33)	- 0.202	47
CMTH (2)	79	145–147	60.90 (61.20)	7.10 (7.30)	17.4 (17.8)	13.60 (13.60)	-	37
Co(CMT)₃(2a)	86	208-211	55.02 (56.75)	6.18 (6.35)	15.08 (16.54)	11.89 (12.62)	- 0.553	37
CETH (3)	59	115–118	62.90 (62.60)	7.60 (7.70)	16.9 (16.8)	14.10 (14.10)	-	-
Co(CET)₃ (3a)	54	220-222	57.71 (58.27)	6.48 (6.77)	15.02 (15.68)	12.88 (11.97)	- 0.561	26
CPTH (4)	45	140–142	68.50 (68.60)	6.20 (6.40)	13.90 (14.10)	10.70 (10.80)	-	-
Co(CPT)₃(4a)	61	155-157	64.65 (64.60)	5.55 (5.74)	12.34 (13.29)	10.80 (10.15)	- 0.324	38

#cgs units   *Ohm⁻¹ cm-2mol⁻¹

These bands (from low energy to high energy) are assigned to $^1A_{1g}\rightarrow{}^1T_{2g}$ and $^1A_{1g}\rightarrow{}^1T_{1g}$ transitions in favour of octahedral low spin cobalt(III) complexes.

Table 2. Electronic spectral data of cobalt(III) complexes.

Complex	CT transition	$^1A_{1g}\rightarrow{}^1T_{2g}$	$^1A_{1g}\rightarrow{}^1T_{1g}$
1a	30769 (88.4)	24390 (82.9)	**15923 (41.9)**
2a	30303 (74.9)	23809 (40.8)	**16393 (42.1)**
3a	28571 (79.8)	22988 (60.1)	**16129 (45.9)**
4a	**27778 (87.8)**	**25641 (76.3)**	16449 (56.9)

Table 3. Selected I.R. bonds (cm^{-1}) of Cobalt (III) complexes with tentative assignment.

Ligand / Complex	υ(NH$_2$/ NHR)	υ(NH)	υ(C=N)	υ(C=S)	υ(C-S)	υ(Co-S)	υ(Co-N)
1	3412(s) 3280(s)	3156(s)	1590(s)	1181(m)	-	-	-
2a	3416 (s) 3261 (br)	-	1554 (s)	-	692 (s)	349 (s)	**538 (br)**
2	3316(br)	3159(m)	1608(s)	1177(m)	-	-	-
2a	3319 (br)	-	1600 (s)	-	690 (s)	-	**544 (br)**
3	3300(br)	3143(br)	1607(s)	1178(m)	-	-	-
3a	3315 (s)	-	1533 (s)	-	697 (s)	398(s)	**537 (br)**
4	3306(s)	3133(br)	1596(s)	1196(s)	-	-	-
4a	**3315 (s)**	-	**1598 (s)**	-	**664 (s)**	**368 (s)**	**542 (s)**

s=sharp, m=medium, br=broad

Infrared Spectra

Important IR spectral bands of the complexes are compared with the ligands spectra. Important infrared spectral data and their tentative assignments are presented in Table 3. The IR spectra of the ligands show bands in 3412–3280 cm^{-1} region assigned to terminal NH$_2$/NHR group vibrations. These bands are not affected in complexes suggesting non-participation of terminal -NH$_2$/NHR group in coordination. In the spectra of the ligands a strong band is observed in 1177–1196 cm^{-1} region due to υ(C=S) stretching vibration, no band is observed near 2575 cm^{-1} suggesting that the ligands remain in thione form at least in solid state. In IR spectra of complexes this band is disappeared indicating the bond formation between cobalt and enolic sulphur.

This is confirmed by the presence of new band in 664–697 cm^{-1} region assignable to υ(C-S). This band is possible when sulphur binds to metal in the thiol form [33, 34]. A strong band is observed in 1590-1608 cm^{-1} region is due to υ(C=N) stretching vibration of azomethine nitrogen atom in ligands. This band is shifted lower/higher frequencies ($\Delta\upsilon = \pm$ 8-17 cm^{-1}), suggesting coordination of azomethine nitrogen atom in complexvformation [35, 36]. Thus the ligands act as mono anionic bidentate ligands.

Based on the molar conductance, magnetic moment, electronic and I.R. data, it is suggested that all the cobalt complexes are non-electrolytes and have octahedral structure (Fig.2).

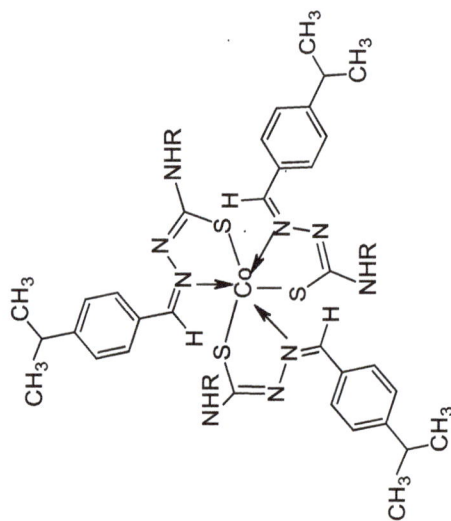

Figure 2. A General and tentative structure for cobalt(III) complexes of ligands.

Electrochemical studies

Electrochemical behavior of cobalt complexes is studied by using cyclic voltammetry. Cyclic voltammograms of the cobalt(III) complexes were recorded in DMF and ethanol in tetra butyl ammonium perchlorate (0.1M) as supporting electrolyte. The electrochemical data of all complexes obtained at the glassy carbon electrode are given in Table 4. Cyclic voltammograms of cobalt(III) complexes showed two active responds in ethanol and DMF medium. In ethanolic medium, two $E_{1/2}$ values are obtained in the potential regions of -1.235 to -1.385 V and -0.545 to -0.715 V vs Ag/AgCl reference electrode.

Table 4. Cyclic voltommetric data of cobalt complexes.

Complex	Redox couple	Ep_c/V		Ep_a/V		ΔEp/mV		$E_{1/2}$		log K_c[a]		$-\Delta G^{o}$ [b]	
		EtOH	DMF	EtOH	DMF	EtOH	DMF	EtOH	DMF	EtOH	DMF	EtOH	DMF
1a	III/II	-0.08	-0.23	-1.01	-0.86	930	630	-0.54	-0.54	0.035	0.052	207	306
	II/I	-1.16	-1.08	-1.31	-1.44	150	360	-1.24	-1.26	0.218	0.090	1285	530
2a	III/II	+0.02	-0.20	-1.19	-0.82	1210	620	-0.58	-0.51	0.027	0.052	159	306
	II/I	-1.18	-1.09	-1.32	-1.46	140	370	-1.25	-1.27	0.233	0.088	1374	518
3a	III/II	-0.20	-0.19	-1.23	-0.94	1030	745	-0.71	-0.57	0.032	0.043	188	253
	II/I	-1.28	-1.08	-1.49	-1.45	210	325	-1.38	-1.26	0.155	0.087	974	513
4a	III/II	+0.10	-0.24	-1.23	-0.97	1240	725	-0.61	-0.61	0.026	0.045	152	265
	II/I	-1.30	-1.10	-1.42	-1.46	120	365	-1.36	-1.28	0.272	0.089	1604	524

[a] log K_c=0.434ZF/RTΔEp, [b] ΔG^{o}= -2.303RT logK_c

These are respectively assigned to Co(II)/ Co(I) and Co(III)/ Co(II) redox couples. Repeated scans as well as various scan rates showed that dissociation does not takes place in these complexes. The non-equivalent current intensity of cathodic and anodic peak difference (ΔEp =120-1240 mV) indicates quasi reversible behaviour of these complexes. The ΔE_p values are greater than the Nernstian values ($\Delta E_p \approx$ 59mV) for one electron redox system. This indicates a considerable reorganization of the coordination sphere during electron transfer has been observed for a number of other cobalt (III) complexes. From Table 4, $E_{1/2}$ values of the complexes in DMF medium are slightly higher than the values obtained in ethanol.

A comparison of the $E_{1/2}$ values of this redox couple of the present complexes with other analogous nitrogen donor macro cycles reveal that these complexes undergo more facile redox change which seem to be a requirement to the DNA cleavage [37].

DNA binding studies

The interaction of cobalt complexes with calf thymus DNA was studied by absorption titrations using spectrophotometer. The absorption titrations were carried out with increasing amount of CT-DNA in 363-367 nm regions. With addition of DNA all cobalt complexes showed hypercromic shift.

Table 5. Electronic absorption data upon addition of CT-DNA to nickel complex.

Complex	$\lambda_{max/nm}$		$\Delta\lambda$ /nm	H (%)	K_b (M^{-1})
	Free	Bound			
1a	365	364	1.0	-25.85	9.58 x 10^7
2a	365.5	364	1.5	-33.29	2.84 X10^7
3a	365	366	1.0	-28.35	4.54 x 10^7
4a	365	366	1.0	-24.92	1.21 x 10^8

From Table 5, it is revealed that in the presence of increasing amount of CT-DNA absorption spectra of complexes show either red-shift or blue-shift ($\Delta\lambda_{max}$:1.0-1.5 nm) and hypercromism [hypercromism: -25.85 % for **1a**, -33.29 % for **2a**, -28.35 % for **3a**, and -24.92 for **4a**. The orders of binding constants of complexes are 2a < 3a < 1a < 4a.

DNA nuclease activity

Gel electrophoresis experiments using pUC18 plasmid DNA were performed with cobalt complexes in presence/absence of H_2O_2 as oxidant. At micro molar concentration for 30 min incubation time all complexes show significant cleavage activity in absence and presence of oxidant. Fig. 3 shows the cleavage pattern of cobalt complexes at physiological conditions. (Compare to lanes 1 and 2, lanes 3-10, shows the linear form in addition to super coiled and nicked forms). To know the cleavage mechanism, control experiments were performed in presence of hydroxyl free radical scavenger, DMSO and singlet oxygen quencher, azide ion (Fig. 4). In the presence of DMSO, azide, there is no significant cleavage activity for complexes. This observation suggests that the complexes produce hydroxyl free radicals (in the presence of oxidant) that cleave DNA.

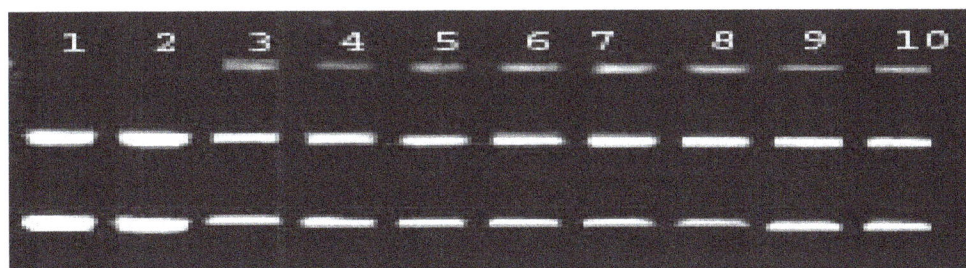

Figure 3. Agarose gel (0.8%) showing results of electrophoresis of 3 μL of pUC18 DNA; 2 μL 0.1M TBE buffer (pH 8); 2 μL complex in DMF (10^{-3}M); 10 μL water, 2 μL H_2O_2 (Total volume 20 μL) were added respectively, incubated at 37 ^{0}C (30 min): Lane 1: DNA control, Lane 2: DNA+H_2O_2, Lane 3:**1a**+DNA, Lane 4: **1a**+DNA+H_2O_2, Lane 5: **2a**+DNA, Lane 6: **2a**+DNA+H_2O_2, Lane 7: **3a**+DNA, Lane 8:**3a**+DNA+H_2O_2, Lane 9:**4a**+DNA, 10.**4a**+DNA+H_2O_2.

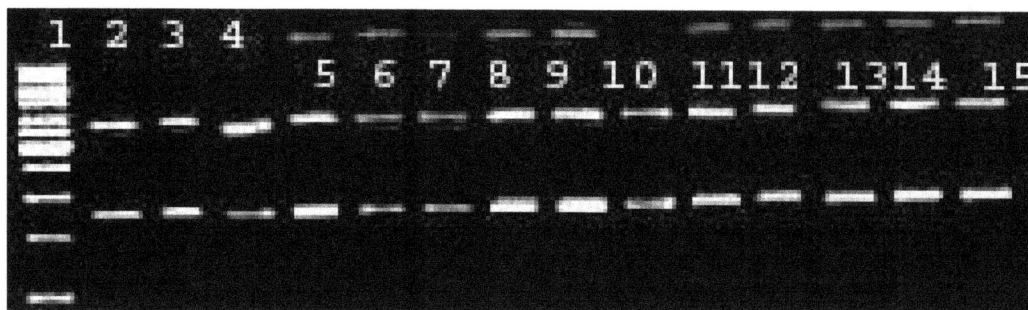

Figure 4. Lane 1: Marker, Lane 2: DNA control; Lane 3: DNA+CoCl₂; Lane 4: 1a+DNA; Lane 5: 1a+DNA+DMSO; Lane 6: **1a**+DNA+azide; Lane 7: **2a**+DNA; Lane 8: **2a**+DNA+DMSO; Lane 9: **2a**+DNA+azide; Lane 10: **3a**+DNA; lane 11: **3a**+DNA+DMSO; Lane 12: **3a**+DNA+azide; Lane 13: **4a**+DNA; Lane 14: **4a**+DNA+DMSO; Lane 15: **4a**+DNA+azide.

References

1-   Elemike E.E., Oviawe A.P. and Otuokere I.E., Potentiation of the Antimicrobial Activity of 4-Benzylimino-2, 3- Dimethyl-1-Phenyl-pyrazal-5-One by Metal Chelation, Res. J. chem. sci., **2011**, 1(8), 6-11.

2-   Opletalová V, Kalinowski DS, Vejsová M, Kunes J, Pour M, Jampílek J, Buchta V, Richardson DR, Identification and characterization of thiosemicarbazones with antifungal and antitumor effects: cellular iron chelation mediating cytotoxic activity, Chem Res Toxicol. **2008**, 21(9), 1878-89.

3-   Laila H Abdel-Rahman, Rafat M El-Khatib, Lobna AE Nassr, Ahmed M Abu-Dief, Fakhr El-Din Lashin, Design, characterization, teratogenicity testing, antibacterial, antifungal and DNA interaction of few high spin Fe (II) Schiff base amino acid complexes, Spectrochimica Acta Part A: Molecular and Biomolecular Spectroscopy, **2013**, 111, 266-276.

4-   Laila H Abdel-Rahman, Rafat M El-Khatib, Lobna AE Nassr, Ahmed M Abu-Dief, Mohamed Ismael, Amin Abdou Seleem, Metal based pharmacologically active agents: synthesis, structural characterization, molecular modeling, CT-DNA binding studies and in vitro antimicrobial screening of iron (II) bromosalicylidene amino acid chelates, Spectrochimica Acta Part A: Molecular and Biomolecular Spectroscopy, **2014**, 117, 366-378.

5-   Laila H. Abdel Rahman, Ahmed M. Abu-Dief, Nahla Ali Hashem and Amin Abdou Seleem, Recent Advances in Synthesis, Characterization and Biological Activity of Nano Sized Schiff Base Amino Acid M(II) Complexes Int. J. Nano. Chem.**2015**, 1(2), 79-95.

6-   Laila H. Abdel-Rahman, Ahmed M. Abu-Dief*, Mohammed Ismael, Mounir A.A. Mohamed, Nahla Ali Hashem, Synthesis, structure elucidation, biological screening, molecular modeling and DNA binding of some Cu(II) chelates incorporating imines derived from amino acids, Journal of Molecular Structure, **2016**, 1103, 232–244.

7-   Laila H. Abdel-Rahman, Ahmed M. Abu-Dief ⇑, Rafat M. El-Khatib, Shimaa Mahdy Abdel-Fatah, Some new nano-sized Fe(II), Cd(II) and Zn(II) Schiff base complexes as precursor for metal oxides: Sonochemical synthesis, characterization, DNA interaction, in vitro antimicrobial and anticancer activities, Bioorganic Chemistry, **2016**, 69, 140–152.

8-   Laila H. Abdel-Rahman, Nabawia M. Ismail, Mohamed Ismael, Ahmed M. Abu-Dief, and Ebtehal Abdel-Hameed Ahmed, "Synthesis, characterization, DFT calculations and biological studies of Mn(II), Fe(II), Co(II) and Cd(II) complexes based on a tetradentate ONNO donor Schiff base ligand," Journal of Molecular Structure, **2017**, 1134, 851–862.

9-   P. Murali Krishna, Hussain Reddy K, Pandey J P, Dayananda Siddavattam DNA binding and Cleavage activity of Binuclear copper(II) complexes of cuminaldehyde thiosemicarbazones, Transition Metal Chemistry, **2008**, 33(5), 661-668.

10- P. Murali Krishna, Hussain Reddy K "Synthesis, single crystal structure and DNA cleavage studies on first 4N-ethyl substituted three coordinate copper(I) complex of thiosemicarbazone", Inorganica Chimica Acta 2009, 362, 4185-4190.

11- D. Pandiarajan, R. Ramesh, Catalytic transfer hydrogenation of ketones by ruthenium (II) cyclometallated complex containing para-chloroacetophenone thiosemicarbazone, Inorganic Chemistry Communications, **2011**, 14 (5), 686-689.

12- 12- Nabanita Saha Chowdhury, Dipravath Kumar Seth, Michael G.B. Drew, Samaresh Bhattacharya, Ruthenium mediated C-H activation of benzaldehyde thiosemicarbazones: synthesis, structure and spectral and electrochemical properties of the resulting complexes, Inorganica Chimica Acta, **2011**, 372, 183-190.

13- N.C Patel,BhaveshA Patel, Spectrophotometric Method for determination of Copper (II) using pChlorobenzaldehyde-4-(2'-carboxy-5-sulphophenyl)-3-thiosemicarbazone [p-CBCST], Research Journal of Chemical Sciences, **2014**, 4(2), 1-6.

14- Tetsumi T., Sumi M., Taraka M. and Shono T, Direct reaction of meta powders with several sodium dithiocarbamates Polyhedron, **1986**, 5(3), 707-710.

15- Buttrus N.H and Mohamed S.M, Synthesis and Characterization of Ni^{+2}, Cu^{+2} and Zn^{+2} complexes with Benzoxazole-2-thionate, Diphenyl Phosphinomethane and Iodine, Research Journal of Chemical Sciences, **2013**, 3(6), 54-59.

16- N.B. Gopal Reddy, P. Murali Krishna, Nagaraju Kottam, Novel metal–organic photocatalysts: Synthesis, characterization and decomposition of organic dyes, Spectrochimica Acta Part A: Molecular and Biomolecular Spectroscopy **2015**, 137, 371–377

17- P. Murali Krishna, N. B. Gopal Reddy, Dr. Nagaraja K, and Yallur B. C, Design and Synthesis of Metal Complexes of (2E)-2-[(2E)-3-Phenylprop-2-En-1-Ylidene] Hydrazinecarbothioamide and Their Photocatalytic Degradation of Methylene Blue, The Scientific World Journal, Volume **2013**, Article ID 828313, 7 pages http://dx.doi.org/10.1155/2013/828313.

18- G. Pelosi, Thiosemicarbazone Metal Complexes: From Structure to Activity, The Open Crystallography Journal, **2010**, 3(2), 16-28.

19- Franco Bisceglie, Silvana Pinelli, Rossella Alinovi, Matteo Goldoni, Antonio Mutti, Alessandro Camerini, Lorenzo Piola, Pieralberto Tarasconi, Giorgio Pelosi, Cinnamaldehyde and cuminaldehyde thiosemicarbazones and their copper(II) and nickel(II) complexes: A study to understand their biological activity Journal of Inorganic Biochemistry, **2014**, 140, 111-125.

20- Quiroga AG, Perez JM, Montero EI, West DX, Alonso C, Navarro-Ranninger C, Synthesis and characterization of Pd(II) and Pt(II) complexes of *p*-isopropylbenzaldehyde N-protected thiosemicarbazones. Cytotoxic activity against *ras*-transformed cells, Journal of Inorganic Biochemistry, **1999**, 75(4), 293-301.

21- Perez JM, Matesanz AI, Martin-Ambite A, Navarro P, Alonso C, Souza P, Synthesis and characterization of complexes of *p*-isopropyl benzaldehyde and methyl 2-pyridyl ketone thiosemicarbazones with Zn(II) and Cd(II) metallic centers. Cytotoxic activity and induction of apoptosis in Pam-*ras* cells, Journal of Inorganic Biochemistry, **1999**, 75(4), 255-261.

22- 22 P Murali Krishna and K Hussain Reddy, "Synthesis, Structural Characterization and DNA Studies of Nickel (II) Complexes of (2*E*)-4*N*-Substituted-2-[4-(propan-2-yl) Benzylidene]hydrazinecarbothioamide Schiff's Bases" Journal of Chemical and Pharmaceutical Research, **2016**, 8(10):61-68.

23- 23. N.B. Gopal Reddy, P. Murali Krishna, S.S. Shantha Kumar, Yogesh P. Patil, Munirathinam Nethaji, "Structure and spectroscopic investigations of a bi-dentate N'-[(4-ethyl-phenyl)methylidene]-4-hydroxybenzohydrazide and its Co(II), Ni(II), Cu(II) and Cd(II) complexes: Insights relevant to biological properties" Journal of Molecular Structure, **2017**, 1137, 543-552.

24- 24. B. S. Shankar, N. Shashidhar, Yogesh Prakash Patil, P. Murali Krishna and Munirathinam Nethaji, (2E)-2-(2-hydroxy-3-methylbenzylidene)-N-methylhydrazine carbothioamide, Acta Cryst. **2013**, E69, o61

25- 25. P. Murali Krishna, B. S. Shankara, N.Shashidhar, Synthesis, characterization and biological studies of binuclear copper(II) complexes of (2E)-2-(2-hydroxy-3-methoxybenzylidene)-^4N-substituted hydrazinecarbothioamides", International Journal of Inorganic Chemistry, vol. 2013, Article ID 741269, 11 pages, **2013**. (doi:10.1155/2013/741269).

26- 26. P. Murali Krishna, K. Hussain Reddy, "Synthesis, characterization, molecular docking and DNA studies of copper(II) complexes of (2*E*)-3-phenylprop-2-enal thiosemicarbazones", Der Pharma Chemica, **2013**, 5 (5), 258-269.

27- 27. P. Murali Krishna, G. N. Anil Kumar, Hussain Reddy K and M. K. Kokila, (2E)-N-methyl-2-(3-phenylpropylidene) hydrazinecarbothioamide, Acta Cryst., **2012**, E68, o2842.

28- 28. P. Murali Krishna, Hussain Reddy K, Pitchika G Krishna & G H Philip "DNA interactions of mixed ligand copper(II) complexes with sulphur containing ligands", Indian Journal of Chemistry, **2007**, 46A, 904–908.

29- 29. Marmur J, A procedure for the isolation of deoxyribonuclcic acid from microorganisms, Journal of Molecular Biology, **1961**, 3(2), 208-218.

30- 30. M. E. Reichmann, S. A. Rice, C. A. Thomas, Paul Doty, A Further Examination of the Molecular Weight and Size of Desoxypentose Nucleic Acid, Journal of American Chemical Soceity, **1954**, *76* (11), 3047–3053.

31- 31. Wolfe A, Chimer GH, Meechan T, Polycyclic aromatic hydrocarbons physically intercalate into duplex regions of denatured DNA, Biochemistry, **1987**, 26(20), 6392-6396.

32- 32 W. J. Geary, "The use of conductivity measurements in organic solvents for the characterisation of coordination compounds," *Coordination Chemistry Reviews*, **1971**, 7(1), 81–122.

33- 33 J. S. Casas, A. Sanchez, J. Sorda, A. Vazquez–Lopez, E. E. Castellano, J. kermann-Schechter, M. C. Rodriguez–Arguelles, U. Russo, Diorganotin(IV) derivatives of salicylaldehydethiosemicarbazone. The crystal structure of dimethyl- and diphenyl-(salicylaldehydethiosemicarbazonato)tin(IV), Inorganica Chimica Acta, **1994,** 216,169-175.

34- 34. J. S. Casas, A. Castineirras, A. Sanchez, J. Sorda, A. Vazquez–Lopez, M. C Rodriguez–Arguelles, U. Russo, Synthesis and spectroscopic properties of diorganotin(IV) derivatives of 2,6-diacetylpyridine bis(thiosemicarbazone). Crystal structure of diphenyl{2,6-diacetylpyridine bis(thiosemicarbazonato)}tin(IV) bis(dimethylformamide) solvate, Inorganica Chimica Acta, **1994,** 221, 61-68.

35- 35. Michel J.M. Campbell, Transition metal complexes of thiosemicarbazide and thiosemicarbazone, Coordination Chemistry Review, **1975,** 15(2-3), 279-319.

36- 36. D. X. West, A. E. Liberta, S. B. Padhye, R. C. Chikte, P. B. Sonawane, A.S. Kumbhar, R. G. Yerande, Thiosemicarbazone complexes of copper(II): structural and biological studies, Coordination Chemistry Review, **1993,** 123 (1-2), 49-71.

37- 37. C. V. Sastri, D. Eswaramoorthy, L. Giribabu, and B. G. Maiya, "DNA interactions of new mixed-ligand complexes of cobalt(III) and nickel(II) that incorporate modified phenanthroline ligands," Journal of Inorganic Biochemistry, **2003,** 94(1-2), 138–145.

20

Inhibition of aluminum corrosion in 0.1 M Na₂CO₃ by *Ricinus communis* oil

Imane Hamdou, Mohamed Essahli and Abdeslam Lamiri

University Hassan 1, Laboratory of applied chemistry and environment, faculty of science and technology, Settat-Morocco

Abstract: The study was carried out using potentiodynamic polarization and electrochemical impedance spectroscopy (EIS) technique. The inhibition efficiency was found around 87 % with 1600 ppm of inhibitor. The efficiency was accentuated with the increase of the concentration while it decreased with the rise of the temperature. Analysis of the polarization curves revealed that the Ricinus communis oil is considered as a mixed inhibitor. The Influence of the temperature was also studied, the values of the activation energy showed that the inhibition occurred by physisorption.

Keywords: Aluminum; corrosion; inhibition; Na₂CO₃; *Ricinus communis* oil

Introduction

Aluminum is currently the focus of many scientific researches in view of its extensive industrial uses, especially in the food industry, which uses Na₂CO₃ as a food additive (E500i). The solubility of aluminum oxide film formed on its surface increases beyond the limits of a pH range from 4 to 9 [1]. The inhibition of materials corrosion by non-toxic organic inhibitors and heavy metal free is an encouraging solution that can effectively replace the use of chemical inhibitors [2-4]. The mechanism action of organic inhibitors is explained by physical and / or chemical adsorption on the metal surface [3, 5-7]. This inhibitive action depends on the physicochemical properties of the inhibitor atoms such as the functional group and the aromaticity as well as the presence of the heteroatoms [8-10], without forgetting the nature of the surface, the temperature and the pressure of the reaction [11].

In light of the good anticorrosive performance of green inhibitors several plant extracts have been used to protect aluminum and these alloys against corrosion in various aggressive media, S.A.Umoren et al [12] used *Raphia hookeri* gum to inhibit The aluminum corrosion in 2M HCl, they found that the addition of 0.5 g / l ensures an efficiency of 56.3% at 30 ° C, E.E. Oguzie [13] found that *Sansevieria trifasciata* leaf extract protects aluminum against corrosion in two aggressive media (2M HCl and 2M KOH), J. Alambek et al [14] studied the effect of the

Lavandula angustifolia extract on Al-3Mg in 3% NaCl, the polarization results show that this extract provides a good efficiency of 99% with a concentration of 20 ppm, G.O.Avwiri et al [15] tested *Vernonia amygdalina* extract on aluminum alloys in two two acidic mediums (0.1M HCl and 0.1M HNO₃), the extract of this plant gave good results in the medium 0.1 M HNO₃. S. Geetha et al [16] evaluated the inhibitory efficacy of *Vitex Negundo* leaf extract in 1M NaOH, based on the polarization results a concentration of 1.5g / l of the extract ensures an efficacy of 78% at 30 ° C.

The oil of *Ricinus communis* is well known for its benefits and uses in cosmetics, in this work we will concentrate on the evaluation of its ability to inhibit aluminum corrosion in 0.1M Na₂CO₃.

Experimental Section

Extraction of *Ricinus communis* oil:
The oil was obtained by the soxhlet method for 3 hours using hexane as a solvent. The oil samples were stored and protected from light. The analysis of the chemical composition was carried out by high performance liquid chromatography coupled with mass spectrometry.

Polarization and Impedance
The electrochemical tests were carried out in a three electrodes cell with a saturated calomel reference electrode and platinum electrode as an

Corresponding author: Imane Hamdou
Email address: imane.hamdou91@gmail.com

auxiliary. The working electrode was an aluminum disk with a surface area of 1 cm². The polarization tests were carried out in a potential range between -1600 mV and -1200 mV with a scanning speed of 0.5 mV / sec. The corrosion inhibiting efficiency was calculated by equation (1), where I'_{cor} and I_{cor} represent respectively the current density with and without inhibitor.

$$E(\%) = \frac{I_{cor} - I'_{cor}}{I_{cor}} \times 100 \quad (1)$$

The curves were realized by the device potentiostat / galvanostats PGZ100 associated with the software "Volta master 4". Impedance spectroscopy (EIS) measurements were carried out using the same electrochemical system. Frequencies between 100 KHz and 10 Hz were superimposed on the corrosion potential. The inhibitory efficiency by impedance method was obtained by formula (2),

where R_T and R'_T are respectively the charge transfer resistance of aluminum in 0.1 M Na_2CO_3 with and without the presence of the inhibitor.

$$E(\%) = \frac{R_T - R'_T}{R_T} \times 100 \quad (2)$$

Preparation of the solution

The corrosive environment was obtained by dissolving sodium bicarbonate in distilled water. Before each electrochemical test the solutions are prepared.

Results and Discussion

Analysis of *Ricinus communis* oil

Analysis of the composition of *Ricinus communis* oil in table 1 [17] shows that the extract is very rich in fatty acid and especially in ricinoleic acid with a percentage of 75.03%.

Table 1. The constituents of *Ricinus communis* oil.

Compound Name	Retention time (min)	Percentage (%)
Palmitic acid	10.0	2.55
Stearic acid	12.5	2.68
Oleic acid	12.8	2.72
Linoleic acid	13.5	9.73
Ricinoleic acid	30.5	75.03

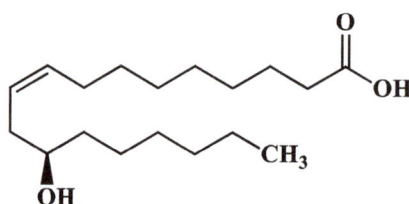

Figure 1. Chemical structure of Ricinoleic acid

Effect of concentration
Polarization

Fig.2 shows the polarization curves with and without addition of the inhibitor. The electrochemical parameters grouped in Table 2 are obtained by extrapolating the linear parts of the anodic and cathodic polarization curves.
The analysis of Table 2 shows that the current intensity decreases from 171.16 µA / cm² to 20.24 µA / cm² in the presence of the inhibitor, which means that the corrosion inhibition of aluminum in Na_2CO_3 is carried out by an adsorption mechanism [18]. The anodic and cathodic slopes of Tafel have greatly

decreased with the gradual addition of the inhibitor, this reduction refers to the mixed type of *Ricinus communis* oil [19]. The polarization curves in the presence and absence of the inhibitor have the same shape, indicating that the mechanisms of the anodic and cathodic reactions do not change. This extract ensures an efficiency of 86, 83 % with a concentration of 1600 ppm. *Ricinus communis* oil also gave good results for other metals such as copper in nitric acid [17] with a 99% efficiency for a concentration of 250 ppm, besides the leaves extract of *Ricinus communis* ensures a good protection of the mild steel in NaCl [20] with an efficacy of 84% for the addition of 300 ppm.

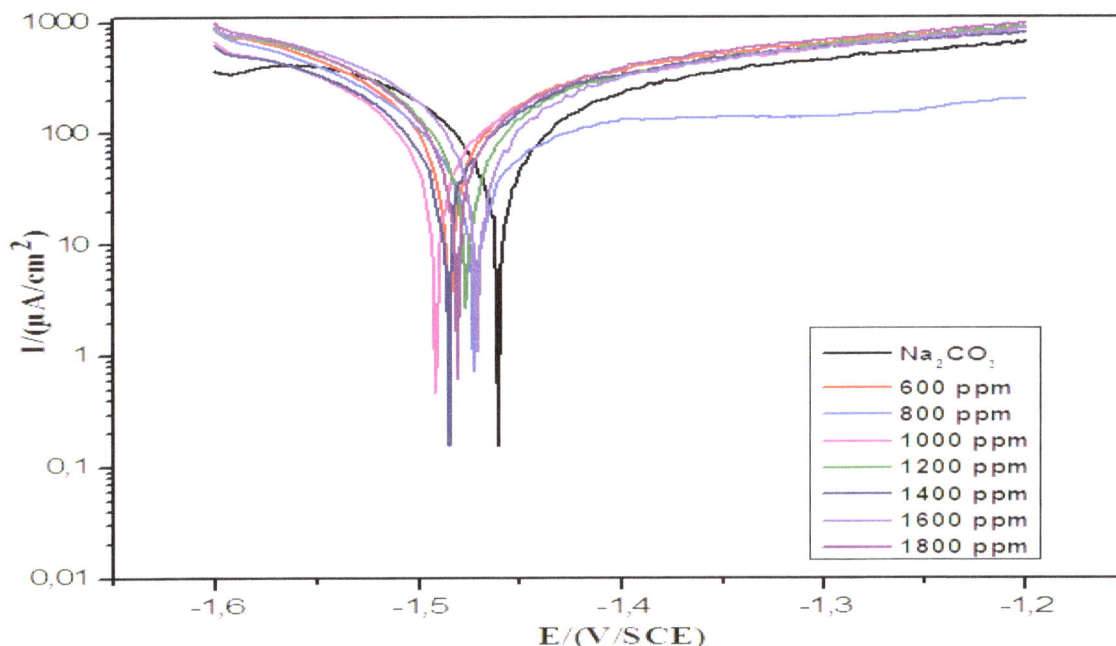

Figure 2. Polarization curves of aluminum in 0.1 M Na_2CO_3 with and without inhibitor at $25 \pm 0.1\,^\circ$ C.

Table 2. Electrochemical parameters of aluminum in 0.1 M Na_2CO_3 with and without addition of *Ricinus communis* oil at different concentrations.

C_{inh} (ppm)	E_{cor} (mV/SCE)	I_{cor} (μA/cm^2)	β_a (mV)	β_c (mV)	E (%)
Blank	-1462	171.16	353.8	-248.6	------
600	-1487	72.30	87.1	-67.4	57.75
800	-1474	58.31	190.6	-83.9	65.93
1000	-1493	46.09	66.6	-53.8	73.07
1200	-1477	30.55	35.9	-33.1	82.14
1400	-1488	26.40	38.3	-33.3	84.57
1600	-1472	22.52	27.6	-23.3	86.83
1800	-1483	20.24	21.4	-19.9	88.17

Electrochemical impedance spectroscopy

The Nyquist diagrams of aluminum in 0.1 M Na_2CO_3 without and with the addition of the inhibitor at different concentrations are given in Fig. 3. The electrochemical impedance parameters are grouped in Table 3.

Most spectra have the same shape (Fig.3) with only one loop, indicating that geometric blocking is the inhibition mode of the oil [21]. From Table 3 the R_T values increase from 45.51 ohm.cm^2 to 203.7 ohm.cm^2 with the gradual addition of the inhibitor while C_{dl} decreases this is due to a decrease in the local dielectric constant and / or an increase in the thickness of the electric double layer [22] due to the formation of a protective layer on the aluminum surface [23]. In the equivalent electrical circuit given in Fig. 4, with R_1 is the resistance of the electrolyte, R_2 is the charge transfer resistor, R_3 is the resistance of passivation layer, Q_1 double layer capacity and Q_2 Passivation layer capacity.

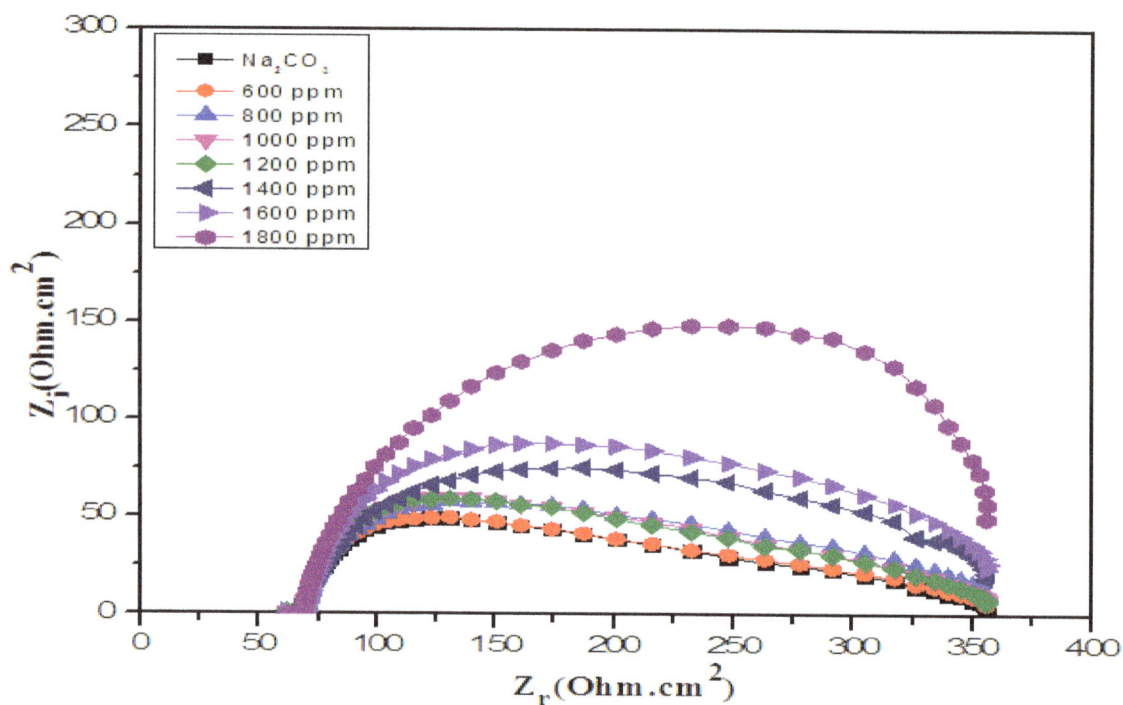

Figure 3. Nyquist curves of aluminum in Na_2CO_3 with and without addition of the inhibitor at $25 \pm 0.1 \,^\circ$ C.

Table 3. Electrochemical parameters of the impedance diagram of aluminum in 0.1 M Na_2CO_3 with and without addition of inhibitor.

C_{inh} (ppm)	R_T (ohm.cm^2)	C_{dl} (μF/cm^2)	E(%)
Blank	46.51	21.62	---
600	113.7	20.49	59.09
800	133.6	18.73	65.18
1000	134.2	18	65.34
1200	139.6	17.63	66.68
1400	173.9	17.49	73.25
1600	203.7	16.66	77.16

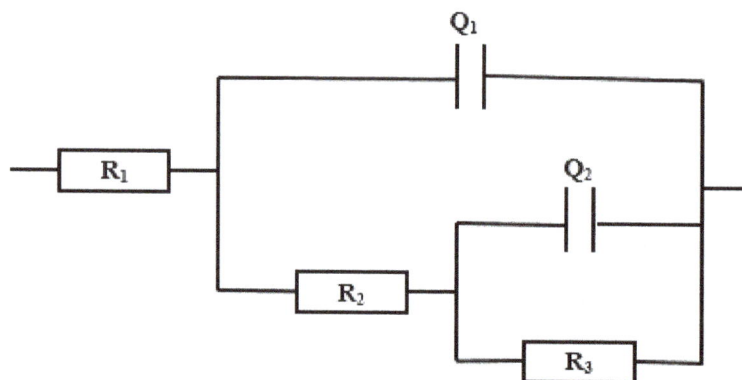

Figure 4. Equivalent circuit model for electrochemical impedance measurements.

Effect of temperature

Polarization

The polarization curves at different temperatures with and without the addition of the inhibitor are given in Figs. 5 and 6. Table 4 shows the temperature effect on the electrochemical parameters taken from the polarization curves with and without addition of 1600 ppm of *Ricinus communis* oil.

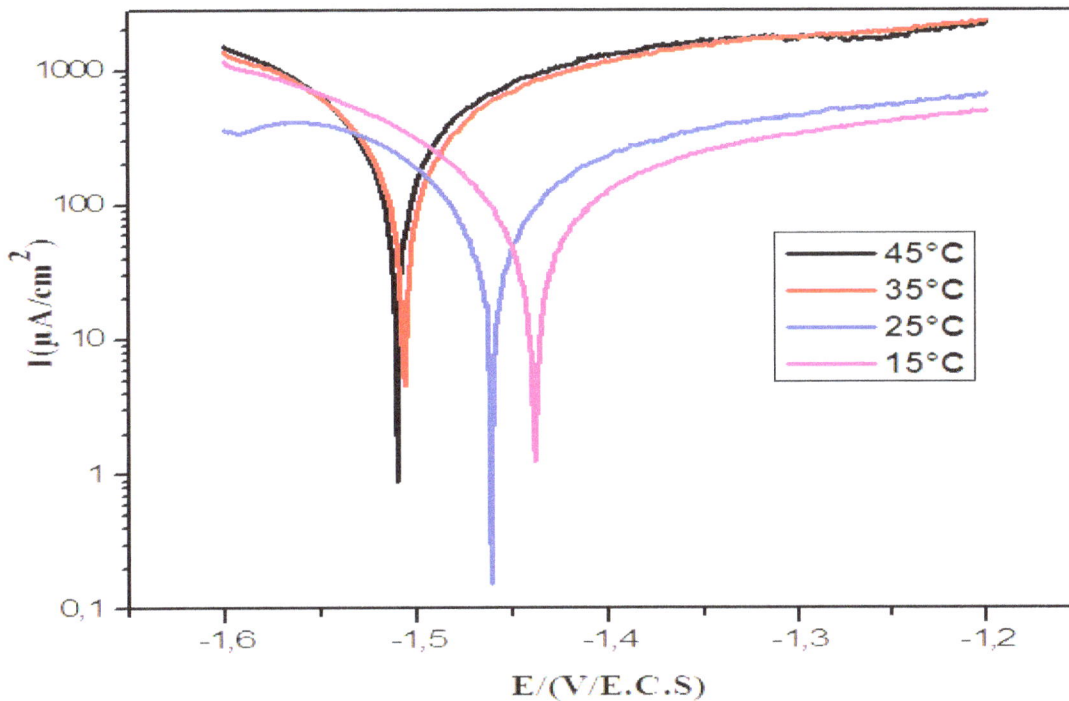

Figure 5. Polarization curves of aluminum in 0.1 M Na$_2$CO$_3$ without inhibitor at different temperatures.

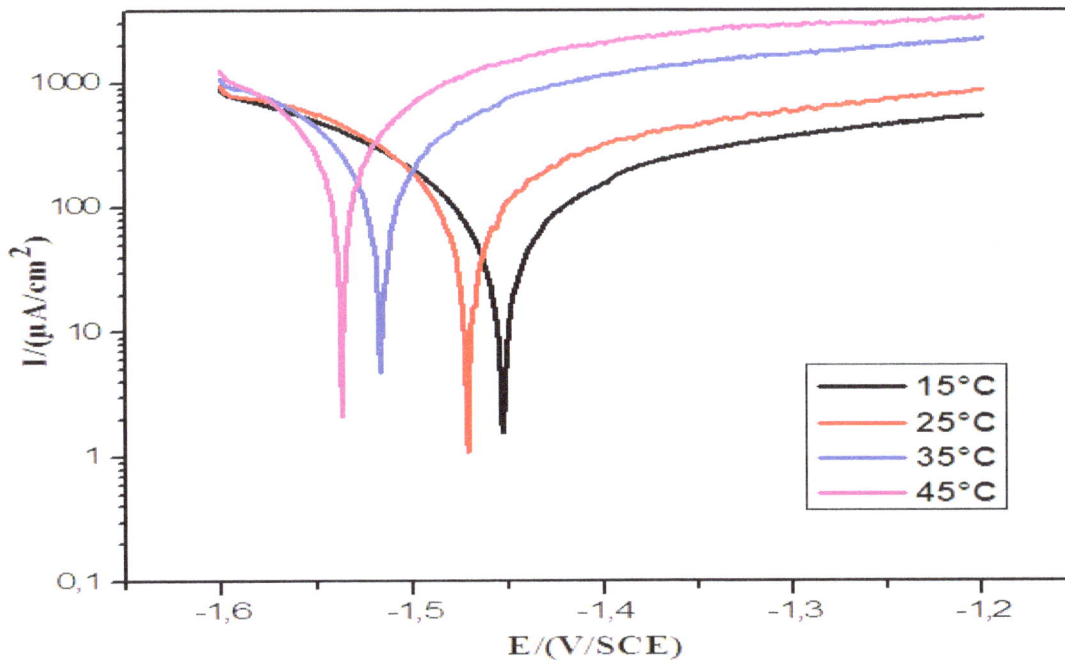

Figure 6. Polarization curves of aluminum in 0.1 M Na$_2$CO$_3$ with the presence of 1600 ppm of the inhibitor at different temperatures.

The results of Table 4 show that as the temperature increases the intensity of current increases with and without addition of 1600 ppm of inhibitor and consequently the inhibition efficiency decreases, this is explained by the physical adsorption of *Ricinus communis* oil molecules on the surface of aluminum [24].

Table 4: Electrochemical parameters of aluminum in a 0.1 M Na$_2$CO$_3$ solution with and without 1600 ppm of the inhibitor at different temperatures.

C$_{inh}$ (ppm)	T(K)	I$_{cor}$ (µA/cm^2)	E$_{cor}$ (mV/SCE)	E (%)
Blank	288	103.32	-1439	----
	298	171.16	-1462	----
	308	197.93	-1509	----
	318	258.03	-1536	----
1600 ppm	288	12.03	-1455	88.34
	298	22.52	-1472	86.83
	308	62.44	-1518	68.44
	318	84.73	-1538	67.15

Electrochemical impedance spectroscopy

The impedance curves obtained at different temperatures with and without the addition of the inhibitor are given respectively in Figs. 7 and 8. The electrochemical impedance parameters are listed in Table 5.

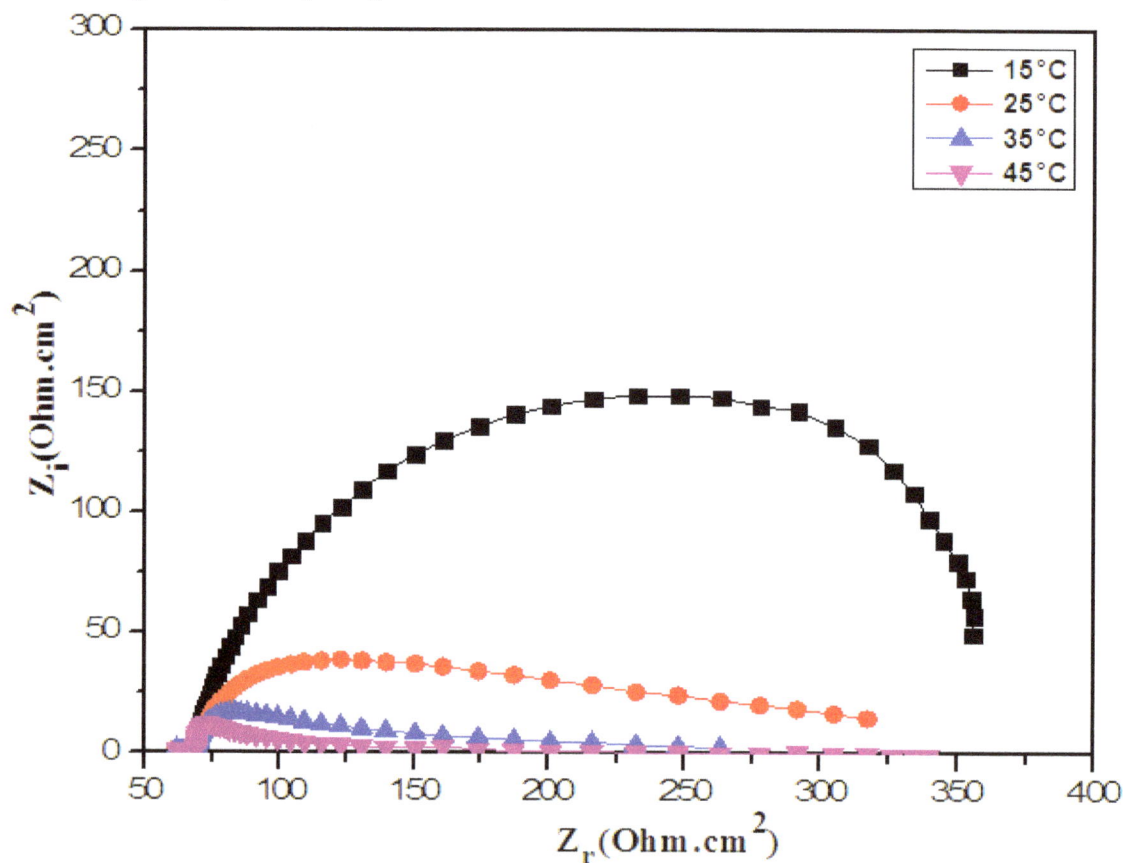

Figure 7. Nyquist curves of aluminum in 0.1 M Na$_2$CO$_3$ without inhibitor at different temperatures.

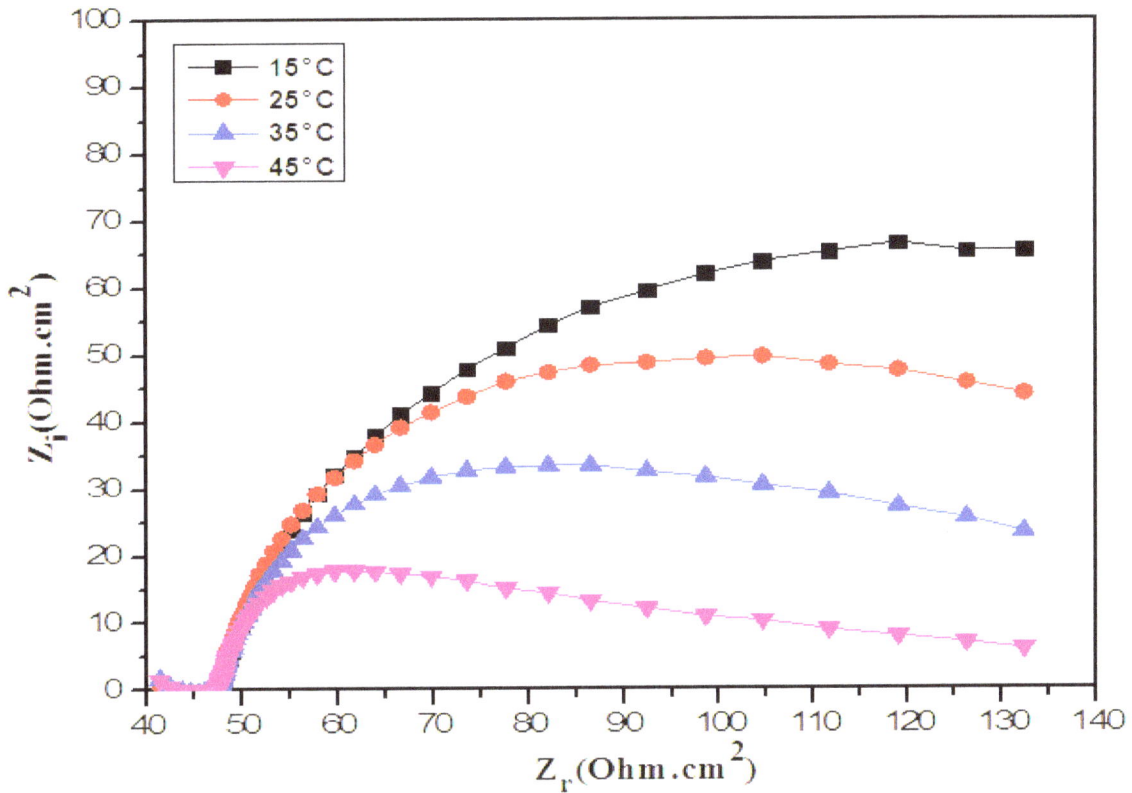

Figure 8. Nyquist curves of aluminum in 0.1 M Na_2CO_3 with 1600 ppm of the inhibitor at different temperatures.

From the inspection of Table 5, the values of the transfer resistance as well as the inhibition efficiency decreased with increasing temperature. These results of impedance correspond to those obtained by polarization ensuring that the increase in temperature negatively influences the inhibition process by *Ricinus communis* oil.

Table 5. Electrochemical parameters of aluminum impedance diagram with and without inhibitor at different temperatures.

C_{inh} (ppm)	T(K)	R_T(ohm.cm^2)	E (%)
Blank	288	55.67	----
	298	46.51	----
	308	39.72	----
	318	22.3	----
1600 ppm	288	353.8	84.26
	298	203.7	77.16
	308	153.9	74.19
	318	74.19	69.94

Determination of activation energy

The activation energy is calculated by equation (4), with K the corrosion rate, A Arrhenius factor and Ea the activation energy. Fig. 9 illustrates the Arrhenius lines in the presence and Absence of the inhibitor. The calculated values of activation energy increase from 21.82 Kj /mol to 52.43 Kj /mol in the presence of the inhibitor, which means that the corrosion inhibition of aluminum in Na_2CO_3 is carried out by physical adsorption [24-27].

$$K = Ae^{-E_a/RT} \qquad (4)$$

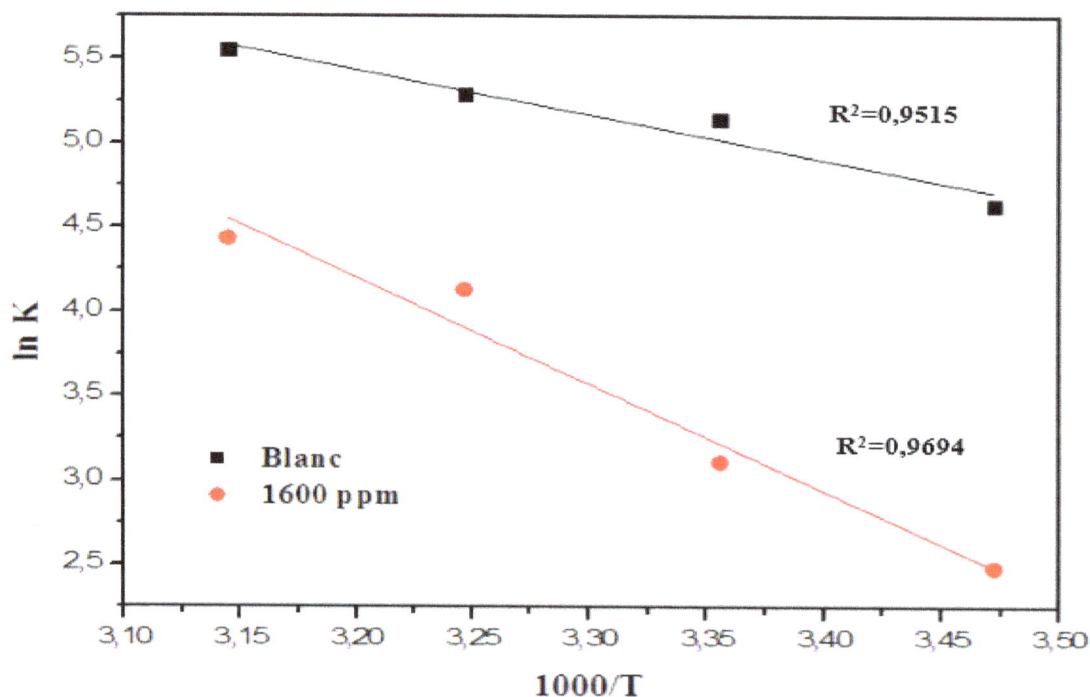

Figure 9. Arrhenius Curves of aluminum in 0.1 M Na_2CO_3 with and without inhibitor

Conclusion

Ricinus communis oil provides a good anticorrosive protection of aluminum in 0.1 M Na_2CO_3, this inhibition is achieved by physical adsorption of its molecules to the aluminum surface. This extract acts as a mixed inhibitor, its inhibitory power increases with the increase of the concentration whereas it decreases with the rise of the temperature.

References

1- K. F. Khaled and M. M. Al-Qahtani, The inhibitive effect of some tetrazole derivatives towards Al corrosion in acid solution: Chemical, electrochemical and theoretical studies, Mater. Chem. Phys, **2009**, 113, 150–158.

2- X. Li and S. Deng, Inhibition effect of Dendrocalamus brandisii leaves extract on aluminum in HCl, H_3PO_4 solutions, Corros.Sci, **2012**, 65, 299–308.

3- S. M. A. Hosseini and A. Azimi, The inhibition of mild steel corrosion in acidic medium by 1-methyl-3-pyridin-2-yl-thiourea,Corros. Sci, **2009**, 51, 728–732.

4- C. M. Goulart, A. Esteves-Souza, C. A. Martinez-Huitle, C. J. F. Rodrigues, M. A. M. Maciel, and A. Echevarria, Experimental and theoretical evaluation of semicarbazones and thiosemicarbazones as organic corrosion inhibitors, Corros. Sci, **2013**, 67, 281–291.

5- W. Li, Q. He, C. Pei, and B. Hou, Experimental and theoretical investigation of the adsorption

behaviour of new triazole derivatives as inhibitors for mild steel corrosion in acid media, Electrochim. Acta, **2007**, 52, 6386–6394.

6- F. Bentiss, M. Lagrenee, M. Traisnel, and J. C. Hornez, The corrosion inhibition of mild steel in acidic media by a new triazole derivative, Corros. Sci,**1999**, 41, 789–803.

7- A. M.Fekry and R. R. Mohamed,Acetyl thiourea chitosan as an eco-friendly inhibitor for mild steel in sulphuric acid medium,Electrochim. Acta, **2010**, 55, 1933–1939.

8- K. F. Khaled, The inhibition of benzimidazole derivatives on corrosion of iron in 1 M HCl solutions, Electrochim. Acta, **2003**, 48, 2493–2503.

9- K. F. Khaled,Experimental, density function theory calculations and molecular dynamics simulations to investigate the adsorption of some thiourea derivatives on iron surface in nitric acid solutions,Appl. Surf. Sci, **2010**, 256, 6753–6763.

10- M. Mehdipour, B. Ramezanzadeh, and S. Y. Arman, Electrochemical noise investigation of Aloe plant extract as green inhibitor on the corrosion of stainless steel in 1M H_2SO_4, J. Ind. Eng. Chem, **2015**, 21, 318–327.

11- A. M. Abdel-Gaber, B. A. Abd-El-Nabey, I. M. Sidahmed, A. M. El-Zayady, and M. Saadawy, Inhibitive action of some plant extracts on the corrosion of steel in acidic media,Corros. Sci., **2006**, 48, 2765–2779.

12- S.A.Umoren, I.B.Obot, E.E.Ebenso, and N.O.Obi-egbedi, The Inhibition of aluminium corrosion in hydrochloric acid solution by exudate gum from Raphia hookeri, Desalination, **2009**, 247, 561–572.

13- E.E.Oguzie, Corrosion inhibition of aluminium in acidic and alkaline media by Sansevieria trifasciata extract, Corros. Sci., **2007,** 49, 1527–1539.

14- J. Halambek, K.Berkovic, and J.Vorkapic-Furac,The influence of Lavandula angustifolia L. oil on corrosion of Al-3Mg alloy, Corros. Sci., **2010**, 52, , 3978–3983.

15- G. O. Avwiri and F.O.Igho, Inhibitive action of Vernonia amygdalina on the corrosion of aluminium alloys in acidic media, Mater. Lett., **2003**, 57, 3705–3711.

16- S. Geetha, S. Lakshmi, and K. Bharathi, Corrosion Inhibition of Aluminium in Alkaline Medium using Vitex Negundo Leaves Extract,Int. J. Adv. Sci. Tech. Res., **2013**, 3, 258–268.

17- Sara Houbairi, Abdeslam Lamiri, Mohamed Essahli,Oil of Ricinus Communis as a Green Corrosion
Inhibitor for Copper in 2 M Nitric Acid Solution, International Journal of Engineering Research & Technology (IJERT), **2014**,3, 698-707.

18- M. N. El-Haddad,Chitosan as a green inhibitor for copper corrosion in acidic medium,Int.J.Biol. Macromol, **2013**, 55,142–149.

19- R.Fuchs-Godec and G.Zerjav,Corrosion resistance of high-level-hydrophobic layers in combination with Vitamin E – (α-tocopherol) as green inhibitor,Corros. Sci., **2015**, 97,7–16.

20- R. A. L. Sathiyanathan, S. Maruthamuthu, M. Selvanayagam, S. Mohanan, and N. Palaniswamy, Corrosion inhibition of mild steel by ethanolic extracts of Ricinus communis leaves, Ind. J. Sci. Technol, 2005, 12, 356–360.

21- A. Khadraoui, A.Khelifa, K. Hachama, R.Mehdaoui, Thymus algeriensis extract as a new eco-friendly corrosion inhibitor for 2024 aluminium alloy in 1 M HCl medium, J. Mol. Liq, **2016**, 214, 293-297.

22- M. Özcan, I. Dehri, and M. Erbil,Organic sulphur-containing compounds as corrosion inhibitors for mild steel in acidic media: Correlation between inhibition efficiency and chemical structure, Appl. Surf. Sci, **2004**, 236, 155–164.

23- K. Boumhara, M. Tabyaoui, C. Jama, and F. Bentiss, Artemisia Mesatlantica essential oil as green inhibitor for carbon steel corrosion in 1M HCl solution: Electrochemical and XPS investigations, J. Ind. Eng. Chem, **2015**, 29, 146–155.

24- M. A. Deyab, Egyptian licorice extract as a green corrosion inhibitor for copper in hydrochloric acid solution,J. Ind. Eng. Chem, **2015**, 22, 384–389.

25- O. K. Abiola and A. O. James, The effects of Aloe vera extract on corrosion and kinetics of corrosion process of zinc in HCl solution, Corros. Sci., **2010**, 52, 661–664.

26- M. Lebrini, F. Robert, A. Lecante, and C. Roos,Corrosion inhibition of C38 steel in 1M hydrochloric acid medium by alkaloids extract from Oxandra asbeckii plant, Corros. Sci., **2011**, 53, 687–695.

27- D. Özkir, K. Kayakirilmaz, E. Bayol, A. A. Gürten, and F. Kandemirli,The inhibition effect of Azure A on mild steel in 1M HCl. A complete study: Adsorption, temperature, duration and quantum chemical aspects, Corros. Sci., **2012**, 56, 143–152.

21

Solubility of phenylboronic compounds in water

Paweł Leszczyński and Andrzej Sporzyński *

Faculty of Chemistry, Warsaw University of Technology, Noakowskiego 3, 00-664 Warsaw, Poland

Abstract: Solubility of six phenylboronic compounds in water was investigated using different methods. The results are consistent with each other, although for particular compounds selected methods should be preferred. The solubility of the investigated compounds is low, with the value of ca. 2 g/100 cm^3 H$_2$O at 20°C for unsubstituted phenylboronic acid. The unsubstituted benzoxaborole is less soluble than phenylboronic acid. Introduction of OiBu, COOH and CF$_3$ groups into the phenyl ring decreases solubility in comparison with unsubstituted phenylboronic acid, especially for the alkoxy substituent.

Keywords: Solubility, solid-liquid equilibrium, boronic, phenylboronic acid, benzoxaborole.

Introduction

Boronic acids are the compounds of growing interest not only for Suzuki coupling reaction [1], but also due to their wide applications in catalysis, medicine and biology [2]. A great interest has been also paid to the supramolecular systems formed by these compounds in which two hydroxyl groups can be involved in various homomeric and heteromeric hydrogen-bonded assemblies, leading to the important compounds applied in materials' chemistry [3]. Recently, rising research activity was directed to benzoxaboroles, internal hemiesters of phenylboronic acids, due to their exceptional properties leading to new applications in biology, medicine and materials' chemistry [4].

The solubility data are crucial for purification and formulation of new biologically active compounds. Many important reactions require the choice of the right solvent and reaction conditions, especially temperature. Suzuki–Miyaura cross-coupling is the most widely used protocol for the formation of the carbon-carbon bond and has become significant method in the synthesis of biaryl compounds, resulting in a wide range of pharmaceuticals. Among the various sustainable media, water is an attractive solvent for this reaction due to its low cost, non-flammability, non-toxicity, and environment friendliness [5].

Taking into account the above-mentioned numerous applications of boronic compounds, there are surprisingly few results of their solubility in water and organic solvents. Even for the known for about 150 years phenylboronic acid, only 2 values of its solubility in water at 0°C and 25°C have been published so far [6]. One can also find some generalized and misleading statements in the literature such as "boronic acids are water soluble" [5] or "since the reactants are insoluble in water" [7]. For benzoxaboroles there are even less information: only general statement like "[benzoxaboroles] show a better solubility profile "without any data given [8].

The aim of the present work is to investigate the solubility of boronic acids and benzoxaboroles in water. Since our earlier research on the solubility of organoboron compounds in organic solvents [9] were focused on different systems, it was necessary to find an appropriate method to determine solubility of these classes of compounds in water. The oldest and frequently used dynamic (synthetic) method with the visual detection of crystals' disappearance [9a] failed for the investigated systems. Several methods (A - C) have been selected and obtained results compared in terms of consistency of obtained data. Advantages and disadvantages of application of the methods in this particular system have been discussed, taking into account possible sources of measurement errors.

Experimental Section

Materials

Phenylboronic acid (**1**) was obtained from Alfa Aesar and was recrystalized from water before use. The following compounds have been synthesized by the procedures described elsewhere: **2** [10], **4** [11]. Compounds **3**, **5** and **6** have been synthesized from the corresponding aryl bromides according to the typical procedure described in [12].

Methods of solubility determination
Method A. Dynamic method with the use of luminance probe [13]

Corresponding author: Andrzej Sporzyński
Email address: *spor@ch.pw.edu.pl*

Mixture of solute and solvent, prepared by weighing pure components, were heated to dissolve the solid. The measurements were conducted using the apparatus described by Hofman et al.[13], in which LED was the source of light and luminance was measured using the LP 471 LUM 2 probe (Photo-

Radiometer HD2102.1). Measured temperatures and light intensities were acquired independently and transformed into the light intensity versus temperature dependence. A typical observed dependence noted for the phenylboronic acid + water system is shown in Figure 1.

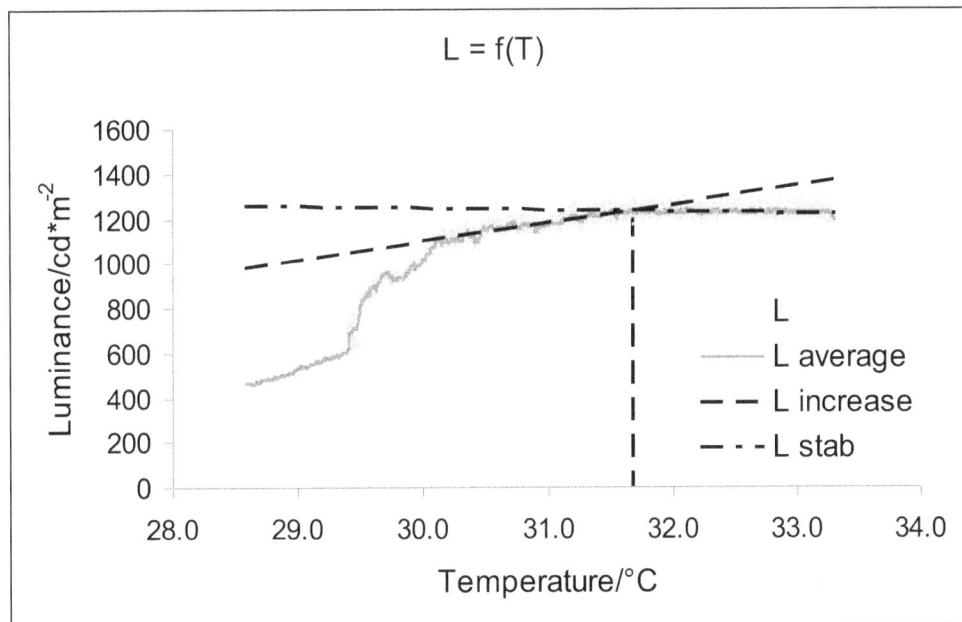

Figure 1. Luminance vs. temperature plot for phenylboronic acid **1** in water.

Method B. Conductometric method [14]

The measurements were conducted by the method originally described by Rogalski et al. for determination of solubility of phenols in water [14]. Weighted amounts of the boronic compound and water were placed in the conductivity sample cell. The cell was designed to minimize vapor volume to avoid the partial vaporization at elevated

temperatures. The amount of solute was in excess with respect to saturation. The solution was stirred for 2 h at a temperature sufficiently high to solubilize all of the solute and cooled to precipitate the solute. The mixture was stirred for 2 h at the measurement temperature before starting conductivity measurements.

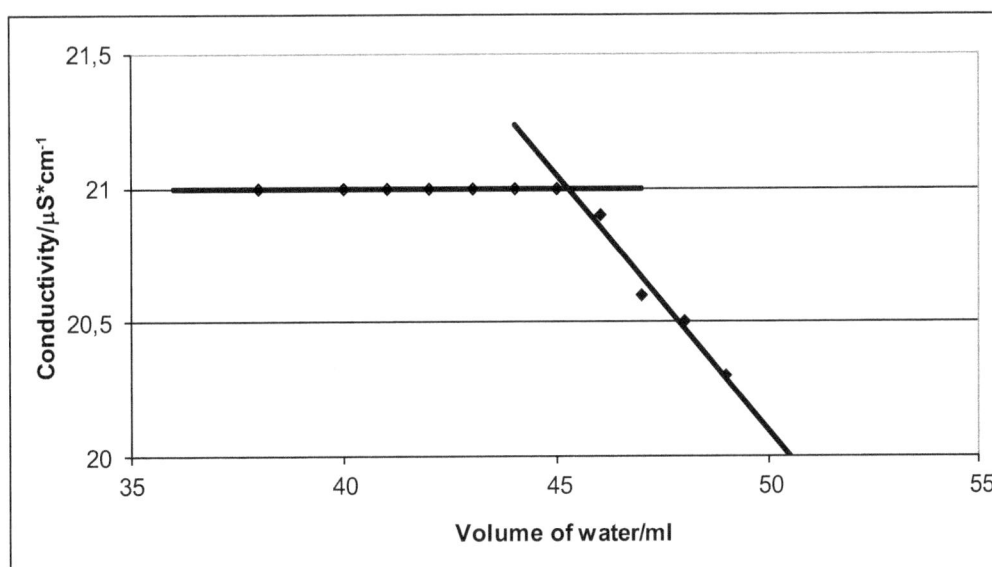

Figure 2. Conductance vs. volume of water plot for the compound **3** at 40°C.

The excess solute was dissolved by injecting known volumes of water (from 0.5 to 2.5 cm³). After each injection, the solution was vigorously stirred for about 20 min to ensure thermal and chemical equilibrium. The stirring was stopped before every measurement. Repeated measurements for each added volume allow to check if the equilibrium was reached (the same conductance values for two measurements). Typical plot of conductance vs. volume of the solvent for a constant temperature is shown in Figure 2.

Method C. Isothermal (static) method

The suspension of solute in water was stirred at constant temperature for about 4 h to reach equilibrium. After sedimentation of the solid a saturated solution was sampled, weighed and evaporated at 90°C to reach a constant mass of the solid. For a given temperature, each determination was repeated to obtain similar results. The presented data are the mean values for each temperature.

Results and Discussion

Solubility of the boronic compounds in water was investigated for the compounds shown in Chart 1.

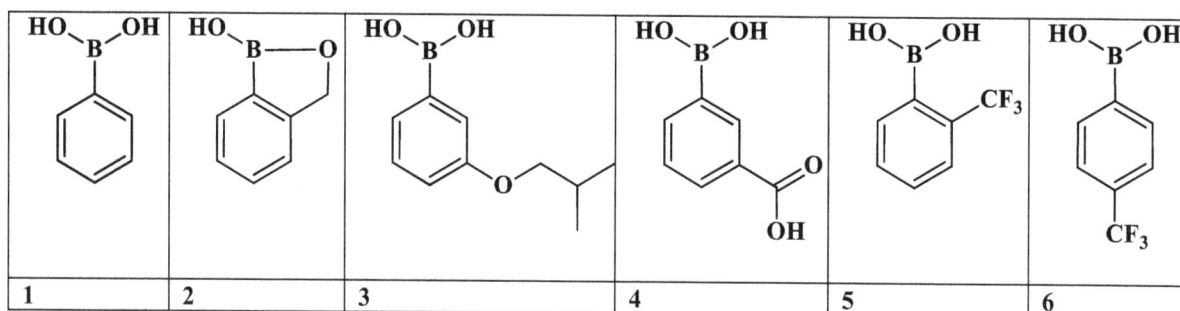

Chart 1. Boronic compounds investigated in the present paper.

Solubility of phenylboronic acid (**1**) determined by methods **A** - **C** is shown in Figure 3.

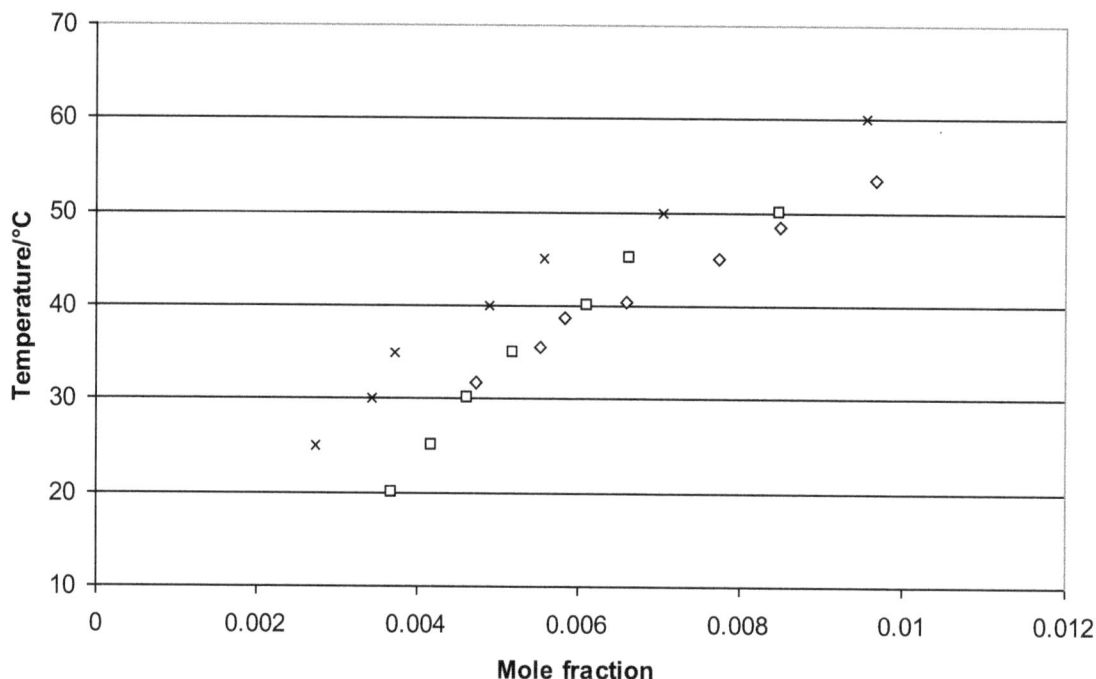

Figure 3. Solubility of phenylboronic acids (**1**) in water. ◊ Method **A**, □ method **B**, ✗ method **C**.

Lower values obtained by method **C** can be caused by the formation of boroxine during evaporation of the solvent at 90°C according to Eq. 1:

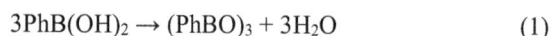

$$3PhB(OH)_2 \rightarrow (PhBO)_3 + 3H_2O \qquad (1)$$

The formation of cyclic anhydrides (boroxines) by boronic acids is a well-known feature of these compounds. The acid/boroxine equilibrium depends on the substituents in the aryl group. DSC measurements of the p-methoxyphenylboronic acid confirm that the fast dehydration starts at about 90°C [15]. However, some compounds dehydrate easily during standing on air at room temperature. DSC curve for unsubstituted phenylboronic acid (**1**) and for the corresponding boroxine is shown in Figure 4.

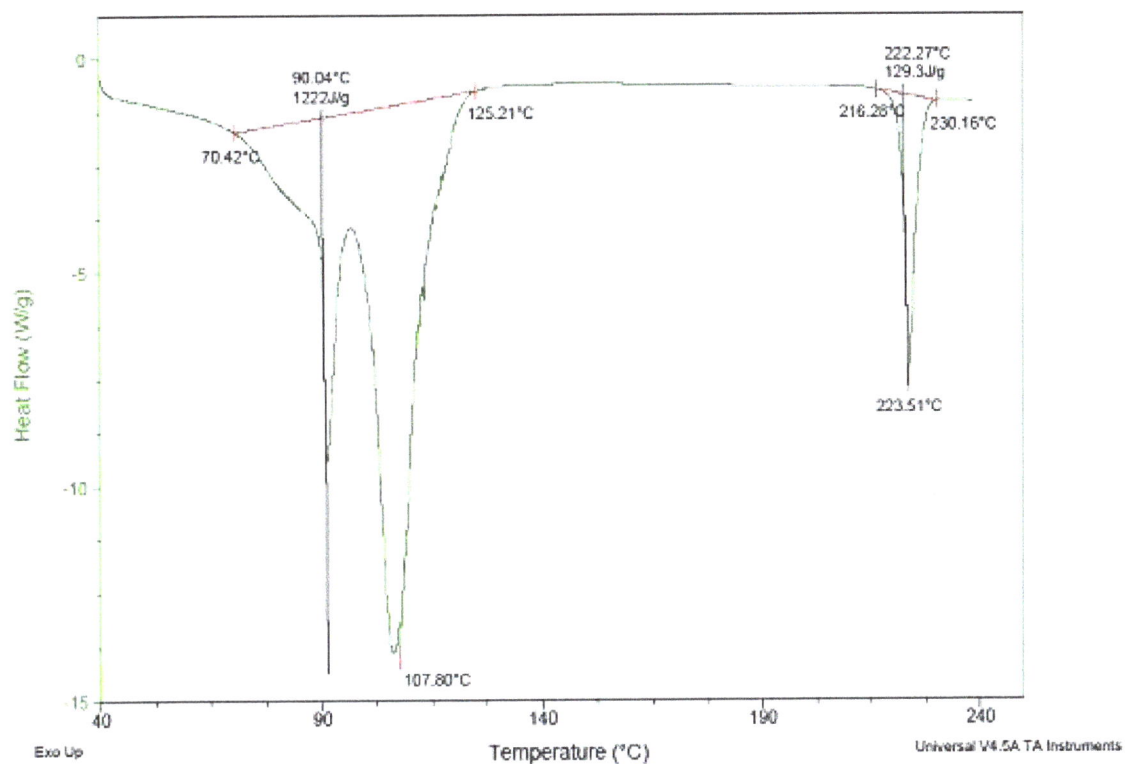

a

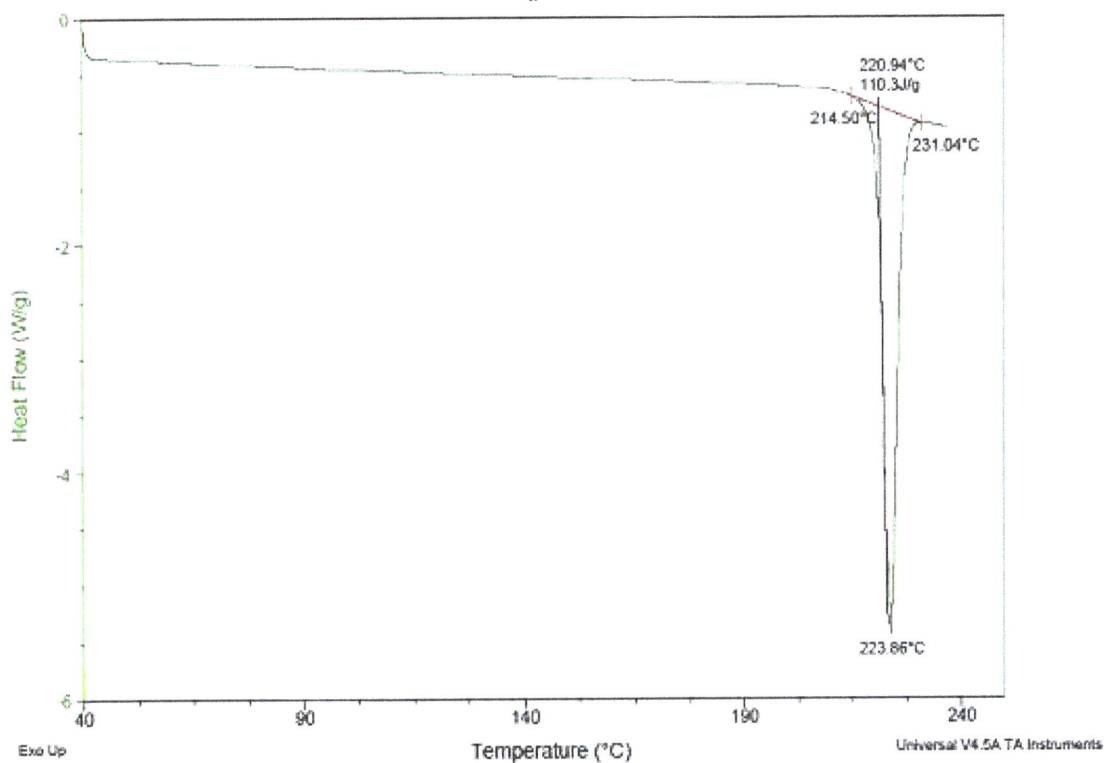

b

Figure 4. DSC plot for phenylboronic acid (**1**) (a) and triphenylboroxine (b).

Solubility of benzoxaborole (**2**) is shown in Figure 5.

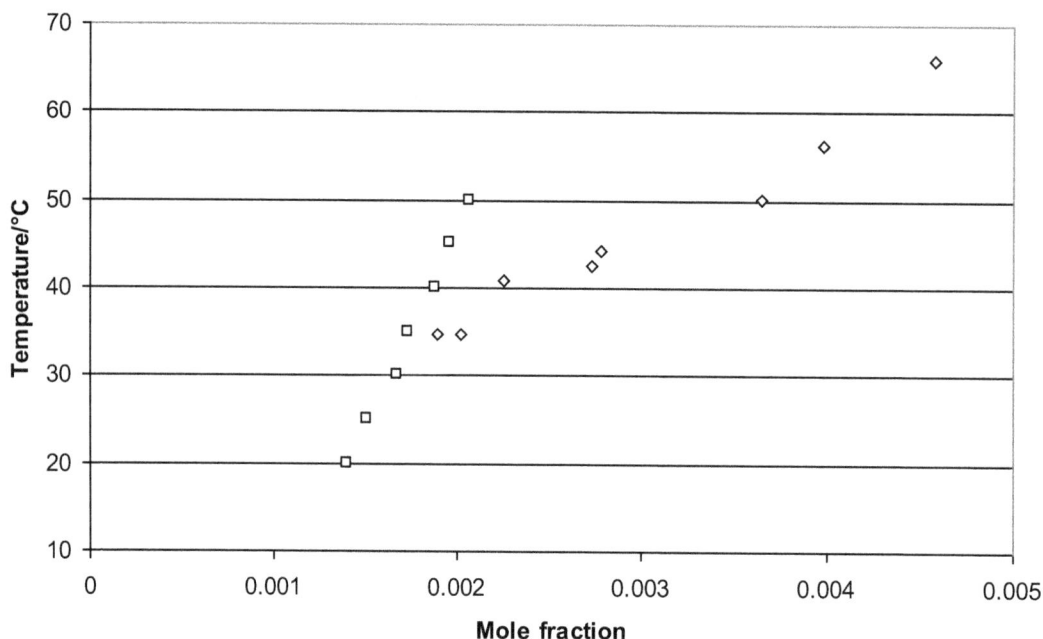

Figure 5. Solubility of benzoxaborole (**2**) in water. ◊ Method **A**, □ method **B**.

In order to investigate the influence of substituents on the solubility of boronic acids in water we investigated compounds **3** - **6**. The results are shown in Figures. 6 - 9.

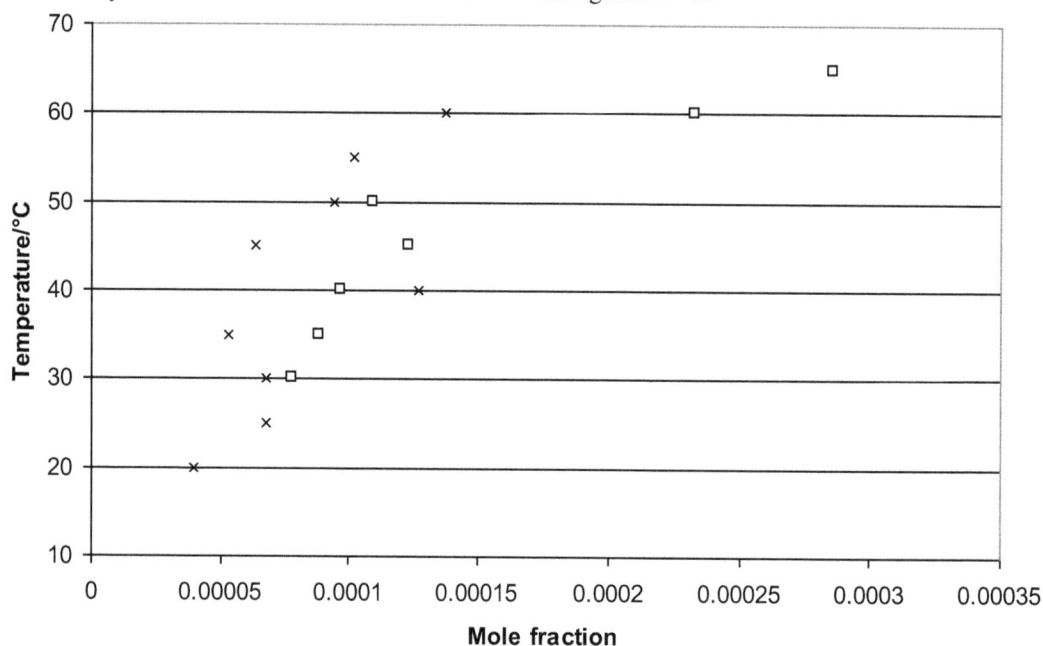

Figure 6. Solubility of 3-isobutoxyphenylboronic acid (**3**) in water. □ Method **B**, x method **C**.

Introduction of alkoxy substituent in compound **3** decreases solubility in water. Moreover, a much larger spread of results is visible for this compound, especially applying the method **C**. The difficulties of solubility determination for this type of compound can be also caused by the possibility of aggregation in aqueous solution into variety of structures like micelles, vesicles, and more complex ones [16].

Solubility of other compounds (**4** – **6**) in water was determined by the method **C**. The results are shown in Figures 7 – 9.

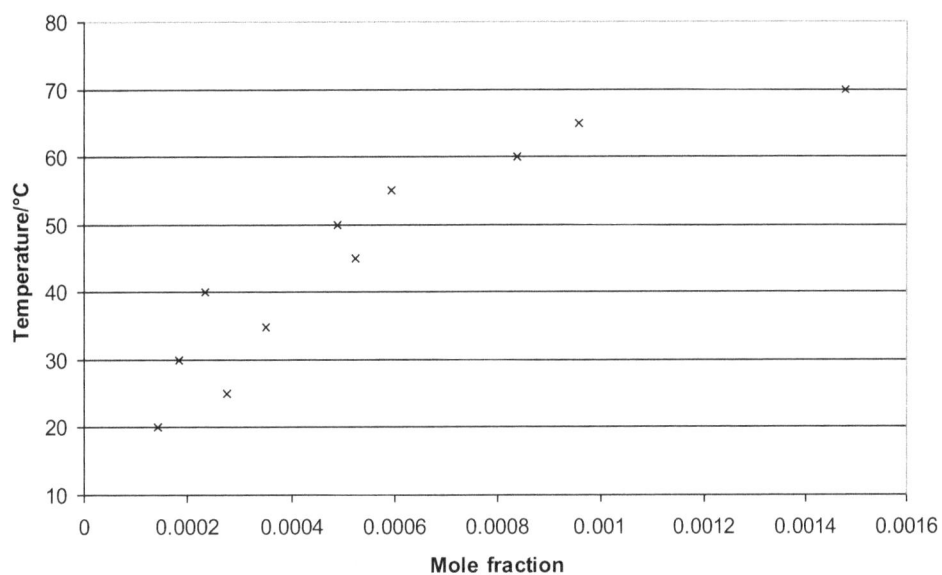

Figure 7. Solubility of 3-carboxyphenylboronic acid (**4**) in water.

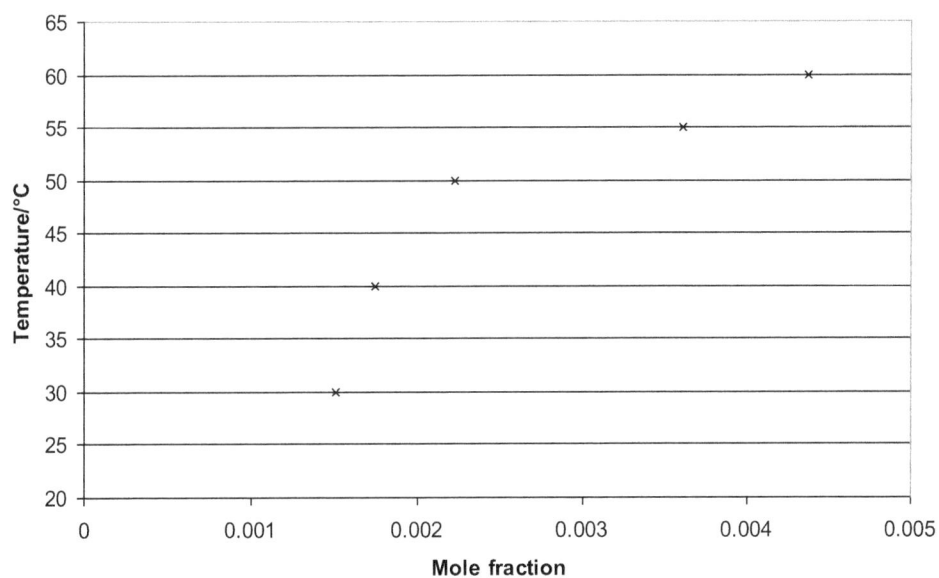

Figure 8. Solubility of 2-(trifluoromethyl)phenylboronic acid (**5**) in water.

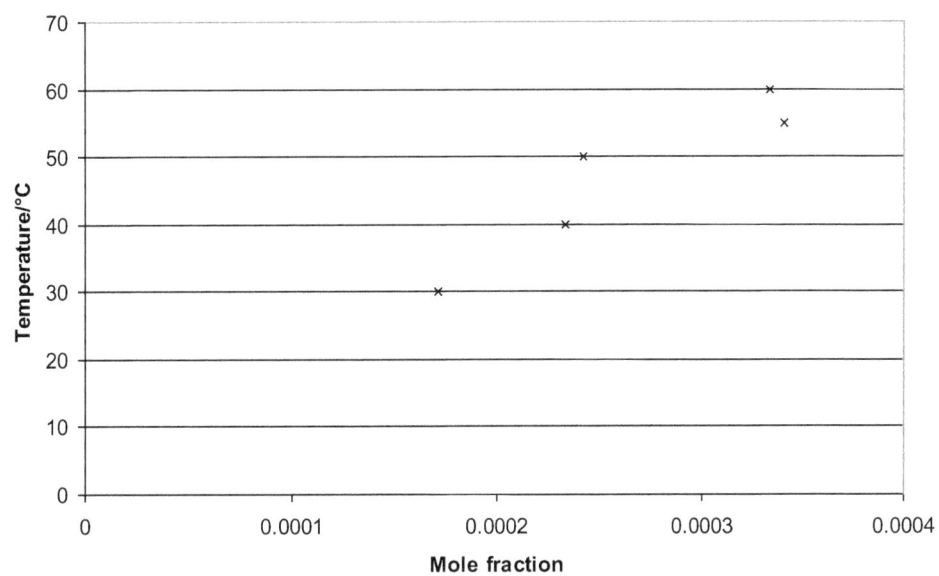

Figure 9. Solubility of 4-(trifluoromethyl)phenylboronic acid (**6**) in water.

Comparison of solubility of investigated compounds is shown in Figure 10.

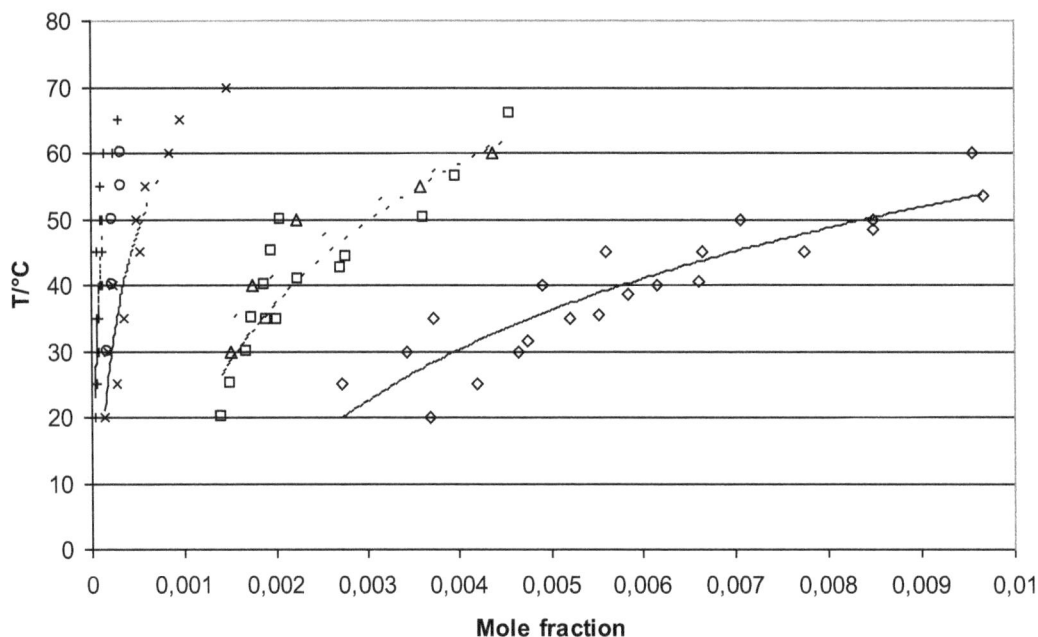

Figure 10. Solubility of investigated compounds in water. ◊ Phenylboronic acid (**1**), □ benzoxaborole (**2**), + 3-isobutoxyphenylboronic acid (**3**), ✕ 3-carboxyphenylboronic acid (**4**), Δ 2-(trifluoromethyl)phenylboronic acid (**5**), ○ 4-(trifluoromethyl)phenylboronic acid (**6**). Solid lines represent logarithmic trend lines.

The solubility of investigated compounds in g/100 g H_2O at 20.0°C and 50.0°C is collected in Table 1.

Table 1. Solubility in water in g/100 g H_2O for selected temperatures, calculated from the curves in Figure 10.

Compound	Temperature/°C		Standard deviation
	20.0	50.0	
1	1.86	5.75	0.025
2	0.84	2.27	0.021
3	0.04	0.15	0.0017
4	0.13	0.51	0.0043
5	0.80	2.67	0.0236
6	0.12	0.26	0.0018

All the primary experimental data are collected in Supplementary Data.

Conclusions

1. All the applied methods of solubility investigations give comparable results. However, the selection of the appropriate method depends on particular compound.

2. Phenylboronic acid (**1**) shows the best solubility in water. Benzoxaborole (**2**) is less soluble than **1**.

3. Introduction of OiBu, COOH and CF_3 groups causes decrease of solubility in comparison with **1**.

4. *ortho*-Isomer of (trifluoromethyl)phenylboronic acid (**5**) is less soluble than *para*-isomer **6**.

5. Significant change of solubility with temperature allows to use water as crystallization solvent.

Acknowledgement

This work was supported by Warsaw University of Technology.

References

1- N. Miyaura and A. Suzuki, *Chem. Rev.*, **1995**, 95, 2457-2483.

2- Boronic Acids. Preparation and Applications in Organic Synthesis, Medicine and Materials, second ed., D.G. Hall (Ed.), Wiley-VCH, Weinheim, **2011**.

3- S. Sene, M.A. Pizzoccaro, J. Vezzani, M. Reinholdt, P. Gaveau, D. Berthomieu, S. Bégu, C. Gervais, C. Bonhomme, G. Renaudin, A. Mesbah, A. van der Lee, M.E. Smith and D. Laurencin, *Crystals*, **2016**, 6, 48.

4- A. Adamczyk-Woźniak, K.M. Borys and A. Sporzyński, *Chem. Rev.*, **2015**, 115, 5224-5247.

5- R.I. Khan and K. Pitchumani, *Green Chem.*, **2016**, 18, 5518-5528.

6- R.M. Washburn, E. Levens, C.F. Albright and F.A. Billig, *Org. Synth. Coll. Vol.*, **1963**, Vol. IV, p. 71.

7- H. Firouzabadi, N. Iranpoor, A. Ghaderi, M. Gholinejad, S. Rahimi and S. Jokar, *RSC Adv.*, **2014**, 4, 27674-27682.

8- M. Dowlut and D.G. Hall, *J. Am. Chem. Soc.*, **2006**, *128*, 4226-4227.

9- (a) U. Domańska, J. Serwatowski, A. Sporzyński, M. Dąbrowski, *Thermochim.Acta*, **1993**, 222, 279-290; (b) U. Domańska, J. Serwatowski, A. Sporzyński, M. Dąbrowski, *Thermochim. Acta*, **1995**, 249, 127-142; (c) M. Dąbrowski, U. Domańska, J. Serwatowski, A. Sporzyński, *Thermochim. Acta*, **1995**, 250, 19-27; (d) M. Dąbrowski, J. Serwatowski, A. Sporzyński, *Thermochim. Acta*, **1996**, 288, 191-202; (e) M. Dąbrowski, J. Serwatowski, A. Sporzyński, *J. Chem. Eng. Data*, **1997**, 42, 1111-1115; (f) M. Dąbrowski, J. Serwatowski, A. Sporzyński, *Pol. J. Chem.*, **1998**, 72, 2423-2431.

10- A. Adamczyk-Woźniak, I. Madura, A. Pawełko, A. Sporzyński, A. Żubrowska, J. Żyła, *Cent. Eur. J. Chem.*, **2011**, 9, 199-205.

11- P. Kurach, S. Luliński, J. Serwatowski, *Eur. J. Org. Chem.*, **2008**, 18, 3171-3178.

12- A. Adamczyk-Woźniak, Z. Brzózka, M. Dąbrowski, I. D. Madura, R. Scheidsbach, E. Tomecka, K. Żukowski, A. Sporzyński, *J. Mol. Struct.*, **2013**, 1035, 190-197.

13- T. Hofman, M. Reda and M. Gliński, *Fluid Phase Equil.*, **2013**, 356, 271-276.

14- C. Achard, M. Jaoui, M. Schwing and M. Rogalski, *J. Chem. Eng. Data* **1996**, 41, 504-507.

15- W. A. Marinaro, L.J. Schieber, E.J. Munson, V.W. Day and V. J. Stella, *J. Pharm. Sci.*, **2012**, 1010, 3190-3198.

16- A. Matuszewska, M. Uchman, A. Adamczyk-Woźniak, A. Sporzyński, S. Pispas, L. Kovacik, M. Stepanek, *Biomacromolecules*, **2015**, 16, 3731-3739.

22

Distribution of some heavy and essential metals Cd, Pb, Cu, Fe and Zn in Mango fruit (Mangifera Indica L.) cultivated in Different Regions of Pakistan

Nusrat Jalbani*, Shahid Bhutto, Suhail Rahujo and Farooq Ahmed

Pakistan Council Scientific and Industrial Research Laboratories Complex, Karachi, Pakistan.

Abstract: In the present study, the concentrations of Cd, Pb, Cu, Zn and Fe from 50 mango samples (Mangifera indica L.) were detected by electro-thermal/flame atomic absorption spectrometry (ETAAS/FAAS). In this study three varieties such as Dusahri, Langra and Chaunsa were collected from different regions (Multan, Sadiqabad, Rahimyar Khan and Mirpurkhas) of Pakistan. The effect of different varieties and environmental conditions on metal accumulation was also investigated. The aim of this study is to compare the level of essential and toxic metal in different regions and correlate the differences observed in metal accumulation to environmental conditions. The accuracy of the conventional acid digestion (CAD) method was checked by analyzing Certified Reference Materials (CRM) i.e. fortified water (TMDA-70), apple leaves (1515) and standard addition technique. The limit of detections (LODs) of the method were found to be 0.113, 2.0, 22.7, 3.85 and 3.05 $\mu g\ L^{-1}$ for Pb, Cd, Cu, Fe, and Zn, respectively.

Keywords: Mangifera indica; Heavy metals; atomic absorption specterometry; Pakistan regions.

Introduction

Fruits are an important fraction of human diet since they contain polysaccharides, sugars, vitamins, minerals, macro and micronutrients which provide wonderful taste and excellent health properties.[1] Mango (Mangifera indica) is the national fruit of Pakistan[2] and is named as "King of the Fruits". It is commonly cultivated in many tropical and sub-tropical regions[2, 3]. Heavy metals are naturally present in environment by natural sources and also from anthropogenic activity or contamination during industrial processes, preservation and cooking[4-10]. Heavy metals are becoming part of the active components of pesticides thus, the continued use of fertilizer and pesticides was identified as the primary pathway from cultivated areas into agricultural soil [11, 12]. Under various conditions, heavy metals are potentially toxic to human health and to the environment even at low concentrations when ingested over a long time period[13-15]. The determination of metal ions at trace level by flame/electrothermal atomic absorption spectrometry (FAAS/ETAAS) has several advantages such as high selectivity, speed and fairly low operational cost. Direct determination of trace elements at extremely low concentration is often very difficult due to the insufficient sensitivity of the methods and the matrix interferences[16, 17].

Pakistan is the 5th largest mango producer with a production capacity of nearly one million tons per year, contributing a share of 7.6% in the world market. Multan, Rahim Yar Khan and Mir Pur Khas are the main mango producing regions of Pakistan. The aim of the present study is to examine the level of toxic elements in mango fruit grown in three major regions of Pakistan and to evaluate most commonly cultivated mango export varieties (Dusahri, Chaunsa, Ratol and Langra) for their nutritional and safety status. We examined the dry matter level of toxic elements in mango fruit, produced from the agricultural land and in order to evaluate the potential corresponding health risk. We then, determined the concentrations of toxic metal contamination and essential metals in mango fruits available in Pakistan super markets and assess how the metal contamination of fruits might have impacted food safety standards. This data will help to provide the status of toxic metals pollution in Pakistan and also to assure food safety and protect the end user health.

Materials and methods

Reagents and glassware

Chemicals and reagents were used of high purity analytical grades (Merck, Darmstadt, Germany). Nitric acid (65%) and hydrogen peroxide

Corresponding author: Nusrat Jalbani
E-mail address: nusratjalbani_21@yahoo.com

(30%) used were obtained from (Merck, Darmstadt, Germany). Stock standard solutions (1000 µg mL^{-1}) of studied metal ions were purchased from Merck (Darmstadt, Germany). While working standards of corresponding metal ions were prepared freshly on daily basis. The laboratory glass wares were kept overnight in 10% (v/v HNO$_3$) and washed with water and rinse with distilled water before use. After washing and rinsing, the glasswares were dried in an oven at 80 ^{0}C for 1 h.

Instrumentation

Atomic absorption spectrometer of Hitachi Ltd. (Model 8000 Z) was used, Electro-thermal atomic absorption spectrometer (ETAAS) was used for determination of Pb, Cd, and flame atomic absorption spectrometer (FAAS) was used for determination of Cu, Fe and Zn. Calibration curves were as following, (10-30 µg L^{-1}) for Pb and (0.5-2 µg L^{-1}) for Cd, while for essential elements (0–1.2 µg mL^{-1}) for Cu, Fe and Zn. Instrumental parameters are listed in Table 1

Table 1. Measurement Conditions for Atomic Absorption Spectrometry (AAS)

Measurement Conditions for electro-thermal Atomization Atomic Absorption Spectrometry

Operating Parameters	Cd					Pb
Lamp current (mA)	7.5					7.5
Wavelength	228.8					283.3
Slit width (nm)	1.3					1.3
Cuvette	Tube					Tube
Temperature Programming						
Drying	80	120	30 Sec	80	120	30 Sec
Ashing	300	300	30	400	400	30
Atomization	1500	1500	10	200	200	10
Cleaning	1800	1800	3.0	2400	2400	3.0
Common Parameters						
Sample volume	10µL analyte					
Backgrund Correction	D^2 Lamp					
Carrier gas Argon	200 mL/mint					

Measurement conditions for flame atomic absorption spectrometer

Elements	Wavelength (nm)	Slit width (nm)	Lamp Current (mA)	Oxidant Pressure	Fuel
Cu	324.8	0.7	30	1.6	Air-acetylene
Fe	372	0.2	5.0	1.6	Air-acetylene
Zn	213.9	0.7	30	1.6	Air-acetylene

Table 2 show the Linearity range of concentration versus absorbance.

Table 2. The range of linearity of concentration versus absorbance graph

Metals	Correlation Coefficients
Cd	Y=0.0059X+0.0017, R^2= 0.9996
Pb	Y=0.0004X+0.0001, R^2= 0.9993
Cu	Y=0.0144X+0.0021, R^2= 0.9985
Fe	Y=0.0285X+0.0003, R^2= 0.9989
Zn	Y=0.0373X+0.0002, R^2= 0.9998

Sampling

50 mango samples were collected from different regions of Pakistan (Multan, Sadiqabad, Rahimyar Khan and Mirpurkhas). From each selected regions (n=12) about three varieties of mango samples (Chaunsa, Dushari and Langra) were randomly collected in the year of 2010-2011. Mango fruit were well blended and homogenized then properly labeled and kept separately in a plastic bag. Fortified water (TMDA-70) and apple leaves (15151) were used as certified reference material (CRM).

Sample preparation

Wet acid digestion method

Duplicate of 10-15 gram of mango fruit (fresh weight) samples were weighed into 250 mL of conical flask. Fifty mango samples were analyzed. 20 to 40 mL of HNO_3 (65%) and 5 mL of H_2O_2 were added to the samples heating on hot plate at 90 ^0C and then mixture was evaporated to dryness. After evaporation and cooling the samples were filtered through Whatman 42 filter paper and then collected into volumetric flasks and diluted with distilled water. Blanks were also treated in the same manner as samples and both were subjected into ETASS/FAAS for determination of the above

mentioned metal ions.

Statistical analysis

Basic statistics on analytical data (mean values with standard deviation) were carried out with Microsoft Excel 2003.

Results and Discussion

The present work was focused on establishing trace elemental levels in different region of three types of samples in order to find ecological and environmental relationship between trace elements. Mango fruit is a summer seasonal commodity growing on agricultural land containing high levels of Cu, Fe, and Zn, with an important role in human nutrition. The analyzed samples were categorized on the basis of variety and region. The mean results of three varieties for Pb, Cd, Cu, Fe, and Zn obtained by proposed procedure of metal ions determination are given in Table 3. The results indicated low levels of Pb (5.28±0.28) in Shujabad, of Cd (2.02±0.11) in Tandoallah Yar, of Fe (0.842±0.02) in Multan whereas, lower level of Cu (0.697±0.04) and Zn (1.01±0.05) were found in Muzfar Garih and Sadiqabad regions (Table 3).

Table 3. Elemental contents in mango fruits of different region of Pakistan (Fresh wt)   (n=12)

Regions	Cd^a	Pb^a	Cu^b	Fe^b	Zn^b
Qitalpur	3.99±0.12	5.78±0.32	1.56±0.05	0.885±0.06	**1.27±0.07**
Tandoallah Yar	2.02±0.11	9.75±0.46	1.78±0.07	1.23±0.08	**1.86±0.09**
Muzfr Garh	2.89±0.21	8.46±0.49	0.697±0.04	1.17±0.06	**1.74±0.08**
Khanewal	2.57±0.22	12.51±0.73	1.63±0.01	1.32±0.07	**1.83±0.11**
Hydrabad	2.07±0.11	12.1±0.77	1.82±0.5	1.25±0.08	**1.86±0.12**
Multan	2.45±0.09	9.53±0.61	0.898±0.09	0.842±0.02	**2.72±0.18**
Shujabad	2.32±0.08	5.28±0.28	0.887±0.07	2.68±0.12	**1.79±0.19**
Sadiqabad	3.38±0.23	11.2±0.64	1.15±0.09	2.35±0.15	**1.01±0.05**
Rahim Yar Khan	5.73±0.33	8.08±0.38	0.953±0.04	2.71±0.18	**1.25±0.06**
Mirpurkhas	**2.37±.11**	**9.33±0.42**	**0.821±0.03**	**2.19±0.12**	**1.63±0.08**

$^a\mu g\ Kg^{-1}$, $^b mg\ Kg^{-1}$

In the view of results, no significant difference was observed among the varieties of different regions at 95% confidence level as indicated in Table

4 where three essential metals Cu, Fe, and Zn were found at low level in the variety of Dushari in most of the regions.

Table 4. Distribution of selected trace and toxic elements in mangoes varieties grown in Pakistan

Regions		Cd^a	Pb^a	Cu^b	Fe^b	Zn^b
Qitalpur	A	2.32±0.15	6.12±0.46	1.75±0.013	0.78±0.065	**1.17±0.055**
	B	3.73±0.21	5.22±0.46	1.61±0.13	1.47±0.02	**1.32±0.120**
	C	5.92±0.36	6.01±0.56	1.32±0.045	0.405±0.021	**1.31±0.055**
Tando allah yar	A	1.90±0.086	9.81±0.55	2.01±0.05	1.33±0.065	**3.02±0.083**
	B	2.11±0.16	9.75±0.62	1.98±0.12	1.49±0.06	**1.3±0.086**
	C	2.05±0.06	9.71±0.66	1.71±0.055	0.865±0.055	**1.27±0.052**
Muzfr garh	A	1.78±0.091	8.41±0.51	0.44±0.023	1.58±0.036	**2.06±0.076**
	B	0.912±0.06	8.50±0.51	1.38±0.02	1.12±0.05	**1.92±0.072**
	C	6.00±0.11	8.48±0.51	0.27±0.01	0.805±0.061	**1.24±0.051**
Khanewal	A	1.56±0.085	13.0±0.73	1.57±0.051	1.36±0.025	**2.70±0.085**
	B	1.75±0.03	12.5±0.91	1.95±0.06	1.635±0.04	**1.65±0.045**
	C	4.40±0.12	12.03±0.68	1.29±0.052	0.956±0.073	**1.33±0.046**
Hydrabad	A	1.17±0.071	12.6±0.61	1.95±0.075	1.63±0.045	**1.45±0.042**

	B	1.23±0.04	11.5±0.82	1.67±0.045	1.49±0.03	**2.89±0.076**
	C	3.83±0.23	12.4±0.71	1.49±0.061	0.63±0.042	**1.25±0.048**
Multan	A	2.11±0.13	9.56±0.62	1.049±0.079	1.07±0.078	**3.01±0.081**
	B	1.63±0.07	9.48±0.43	1.124±0.02	0.615±0.02	**3.25±0.160**
	C	3.61±0.13	9.55±0.58	0.522±0.034	0.84±0.056	**1.92±0.078**
Shujabad	A	2.08±0.11	5.20±0.23	0.8±0.065	3.0±0.125	**2.07±0.078**
	B	2.33±0.06	5.50±0.23	1.2±0.03	1.65±0.071	**1.78±0.032**
	C	2.56±0.15	5.15±0.45	0.66±0.043	3.45±0.25	**1.53±0.036**
Sadiqabad	A	3.44±0.25	13.2±0.76	0.678±0.044	2.56±0.16	**1.48±0.048**
	B	2.62±0.22	10.1±0.78	1.3±0.05	2.66±0.15	**1.14±0.043**
	C	4.08±0.27	9.78±0.61	1.48±0.086	1.85±0.18	**0.413±0.011**
Rahim yar Khan	A	3.09±0.25	8.67±0.57	0.826±0.067	2.16±0.11	**1.23±0.041**
	B	2.41±0.21	8.12±0.31	0.629±0.04	3.19±0.26	**1.07±0.036**
	C	11.70±0.21	7.45±0.72	1.403±0.076	2.75±0.16	**1.46±0.034**
Mirpurkhas	A	0.73±0.03	11.3±0.73	0.557±0.034	2.11±0.12	**1.60±0.073**
	B	2.15±0.19	8.14±0.33	1.31±0.05	1.75±0.13	**1.74±0.025**
	C	**4.24±0.11**	**8.56±0.53**	**0.598±0.033**	**2.71±0.17**	**1.55±0.038**

A = Langra,   B = Chonsa, C = Dusheri. ^{a}μg Kg^{-1} ^{b}mg Kg^{-1}

The level of Cu in mango fruits ranged from 0.697 to 1.82 mg Kg^{-1}(Table 3). This value was not significantly different between the different regions under study. It is noteworthy that higher level of Cu was observed in the city of Hyderabad while low level was found in Muzfar Garh. An elevated level of Fe (2.71±0.18) and Zn (2.72±0.18) mg Kg^{-1} was found in Rahim Yar Khan and Multan respectively (Table 3). Zinc is an essential element for proper metabolism. The significant differences were found among the different regions of Pakistan when comparing the obtained values at a confidence interval of 95%. In this work the lower level of Fe, Zn, and Cu were found in Multan, Sadiqabad and Muzfr Garh (Table 3). On contrary, the concentration of Cd was higher in Dushari variety and of Pb was higher in most of the studied regions[18]. However, in eight samples level of Pb was found higher in Langra (L) variety while 2nd and 3rd highest level of Pb was obtained in chaunsa (B) and Dushari (C), accordingly (Table 4). The Pb content of mango fruit collected from different regions was found to be within the permissible limit based on EU regulations (Table 7). The highest average concentration of Cd was 5.73±0.33 mg Kg^{-1} in the Rahim Yar Khan and the level of Pb was 12.51±0.73 μg Kg^{-1} in Khanewal (Tabel 3). The concentration of Cd 5.73 μg Kg^{-1} was found at maximum level in all selected variety, while the levels of Cd and Pb were relatively lower in Multan with 3.81±0.21 μg Kg^{-1} and 4.06±0.32 μg Kg^{-1} respectively. From the results it was clear that the Dushari and Langra contained higher concentration of Cd and Pb as compared to variety of chounsa.

Discussion

Fruits are considered key factor for providing nutritional basis in human diet and as well as for their bio-functional components[19]. Hence, the objective of our study was to evaluate the heavy metals present in three varieties of Mango fruit from different regions of Pakistan. Macro and trace elements play a significant role in maintaining health in humans. Heavy metals i.e. Cd and Pb are regularly present in foods and considered dangerous elements even at low contamination[20-21]. These metals are concentrated particularly in kidneys, liver, blood forming organs and lungs. They most frequently result in kidney damage (necrotic protein precipitation) and metabolic anomalies caused by enzyme Inhibitions[22, 23]. These heavy metals may replace the essential nutrients in human body adjoining their sites to the vital organs. Their ingestion could lead to severe liver damage, symptoms of chronic toxicity in kidney, pancreas or Alzheimer etc. diseases[24-26]. Nutrients are important and beneficial components for human health[27, 28]. The maximum permitted level of Fe for food is 15mg Kg^{-1} according to WHO[29], the concentration of essential and non-essential elements in mango fruit are found within acceptable limits except Cd [30, 31]. These results indicated that there are common anthropogenic contamination sources regarding heavy metals.

Analytical figure of merit

Under the optimized conditions (Table 1), the analytical performance of the CAD was evaluated. The linear range was 10-30, 0.5-2.0 μg L^{-1} for Pb and Cd, whereas 0.2, 2.0 μg mL^{-1} for Cu, Fe and Zn respectively with the correlation coefficient (R) of 0.998 and 0.9999. The limit of detection (LOD) was calculated as equivalent of three times of standard deviation of blank readings, the calculated LODs are: 0.113, 2.0, 22.7, 3.85, and 3.05 μg L^{-1} for Pb, Cd, Cu, Fe, and Zn respectively.

Validation of proposed methodology

In order to validate the methodology, the accuracy of the proposed method was evaluated by the analysis of CRM; fortified water (TMDA-70) apple leaves (1515) and standard addition recovery/test method to determine trace and toxic metals in mango fruit. The results are given in Tables 5 and 6. The proposed procedure was applied on four replicates of each samples as discussed in "digestion procedure" section. A recovery of the spiked metals is close to (97.9-102%) which show quantitative recovery of the understudied metals as seen in Table 5. A good agreement exists between the results of the proposed method and can be applied successfully to real samples.

Table 5. The results of certified reference material (CRM), TMDA 70 (Fortified water) µg L⁻¹ and Apple leaves (NIST-1515), mg Kg⁻¹.

	Fortified water (TMDA 70)			Apple leaves (NIST-1515)		
Analyte	Certified	CAD	% Recovery	Certified	CAD	% Recovery
Cd	145	144±5.3	99.3	0.013	0.0131±0.001	101
Pb	444	445±12	100	0.47	0.46±0.02	97.9
Cu	398	399±9.8	100	5.64	5.63±0.35	99.8
Fe	368	367±11	99.7	83	84±3.21	101
Zn	477	476±18	99.8	12.5	12.8±0.55	102

Table 6. The results of standard addition/recovery for Cd, Pb, Cu, Fe and Zn determination in mango fruit samples (Multan) (*n*=6).

	Added	CAD $\bar{x} \pm \dfrac{ts}{\sqrt{n}}$ [a]	% Recovery
Cd[a]			
	0.0	3.81±0.18	99.5
	0.5	4.29±0.13	99.8
	1	4.8±0.18	99.7
	2	5.79±0.19	99.5
Pb[a]			
	0.0	3.45±0.15	99.7
	10	23.43±0.97	99.9
	20	33.4±0.81	99.8
	30	13.41±0.78	99.7
Cu[b]			
	0.0	2.21±.09	98.9
	0.4	2.58±0.09	98.9
	0.6	2.78±0.078	98.7
	0.8	2.97±0.016	98.9
Fe[b]			
	0.0	3.01±.0.16	99.4
	0.5	3.49±0.15	99.3
	1.0	3.98±0.23	99.2
	2.0	4.97±0.19	99.4
Zn[b]			
	0.0	2.21±0.12	99.2
	0.4	2.59±0.12	98.9
	0.6	2.78±0.16	99.3
	0.8	2.99±0.15	99.2

[a]Average value ± confidence interval ($P = 0.05$) µg L⁻¹, [b]Average value ± confidence interval ($P = 0.05$) µg mL⁻¹

Applications

In order to evaluate the accuracy of the CAD method, certified reference materials fortified water (TMDA-70) apple leaves (1515) were analyzed and the results are presented in Table 4. Different mango fruit samples were subjected to the CAD procedure for determination of concentrations of Pb, Cd, Cu, Fe and Zn respectively. The results are given in Table 6. The daily intake of all studied metals were calculated on the basis of consumption of a minimum 300 gm/person/day of mango and compared with permissible limits by WHO/RDA as represented in Table 7.

Table 7. Concentration of trace and toxic elements in mango fruit (fresh weight) grown on agricultural land of different regions of Pakistan

Metals	Permissible level ($mgKg^{-1}$)	daily consumption/person/day (mg/person/day)
Cd	0.05[32]	**0.015**
Pb	0.10[32]	**0.03**
Cu	5.0[33]	**1.5**
Fe	5.0[33]	**1.5**
Zn	**5.0[33]**	**1.5**

Average daily consumption rate of fruits per person (fresh weight) = 300 g

In the present study it is important to remark that the concentration of all studied macro and micronutrient were found within permissible limits except Cd.

Conclusion

Pakistan is an agricultural country and production of mango fruits is an important part of this sector. Trace and toxic metals may get into natural water and soil and thus into fruit or plant through polluted air. This can result in the contamination of food chain causing serious problems in human health which is directly affected by ingestion of contaminated food and water.

The uptake of elemental contents Cd, Pb, Fe, Cu and Zn by three varieties of Mango growing in different regions of Pakistan was studied. The variations in uptake of elemental contents by different varieties were evaluated to check higher tolerance limit for toxic elements present in mango fruit. In view of these results, it was concluded that the variety of Mirpur Khas is the best because the trace and toxic metals were found within safe limits as compared to other regions.

References

1- S. Akhtar, N. Safina, M.T. Sultan, S. Mahmood, M. Nasir and A. Ahmad, *Pak. J. Bot.* **2010,** *42(4)*, 2691-2702.

2- M. B. Usman, Fatima, M. M. Khan and M. I. Chaudhry, *Pak. J. Agri. Sci.* **2003**, 40 (3-4).

3- F. A. Jam, S. Mehmood and Z. Ahmad. Academy of Contemporary Research Journal V II (I), 10-15, ISSN: 2305-865X, 2013.

4- O. Mumzuroglu, F. Karatas and H. Geekil, *Food Chem.* **2003**, *83*, 205-212.

5- S.G. Ozcan, N. Satiroglu and M. Soylak, *J Food Chem. Toxicol.* **2010**, *48*, 2401-2406.

6- M. Tuzen, E. Sesli and M. Soylak, *Food Cont.* **2007a,** *18*, 806-810.

7- M. Tuzen, S. Silici, D. Mendil and M. Soylak, *Food Chem,* **2007b.** *103*, 325-330.

8- S. Singh, M. Zacharias, S. Kalpana and S. Mishra. *J. Environ. Chem. Ecotoxicol.* **2012**, 4(10), 170-177.

9- A. Duran, M. Tuzen and M. Soylak, *Int. J. Food Sci. Nutr.* **2008,** *59*, 581-589.

10- M. Tuzen, M. Soylak, D. Citak, H.S. Ferreira, M.G.A. Korn and M.A. Bezerra, *J. Hazard. Mater.* **2009**, *162*, 1041-1045.

11- I. Narin, M. Soylak, K. Kayakirilmaz, L. Elci, and M. Dogan, *Anal. Lett.* **2003**, *36*, 641-658.

12- J.E.V. Nunez, N.M.B.A. Sobrinho and N. Mazur, *Ci. Rural.* **2006**, *36*, 113-119.

13- Z. Parveen, M.I. Khuhro and N. Rafiq, *Bull. Environ. Contam. Toxicol.* **2003,** *71*, 1260-1264.

14- S. Baytak and A.R. Turker, *Talanta,* **2005**, *65*, 938–945.

15- P.C. Aleixo, D.S. Junior, A.C. Tomazelli, I.A. Rufini, H. Berndt and F. J. Krug, *Anal. Chim. Acta.* **2004**, *512*, 329-337.

16- M.K. Jamali, T.G. Kazi, M.B. Arain, H.I. Afridi, N. Jalbani and G.A. Kandhro, *J. Hazard. Mater.* **2008**, 158, 644-651.

17- N. Jalbani, F. Ahmed, T.G. Kazi, U. Rashid, A.B. Munshi and A. Kandhro. *Food Chem. Toxicol.* **2010,** *48*, 2737-2740.

18- D. Petit, F. Claeys, C. Sykes and Y. Noefnet, *J. Physique. IV France*, **2003**, *107* (1053), 1053 - 1056.

19- M.A.M. Sajib, S Jahan, MZ Islam, TA Khan and B.K. Saha, *Int. Food Res. J.* **2014,** *21*(2), 609-615.

20- X. Bi ,L. Ren, M. Gong, Y. He, L. Wang, Z. Ma. The Global J. Soil Sci. 2010, 155 (1-2), 115-120.

21- A.0. Igwegbe, H.M. Belhaj, T.M. Hassan and A.S. Gibali, *J. Food Saf.* **1992.** 13, 7-18.

22- N. Jalbani, T.G. Kazi, N. Kazi, M.K. Jamali and M.B. Arain, *Biol.Trace Elem. Res.* **2008,** 1-12.

23- S. I. Khan, A.K. M. Ahmed, M. Yunus, M. Rahman, S. K. Hore, M. Vahter and M.A. Wahed. *J. Health Popul. Nutr.* **2010,** 28(6), 578-584.

24- M. Yaman, *Anal. Biochem.* **2005.** *339* (1), 1-8.

25- M. Tuzen and M. Soylak, *J. Hazard. Mater.* **2009,** *164*(1), 1428-132.

26- Tu. Zhifeng, He. Qun, X. Chang, Hu. Zheng, Ru. Gao, L. Zhang and Li. Zhenhua, *Anal. Chim. Acta,* **2009,** *649* (2), 252-257.

27- A. Waheed, M. Jaffar and K. Masud, *Nutr. food Sci.* **2003,** *33*, 6.

28- E. Zahir, I.I. Naqvi and S. Mohi Uddin, *J. Basic Appl. Sci.* **2009,** *5* (2), 47-52.

29- WHO. Evaluation of certain food additives and contaminants: 46th Report of the joint FAO/WHO expert committee on food additives, WHO. Technical report series No. **1997,** *868,* pp i-viii.).

30- G. Herrick and T. Friedland, *Water Air. Soil Poll.* **1990,** *53*, 151-157.

31- FAO/WHO, Toxicological evaluationof certainfoodadditives and contaminants. Forty-first report of the, Joint FAO/WHOExpert CommitteeonFoodAdditives (WHO Technical Report Series, No. 837, WHO, Geneva, **1993**.

32. Commission Regulation (EC) No 629/2008. Amending Regulation (EC) No 1881/2006 setting maximum levels for certain contaminants in foodstuffs, **2008.**

33. A. S. M. Hassan, T. A. Abd-El-Rahman, A. S. Marzouk. *Int. J. Food Sci. Nutr. Eng.* **2014,** 4(3), 66-72.

23

Dielectric relaxation behavior of nematic liquid crystal cell using β-cyclodextrin as an alignment layer

Marwa Sahraoui[1,*], Asma Abderrahmen[1], Rim Mlika[1], Hafedh Ben Ouada[1] and Abdelhafidh Gharbi[2]

[1]Laboratoire des Interfaces et Matériaux Avancés, Université Monastir, Faculté des Sciences de Monastir, Département de Physique, Avenue de l'environnement 5000 Monastir, Tunisia
[2]Laboratoire de la Matière molle, Faculté des Sciences de Tunis, Tunisia

Abstract: In the present investigation, we report the dielectric properties of a symmetric Nematic Liquid Crystal (NLC) cell using Beta Cyclodextrins (β-CD) as alignment layers. These layers were deposited onto Indium Tin Oxide (ITO) surface by thermal evaporation and then characterized using contact angle measurement. This revealed a hydrophilic character attributed to the presence of hydroxyl groups. Morphological study was carried out by Scanning Electronic Microscopy (SEM).

The dynamic impedance study of the Liquid Crystal (LC) cell in a wide frequency range from 1mHz to 13MHz was reported. It was found that the β-CD alignment layer had a blocking effect on the NLC cell at a high frequency range. We also report the relaxation mechanism of NLC cell which is modeled by an appropriate equivalent circuit in order to understand the electrical properties of the liquid crystal cell and to investigate the processes taking place at different interfaces.

Keywords: Liquid crystals; molecular dynamics; dielectric properties; interfaces.

Introduction

Liquid crystals (LCs) have been extensively investigated thanks to their increasing importance in technology, industry, science and medical applications. They have received considerable attention from many scientists over the last few decades [1]. These materials are used in electro-optical devices due to their anisotropic molecular order which can be controlled by external electric field [2,3]. It is well known that the dielectric anisotropy of the liquid crystals has an important impact on the image quality of Liquid Crystal Displays (LCDs) [4,5]. Moreover the high birefringence of liquid crystals is important for fast switching display [6]. For more advanced applications of LCs, it is necessary to understand the orientation of LC molecules and the effect of the surface on their alignment [7]. The anchoring properties of NLCs and their alignment on solid surfaces are the heart of most LC device applications [8,9,10].

Nowadays, there is a considerable interest in the use of macromolecules as components of LC devices. The particular features of these materials, which make them attractive for these applications, are their conformability, their low cost of fabrication, and the extreme ease of their deposition under thin film [11].

Previous works discussed the possibility of using these macromolecules as potential materials to display devices. Peralta et al. have recently reported a reorientation of a liquid crystal induced by such a macrocyclic molecule like a calixarenes[7]. Cyclodextrins are a category of macrocyclic platform having a rigid natural cavity [12,13], a good physical and chemical stability and compatibility with a variety field applications that can lead to low-cost devices. Up to now, they have been studied mostly for use in sensor devices and electronic device applications [11,14]. So, in this work, in order to develop a new liquid crystal device and to improve the dielectric properties and the alignment effect, Beta cyclodextrins are used for the first time as alignment layer. They are considered as an ideal candidate for this purpose due to their well defined structure [13].

In this context, the NLC cell was fabricated using nematic liquid crystal (5CB) and Beta cyclodextrin (β-CD) as alignment layer. Before the manufacturing of NLC cell, the morphology of the films was characterized using Scanning Electron Microscopy (SEM). The relaxation dynamics and the dielectric properties of NLC cell were investigated.

Corresponding author: Marwa Sahraoui
E-mail address: sahraoui.marwa@yahoo.fr

Exprimental Section
Materials

The NLC used in this work is the polar liquid crystal (4-*n*-pentyl-4'-cyanobiphenyl) (commonly known as 5CB) which has a positive dielectric anisotropy. The 5CB is one of the best known liquid crystalline materials. It is in the nematic phase between 22.5°C and 35.5°C [15,16].

As an alignment layer, we chose the native Beta Cyclodextrin (β-CD) (99 °/° Sigma Chemical Co, St Louis, Mo, USA). The native β-CD is a bucket-shaped oligosaccharide consisting of seven glucose units with a hydrophobic cavity and a hydrophilic exterior [12,17]. The chemical structure of the molecule is presented in Fig. 1.

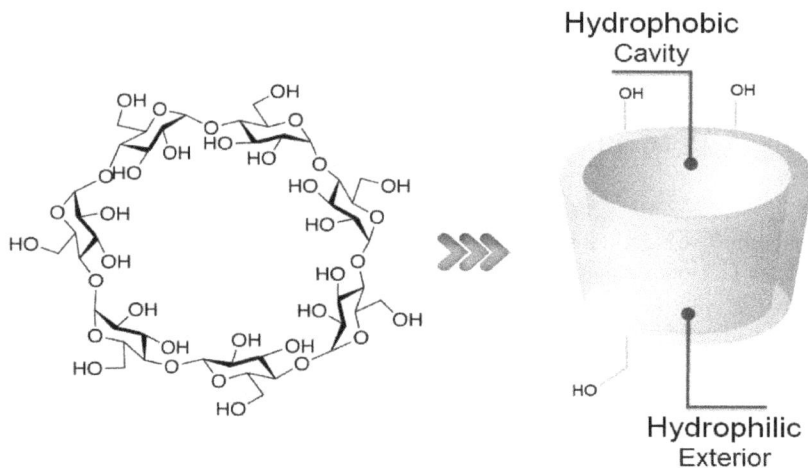

Figure 1. The molecular structure of β-Cyclodextrin.

Indium Tin Oxide (ITO) is usually used as an electrode material because of its transparency, high conductivity, good etchability, hardness and good adherence on many types of substrates. These characteristics allow its use in a wide range of optoelectronic devices, particularly LCDs [9].

NLC cell fabrication

The NLC cell is prepared by two ITO-covered glass plates separated by Mylar sheets [4,5]. Firstly, before the deposition of the alignment layer, the native ITO-covered glass plates were pre-cleaned in acetone and methanol in ultrasonic bath for 20 min for each solvent [9,18].

The substrates were finally dried under nitrogen flow to remove contaminants from the surface, improve adhesion of coating and prevent the humidity. The alignment layers of β-CD were deposited onto ITO by thermal evaporation under secondary vacuum of 10^{-6} torr [13]. During this process the temperature was maintained at about 200°C [14] (evaporating temperature of the organic material), the evaporation rate and the deposited thickness were controlled by a piezoelectric quartz crystal microbalance. The measured deposited alignment layer thickness was about 300 nm, measurements were carried out using dektak-3010 surface profilometer [13]. The liquid crystal 5CB was injected into the sample cells by capillary action [4,5] at 25°, all samples were filled at room temperature. It is well known that 5CB liquid crystal has a nematic phase at room temperature [15]. The area and the thickness of the cell were 2.25 cm² and 20 μm, respectively. The deposited alignment layer β-CD favors a planar orientation of the LC molecules. This was checked through observation with a polarized

microscope under a crossed polarizer-analyzer arrangement [9].

Measurements

To characterize the β-CD layers, measurements of contact angle were performed with a model contact instrument (Dig drop) form GBX (Romans, France). The image of the liquid droplet behavior on the surface was taken with a digital camera and then analyzed [13]. Deionised water, formamide (Sigma Chemical Co, St Louis, Mo, USA) and diiodomethane (DIM) (Sigma Chemical Co, St Louis, Mo, USA) were used as the three liquid probes for surface free-energy calculations [9]. Van Oss theoretical model is an approach which has been widely used to measure wettability [9,12], surface energy and its polar and dispersive components. A software based on Van Oss equation was used to calculate surface energies [20].

A scanning electron microscopy (SEM), (JSM-5400 Scanning Microscope) was used to study the surface topography of alignment layers.

The principle of Impedance Spectroscopy (IS) is to measure the electrical characteristics of a material as a function of the frequency of a small-amplitude AC signal. Changes in impedance characteristics with frequency provide information about conduction mechanisms, charge accumulation, dipole behavior and the dielectric relaxation of a material [9, 21].

In our study the dielectric measurements were carried out at ambient temperature and in a frequency range from 1mHz to 13MHz. In the lower frequency range (1mHz to 1MHz), an impedance analyzer (Autolab PGSTAT) was used. The measurements of impedance in the frequency range from 5 Hz to 13MHz were taken using an HP 4192A impedance analyzer. For all the measurements, the amplitudes of the AC were 100 mV [9,13].

Results and discussions

Contact angle measurement

The wetting of the solid phase is always the first step in the preparation of optoelectronic devices. A wettability study was carried out before and after thin film deposition in order to check the non-

degradation of the deposited alignment layer after the deposition technique [20]. The results of liquid contact angle and surface energy of the samples are summarized in Table 1. When the water contact angle on a surface is smaller than 90°, the surface is wetted by the water so the surface is hydrophilic [20]. We have found that the value of water contact angle value decreases from (73°) for the cleaned ITO-covered glass plates to (46°) for β-CD alignment layer (Fig. 2). This result proves that the surface became more hydrophilic. The reduction of the surface hydrophobicity can be attributed to the presence of the hydroxyl groups (OH: Hydrophilic functional) on the surface of the membrane [22]. On the other hand, we notice that the basic component γ^- increases after the modification of ITO surface by the β-CD from (**8.3** mJ /m^2) to (**35** mJ /m^2), due to the presence of basic groups (OH) (Table 1). This clearly proves that the alignment layers were well deposited [22].

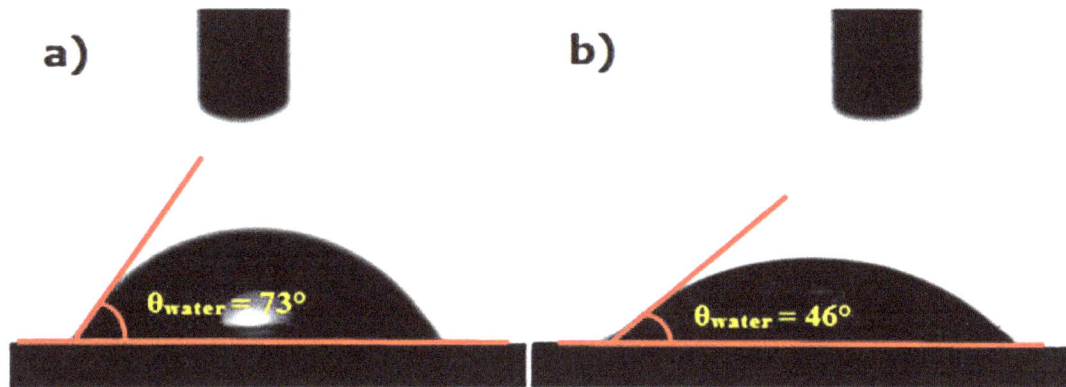

Figure 2. Water contact angle of **a)** Cleaned ITO electrode **b)** β-Cyclodextrin layer.

Table 1: Contact angle and surface energy data

Sample	Θ° water	Θ° For	Θ° DIM	γs (mJ /m^2)	γ^d (mJ /m^2)	γ^p (mJ /m^2)	γ^+ (mJ /m^2)	γ^- (mJ /m^2)
Cleaned ITO	**73**	50	35	**44.6**	**42**	2.6	0,2	**8.3**
β-CD native	**46.4**	40.3	32.0	**45.4**	**43.4**	2.0	0.0	**35.0**

Θ°water : water contact angle (°) ; Θ°For : formamide contact angle (°) ;
Θ°DIM : diiodomethane contact angle (°); $\gamma^s = \gamma^d + \gamma^p$: surface energy; γ^d : dispersive energy; $\gamma^p = (2 \gamma^+ \gamma^-)^{1/2}$: polar energy; γ^+: acid energy component; γ^-:basic energy component [9,12].

Surface Morphology Studies

The realization of the desired orientation of the LC molecules depends on the alignment layers' morphology. So, before to manufacturing of the LC cell, the morphology of the films was characterized using scanning electron microscope. SEM images prove that the alignment layers were well deposited

on the ITO (Fig. 3(a) and (b)). Additionally, the SEM image of the β-CD membrane (Fig. 3(b)) shows a compact structure and homogeneous surface [23]. This is due to the high interaction between ITO substrate and the alignment film [9]. This morphology enormously affects the alignment of liquid crystal molecules [1].

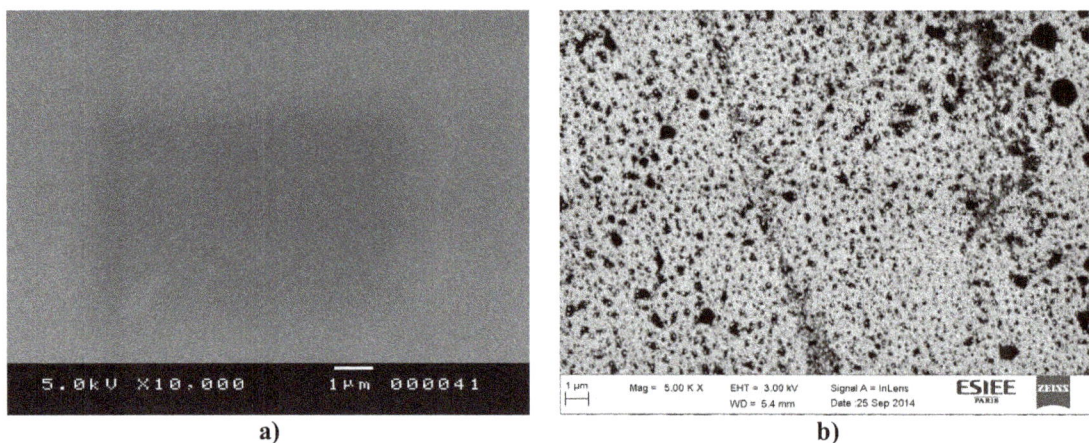

Figure 3. SEM images of **a)** cleaned ITO electrode **b)** β-Cyclodextrin layers.

Impedance spectroscopy

Impedance Spectroscopy (IS) technique is an effective tool for the qualitative and quantitative characterization of the electrical processes occurring conducting NLC cells. Moreover, it can be used to study the interfacial properties of modified alignment layer. In the present work, the applied DC bias voltage was chosen higher than or equal to 2V to avoid the diffusion effect observed at 0V and 1V DC [9].

Fig. 4(a) shows the dependence of the conductance on frequency from 1 mHz to 13 MHz of the ITO/β-CD/5CB/β-CD/ITO cell at 2V. In such materials, the frequency dependent conductivity G (ω) is expressed as the following equation [24]:

$$G(\omega) = G_{dc} + G_{ac}(\omega) \tag{1}$$

Where $G_{ac}(\omega) \approx \omega^s$ [18]   (2)

Where ω is related to the angular frequency of the applied excitation and the exponent "s" is the critical exponent [21,24].

Conductance measurements

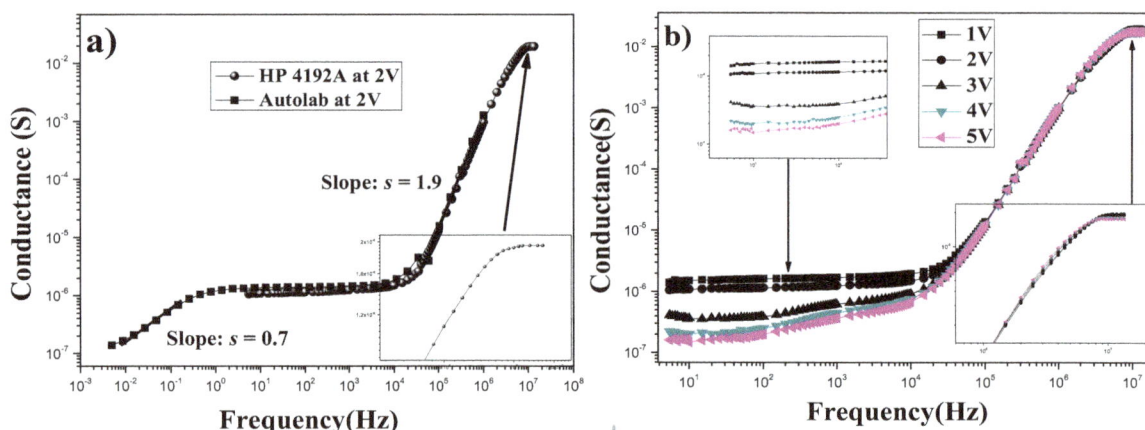

Figure 4. Conductance as a function of frequency of the ITO/β-CD/5CB/β-CD/ITO cell **a)** at the bias voltage (2V). **b)** Conductance as a function of frequency at different applied bias voltages.

It is evident that G (ω) increases with increasing frequency. Fig. 4(a) shows that the conductance is characterized by the presence of two power laws and a plateau region. This latter corresponds to the frequency independent conductance G_{dc} which is related to free charges available in the material. The increase in ac conductance G_{ac} can be attributed to the presence of the charge carriers in the material which can be caused by the reorientation of molecules [2,4]. The first power at low frequency has an exponent close to unity ($0<s<1$) [24]. The second power, above 10^5 Hz, with an exponent s in the order of 2, can be related to the series resistance of the ITO electrodes (Fig. 4(a)) [21,25].

The transition from frequency-independent (dc conductance) to frequency-dependent (ac conductance) behavior indicates the beginning of the relaxation phenomena and is in agreement with the dielectric relaxation behavior studied in the next part. On the other hand, the conductance plot shows a saturated region at high frequencies [21,25] (Fig. 4(a) lower inset), which indicates that the β-CD layer has a blocking effect on the electrodes [9]. The conductance variation of the ITO/ β-CD /LC/ β-CD /ITO cell as a function of frequency at different bias voltages is plotted in Fig. 4(b). Fig. 4(b) shows that the conductance decreases with the applied bias voltage. This can be attributed to the migration of ionic impurities from the volume to the interfaces. Indeed, the mechanism of generation of charges in the electrodes allows the stabilization of the orientation of the molecules [3,19].

Admittance measurements

To better understand the blocking effect, the admittance measurement was carried out. The admittance is proportional to the conductance; it can be written as the following equation [25].

$$Y(\omega) = G(\omega) + i\omega C \qquad (3)$$

Where G (ω) is the equivalent parallel conductance and C (ω) is the equivalent parallel capacitance related to permittivity and both are in general functions of frequency [3].

Fig. 5(a) shows that the entire spectra exhibit a single semicircle. The slope of log plot is ~1/2. This is shown in the lower inset of Fig. 5(b).

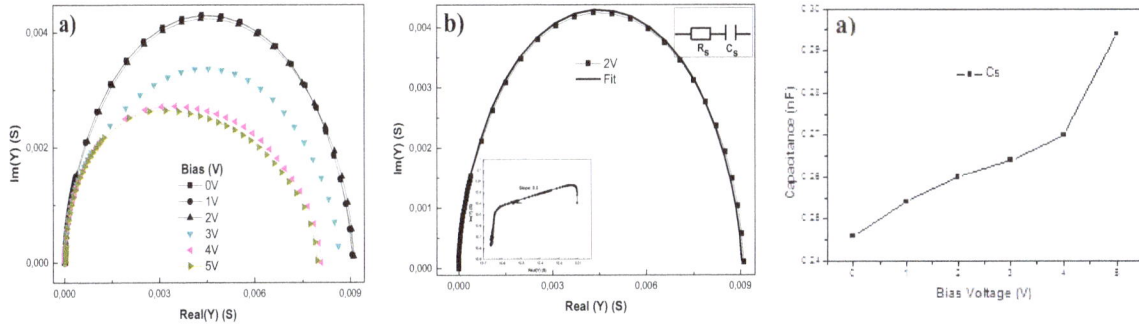

Figure 5. Cole-Cole plots of complex admittance of the ITO/β-CD/5CB/β-CD/ITO cell
a) With variation of the bias voltage **b)** At 2V, Symbols denote experimental results and line show fit according to the equivalent circuit, the upper inset shows the equivalent circuit and lower inset shows the Cole-Cole plot in logarithmic mode **c)** Variation of the series capacitance with bias voltage.

The curvature of the semi-circle decreased with increasing the applied dc bias. This decrease is proportional to the decrease of conductance (previous part Fig. 4(b) upper inset). This result can be explained by the blocking effect of alignment layer on the LC cell and the reaction of reorienting molecules which became more stable [21,25,26]. This result is in good agreement with the measurements of conductance. It is clear that the admittance for different values of the applied bias can provide a useful means for the existence of the blockage phenomena [25,26]. The admittance data of the cell at different bias voltages are fitted by the equivalent circuit given in Fig. 5(b) upper inset [21]. The fitting of the experimental data is in good agreement with the theoretical models as shown in Fig. 5(b) [26]. The resulting parameters of the electrical equivalent circuit are summarized in Table 2. The capacitance Cs increases with the increase of the bias voltage (Fig. 5(c)), it amounts to the barrier capacity at high frequency. However, the series resistance Rs is almost independent of the applied bias (Table 2).

Table 2: Fitting parameters of admittance data at different bias voltages.

Bias (V)	0	1	2	3	4	5
R1(Ω)	201.1	200.8	210.4	216.0	218.5	208.2
C1(nF)	0.246	0.254	0.260	0.264	0.270	0.294
Chi-Square	2.9E-2	8.48E-3	1.86E-2	1.63E-2	2.41E-2	5.32E-2

Impedance spectroscopy

The measurement of the dielectric relaxation at different frequencies gives information about the dynamics of polar groups and molecular motion [3]. In general, the complex impedance Z (ω) under a sinusoidal regime can be expressed as the following equation [3,26]:

$$Z (\omega) = Re(Z) + j\, Im(Z) = Z' + j\, Z'' \qquad (4)$$

Where Re(Z) = Z' and Im(Z) = Z'' represent real and imaginary parts of the impedance Z (ω), respectively [3].

Indeed, the real and imaginary parts of the impedance assigned to the circuit represented in Fig. 6 can be calculated according to the following equations, respectively:

$$Z' = R_s + \frac{1}{1+\omega^2 R_{P0}^2 C_{P0}^2} + \frac{1}{1+\omega^2 R_{P1}^2 C_{P1}^2} \qquad (5)$$

$$-Z'' = \frac{\omega^2 R_{P0}^2 C_{P0}^2}{1+\omega^2 R_{P0}^2 C_{P0}^2} + \frac{\omega^2 R_{P1}^2 C_{P1}^2}{1+\omega^2 R_{P1}^2 C_{P1}^2} \qquad (6)$$

Where ω is related to the angular frequency of the applied excitation of the circuit.

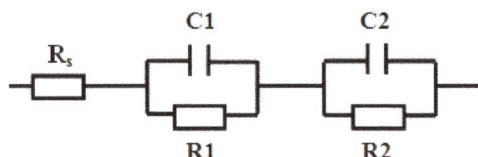

Figure 6. Equivalent electrical circuit model corresponding to the ITO/β-CD/5CB/β-CD/ITO cell. R_s is the resistance of the electrode (ITO). R_1 and C_1 are associated to the cell volume; R_2 and C_2 correspond to the LC/(β-CD) interface.

- *Variation of the real and imaginary parts of the impedance with frequency*

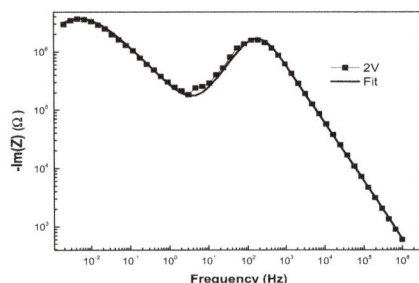

The frequency dependence of the real and imaginary parts of the impedance at 2V is shown in Fig. 7. The imaginary impedance part shows two peaks (Fig. 7).

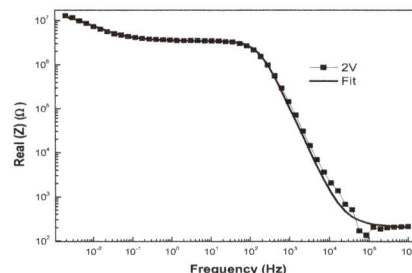

Figure 7. Imaginary and Real parts of the complex impedance as a function of the frequency of the ITO/β-CD/5CB/β-CD/ITO cell at the bias voltage (2V). Symbols are measured data and the solid lines are the fitting ones.

These peaks prove the existence of two kinds of dielectric relaxation in the sample [1,4,9]. Thus, we can propose that the change in dielectric relaxation is related to the molecular reorientation of LC molecules (dipolar orientation) [5,25,27] in the high-frequency range which is triggered by external applied voltage [1,4,8] and an ionic relaxation at the LC-electrodes interfaces in the lower one [9,27]. On the basis of Murakami's study [9,28], we can consider that the relaxation in the low- frequency domain is due to the formation of Helmholtz double layer at the electrode/5CB interfaces. The double layer is formed via the adsorption of impurity ions in the LC 5CB onto the electrode surfaces 28 of the cell and therefore the electrode/5CB interface behaves as a parallel-plate capacitor [8,9]. The dielectric relaxation due to the orientational polarizability was observed in frequency higher than 10^3 Hz. It provided information on the molecular rotation [1,8,28]. In the low frequency regime (10^{-1} -10^3 Hz), electrode polarization was observed in NLCs as well as alignment layers [1,19,28]. The occurrence of this relaxation is related to ionic impurities present in the materials [2,4].

- *Complex impedance analysis*

To understand the dynamic response in the liquid crystal cell, a Cole-Cole plot, is shown in Fig. 8(a). Cole-Cole curves give useful information about the relaxation mechanism of the cells [3]. The arc law is due to a distribution of relaxation times [3,4,28].

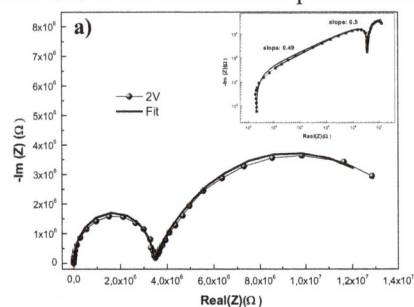

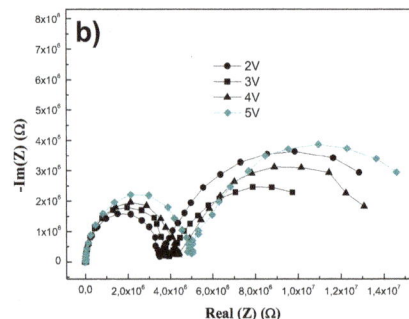

Figure 8. Cole-Cole plots of complex impedance of the ITO/β-CD/5CB/β-CD/ITO cell **a)** at the bias voltage (2V). Symbols are measured data and the solid lines are the fitting ones. The upper inset shows the Cole-Cole plot in logarithmic mode with slope ~ 0.5 **b)** at 2, 3, 4 and 5 bias voltages.

The Cole-Cole plot explains the kinds of dielectric relaxation existing in the frequency

dependent response of the cell. So, the Cole-Cole plot shows the presence of two dielectric relaxations in the complex impedance at 2V. As can be seen on logarithmic mode, the slope of the plots is ~ 0.5 (Fig. 8(a) upper inset) which means that the curvature is semicircle. The first one is a half circle which is supposed to be due to a molecular orientational polarization [5,29], and the second one is at low-frequency response (below 100Hz), which is possibly due to an interfacial polarization [1,29]. The dielectric relaxation in the low-frequency regime was attributed to the ionic space-charge polarization or to the electrical double layer formed at the electrode/5CB interface [1,9]. The ionic space-charge polarization arises from long-range movements of impurity ions in 5CB. The Cole-Cole plots have the same behavior for different bias voltages (2, 3, 4 and 5 V) (Fig. 8(b)). The impedance values change by a high amount when the applied voltage is increased. The impedance is dependent on the applied voltage. This is due to the anisotropy property of the samples [3]. The plot shows that the first semicircle's curvature (size) increases with increasing bias voltage. This suggests that the relaxation process are associated with the reorientation of LC molecules [4,9],

however the second semicircle decreases in size with increasing bias voltage, as illustrated in Fig. 8(b).

To gain more information on the relaxation mechanism, we assimilated the LC cell to an electrical equivalent circuit. This latter allows the analysis of the Cole-Cole plot and includes contributions from the electrodes, the LC alignment layer and the LC in the volume of the cell. On account of the existence of two relaxation times, we assign, in the modeling, a parallel resistor capacitor circuit for LC/β-CD interface (R_2 and C_2) and another one for the volume (R_1 and C_1) [9]. These circuits are linked serially to a resistance R_s associated with the ITO electrode, which was found to be approximately 200 Ω. The series resistance values R_s are small but they remain constant for higher bias voltages [29].

- *Parameters obtained by modeling the impedance spectra*

The impedance data of the cell at different bias voltages are fitted by the equivalent circuit given in Fig. 6. The fitting of the experimental data is in good agreement with the theoretical models as shown in Fig. 8(a) [6].

Table 3: Fitting parameters of impedance data at different bias voltages.

Bias (V)	0	1	2	3	4	5
R_s (Ω)	186	186.2	198.4	198.5	212.6	199.4
R_1 (M Ω)	2.845	2.893	3.42	3.81	4.23	4.85
C_1 (nF)	0.258	0.262	0.266	0.268	0.272	0.298
n_1	0.9927	1	1	1	1	1
R_2 (M Ω)	*	73.8	12.67	9.85	9.51	8.39
C_2 (μF)	0.9673	0.9549	0.925	0.848	0.815	0.811
n_2	0.729	0.671	0.716	0.688	0.75	0.729
Chi-Square	6.387E-2	3.2759E-2	6.87E-2	5.631E-2	1.38E-2	1.01E-2
τ_1 (ms)	0.73401	0.75796	0.90972	1.02108	1.15056	1.4453
τ_2 (s)	*	70.471	11.719	8.3528	7.75065	6.8042
f_1 (Hz)	216.82	210.08	175.03	155.94	138.39	110.17
f_2 (Hz)	*	0.00226	0.01358	0.01906	0.020545	0.0234025

The resulting parameters of the electrical equivalent circuit are summarized in Table 3. The capacitance values C_1 and C_2 are of the order of nF and μF, respectively, which is in agreement with previous works [9]. This is also the case for the high

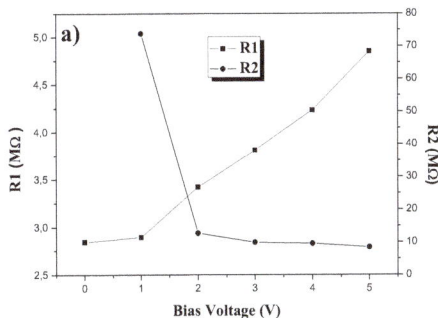

resistance values (order of magnitude of MΩ). It is seen that R_1 values increase with increasing the applied voltage (Fig. 9(a) upper inset). This suggests that the R_1 values related to the relaxation process are associated with the reorientation of molecules [2].

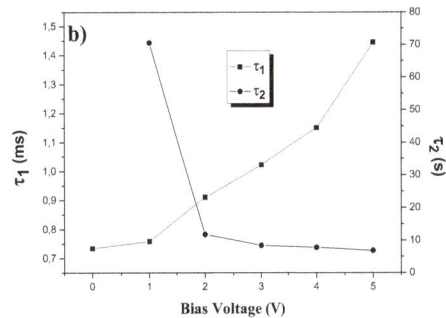

Figure 9. a) Variation of the parallel resistance (R_1, R_2) with bias voltage **b)** Variation of the relaxation times (τ_1, τ_2) with bias voltage.

Moreover, Fig. 9(a) shows the variation of R_2 with the applied bias. The resistance R_2 decreases with the increase of the bias voltage. This is due to the increase in the number of injected carrier into molecules. The carriers may be trapped by structural defects or chemical impurities. The impurity ions in 5CB adsorb on the electrodes and form Helmholtz double layers. Since some of the ions may be incorporated during the synthesis of 5CB and cell manufacturing, many ionic species exist in the 5CB cells.

The Nyquist plot also can give the information about the relaxation time, τ, as in the following equation [1,30].

$$\tau_i = \frac{1}{2\pi f i} = \frac{1}{\omega i} = R_i C_i \qquad \text{Where i=1, 2} \qquad (7)$$

The relaxation time τ_1 (order of ms) and τ_2 (order of s) are suitable for the contribution to the orientationel relaxation and the ionic relaxation, respectively (Table 3) [9].

The relaxation time τ_1 increases with increasing bias voltage (Fig. 9(b)). It suggests that the relaxation process is associated with the rotational motion of the LC molecules. The millisecond values obtained for τ_1 indicate that the relaxation is dipolar [1,3]. On the other hand, the relaxation time τ_2 decreases with the increase in the bias voltage as given in Fig. 9(b). This behavior can be understood as follows: as the bias voltage is increased, a large number of charge carriers are injected into the devices, resulting in a decrease in the dielectric relaxation time [24].

For liquid crystal devices, the relaxation time is of extreme importance, which further depends on the cell gap, the degree of molecular alignment and many other parameters. When a high voltage was applied to the sample cell, the relaxation time was affected.

Conclusion

In this paper, we presented the fabrication and the characteristics of nematic liquid crystal (NLC) cell using Beta Cyclodextrins (β-CD) as alignment layers. We found that the hydrophilic surface and its morphology play an important role in the planar alignment of liquid crystal.

Impedance spectroscopy was used to report the effect of this alignment layer on dielectric properties and the liquid crystal molecular orientation. It was found that this layer has a blocking effect on the LC cell. Therefore, we studied the dielectric relaxation behavior of the cell. Two kinds of relaxation were observed; the first was attributed to the reorientation of the LC molecules in the volume and the second was associated with the ionic relaxation at the LC-(β-CD) interfaces of the cell. We obtained good agreement between the experimental and the fitted curves. We also determined the different component parameters of the electrical equivalent circuit and we obtained the expected values.

References

1- S.S. Parab, M.K. Malik, R.R. Deshmukh, Dielectric relaxation and electro-optical switching behavior of nematic liquid crystal dispersed in poly (methyl methacrylate), J. Non-Cryst. Solids. 358 (2012) 2713-2722.

2- O. Köysal, M. Okutan, M. Gökçen, Investigation of dielectric properties and diffraction efficiency enhancements caused by photothermal effect in DR9 dye-doped nematic liquid crystal, Opt. Commun. 284 (2011) 4924-4928.

3- M. Okutan, F. Yakuphanoglu, S.E. Sana, O. Koysal, Impedance spectroscopy and dielectric anisotropy-type analysis in dye-doped nematic liquid crystals having different preliminary orientations, Physica B. 368 (2005) 308-317.

4- M. Okutan, F. Yakuphanoglu, O. Koysal, M. Durmus, V. Ahsen, Dielectric spectroscopy analysis in employing liquid crystal phthalonitrile derivative in nematic liquid crystals, Spectrochim. Acta, Part A. 67 (2007) 531-535.

5- F. Al-Hazmi, A. Al-Ghamdi, N. Al-Senany, F. Alnowaiser, F. Yakuphanoglu, Dielectric anisotropy and electrical properties of the copper phthalocyanine (CuPc): 4-40-n Heptylcyanobiphenyl (7CB) composite liquid crystals, Composites Part B. 56 (2014) 15-19.

6- R. Manohar, A.K. Misra, D.P. Singh, S.P. Yadav, P. Tripathi, A.K. Prajapati, M.C.Varia, Dielectric, thermal and optical study of an unusually shaped liquid crystal, J. Phys. Chem. Solids. 71 (2010) 1684-1689.

7- S. Peralta, F. Hapiot, Y. Barbaux, M. Wrentham, Alignment of nematic crystal using substituted calixarene Langmuir-Blodgett films, Liq. Cryst. 30 (2003) 436-469.

8- M. Rahman, C.W. Hsieh, C.T. Wang, B.R. Jian, W. Lee, Dielectric relaxation dynamics in liquid crystal-dye composites, Dyes Pigm. 84 (2010) 128-133.

9- A. Abderrahmen, F. F. Romdhane, H. Ben Ouada, A. Gharbi, Investigation of the liquid crystal alignment layer: effect on electrical properties, Sci. Technol. Adv. Mater. 9 (2008) 025001.

10- B.J. Shen, Z. Xie, Y.N. He, Y.Q. Lian, Cinnamate-functionalized hyperbranched polymeras liquid crystal photo-alignment layer, Chin. Chem. Lett.19 (2008) 1131-1134.

11- M. Echabaane, A. Rouis, I. Bannamour, H. Ben Ouada, Optical electrical and sensing properties of β-ketoimine calix[4]arene thin films, Mater. Chem. Phys. 141 (2013) 781-789.

12- A. Gaied, N. Jaballah, S. Teka, M. Majdoub, A water-Insoluble β -Cyclodextrin Derivative for Hydroquinone Sensor Applications, International Peer Reviewed Journal,

2014, 3 (4): 1655-1664.

13- R. Mlika, S. Hbaieb, R. Ben Chaabene, Y. Chevalier, R. Kalfat, H. Ben Ouada, Electrochemical properties of gold electrodes functionalized by new pseudo-polyrotaxanes of polyaniline and chemically modified β-cyclodextrin inclusion complex, Synth. Met. 162 (2012) 186-192.

14- M. Raoov, S. Mohamad, M.R. Abas, Synthesis and characterization of β-cyclodextrin functionalized ionic liquid polymer as macroporous material for the removal of phenols and As(V), Int. J. Mol. Sci. 15 (2014) 100-119.

15- P. Upadhyay, M.K. Rastogi, D. Kumar, Polarizability study of nematic liquid crystal 4-cyano-40-pentylbiphenyl (5CB) and its nitrogen derivatives, Adv. Chem. Phys. 456 (2015) 41-46.

16- P. Jedlovszky, L.B. Pártay, Investigation of the saturated adsorption layer of 5-cyano-biphenyl and 5-cyano-terphenyl at the free water surface by Monte Carlo simulation, J. Mol. Liq. 136 (2007) 249-256.

17- A. Gunaratne, H. Corke, Effect of hydroxypropyl b-cyclodextrin on physical properties and transition parameters of amylase-lipid complexes of native and acetylated starches, Food Chem. 108 (2008) 14-22.

18- A. Rouis, C. Dridi, I. Bonnamour, J. Davenas, H. Ben Ouada, Transport mechanism and trap distribution in ITO/azo-calix[4]arene derivative/Al diode structure, Physica B. 399 (2007) 109-115.

19- P. K. Tripathi, A. Misra, K. Pandey, R. Manohar, Study on dielectric and optical properties of ZnO doped nematic liquid crystal in low frequency region, Chem. Commun. 1(2013)20-26.

20- R. Ploeger, S. Musso, O. Chiantore, Contact angle measurements to determine the rate of surface oxidation of artists' alkyd paints during accelerated photo-ageing, Prog. Org. Coat. 65 (2009) 77-83.

21- M.C. Petty, C. Pearson, A.P. Monkman, R. Casalini, S. Capaccioli, J. Nagel, Application of impedance spectroscopy to the study of organic multilayer devices, Colloids and Surfaces A: Physicochem. Eng. Aspects. 171 (2000) 159-166.

22- R. Mlika, A. Rouis, I. Bonnamour, H. Ben Ouada, Impedance spectroscopic investigation of the effect of thin azo-calix[4]arene film type on the cation sensitivity of the gold electrodes, Mater. Sci. Eng., C. 31 (2011) 1466-1471.

23- Y. Shiraishi, T. Uehara, H. Sawai, H. Kakiuchi, S. Kobayashi, N.Toshima, Electro-optic properties of liquid crystal devices doped withcucurbit(6)uril-protected zirconia nanowires, Colloids and Surfaces A: Physicochem. Eng. Aspects. 460 (2014) 90-94.

24- I. Zahou, R. Ben Chaabane, R. Mlika, S. Touaiti, B. Jamoussi, H. Ben Ouada, Optical and electrical properties of novel peripherally tetra and mono -quinoleinoxy substituted metallophtalocyanines, J Mater Sci: Mater Electron. DOI 10.1007/s10854-014-1955-6.

25- A.K. Jonscher, Alternating Current Diagnostics of Poorly Conducting thin films, Thin Solid Films. 36 (1976) 1-20.

26- R.P. Suvarna, K.R. Rao, K. Subbarangaiah, A simple technique for a.c. conductivity measurements, Bull. Mater. Sci. 25 (2002) 647–651.

27- F.M. Aliev, Z. Nazario, G.P. Sinha, Broadband dielectric spectroscopy of confined liquid crystals, J. Non-Cryst. Solids. 305 (2002) 218-225.

28- S. Murakami, H. Iga, H. Naito, Dielectric properties of nematic liquid crystals in the ultralow frequency regime, J. Appl. Phys. 80 (1996) 6396-6400.

29- J. Ahn, D. Chung, J. Lee, G. Lee, M. Song, W. Lee, W. Han, T. Kim, Equivalent-Circuit Analysis of Organic Light-Emitting Diodes by Using the Frequency-Dependent Response of an ITO/Alq3/Al Device, J. Korean. Phys. Soc. 46 (2005) 546-550.

30- M. Kondo, N. Suzuki, Y. Nakajima, T. Tanaka, T. Muroga, Electrochemical impedance spectroscopy on in-situ analysis of oxide layer formation in liquid metal, Fusion Eng. Des. 89(2014)1201-1208.

24

Phenolic composition and antioxidant, antimicrobial and cytochrome P450 inhibition activities of *Cyperus rotundus* tubers

Ismahen Essaidi [1,2], Zeineb Brahmi [3], Hayet Ben Haj Koubaier [1,2], Ahmed Snoussi [1,2], Hervé Casabianca [4], Naoki Abe [3] and Nabiha Bouzouita [1,2]*

[1] Ecole Supérieure des Industries Alimentaires, 58, Avenue Alain Savary; 1003 Tunis, Tunisia.
[2] Faculté des Sciences de Tunis ElManar, Laboratoire de Chimie Organique Structurale : Synthèse Chimique et Analyse Physico-chimique, Tunis, Tunisia.
[3] Department of Nutritional Science, Faculty of Applied Bio-Science, Tokyo University of Agriculture, 1-1-1 Sakuragaoka, Setagaya-Ku, Tokyo 156-8502, Japan.
[4] Centre National de Recherche Scientifique, Service Central d'Analyse, 5 rue de la Doua 69100, Villeurbanne, France

Abstract: The chemical composition of a *Cyperus rotundus* methanol extract was investigated. The phenol content determined by the Folin-Ciocalteu method was 83.6 ± 5.42 mEGA.g^{-1} and the flavonoid content assayed by the AlCl$_3$ method was 32.65 ± 3.5 mEQ.g^{-1}. The phenolic composition determined by HPLC-DAD showed the presence of phenolic acids (chlorogenic, caffeic, ferulic, syringic acids) and flavonoids (luteolin, quercetin, apigenin, myricetin). The antioxidant activity of the extract was evaluated using the 1,1-diphenyl-2-picrylhydrazyl (DPPH·) and 2,2'-azino-bis(3-ethylbenzothiazoline-6-sulphonic acid)(ABTS·$^+$) radical scavenging assays and the β-carotene bleaching test. The EC$_{50}$ values were 5.76 ± 0.83 µg/mL and 18.08 ± 0.6µg/mL for DPPH· and ABTS·$^+$, respectively. The antioxidant activity coefficient (AAC) was 670.8 ± 4.2. The antibacterial activity determined by the disc diffusion and submerged culture methods against four Gram negative and two Gram positive bacteria showed effects against all strains. Cytochrome P450 (CYPs) inhibitory activity was evaluated against CYP1A2, CYP3A4 and CYP2D6, which was the most inhibited (IC$_{50}$ =11.13 ± 0.04µg/mL).

Keywords: *Cyperus rotundus*, chemical composition, antioxidant activity, antimicrobial activity, cytochrome inhibition

Introduction

For many years, plants have been used as therapeutic resources either as herbal teas or other homemade remedies, as crude extracts, or as standard enriched fractions in pharmaceutical preparations such as tinctures, fluid extracts, powders, pills, and capsules[1,2]. Plants are rich sources of bioactive molecules, most of which probably evolved as chemical defenses against infections[3]. Increasingly, natural products have been studied for their antibacterial antioxidant and anticancer activities[4,5]. Recently, the ability of numerous herbs to act as substrates or inhibitors for different cytochrome P450 (CYPs) has been reported[6]. The relationship between cytochrome P450 and chemical carcinogenesis has been extensively studied. CYP activation of certain anticancer drugs has long been known and the importance of this process as a way of targeting novel anticancer therapy is being explored.

However, certain CYPs involved in hormone and vitamin metabolism and in the metabolic activation of genotoxic substances have been reported to play important roles in tumor formation and development[7].

Cyperus rotundus L., (family Cyperaceae), also known as purple nutsedge or nutgrass, is a common perennial weed with slender, scaly creeping rhizomes, bulbous at the base and arising singly from tubers which are about 1-3 cm long[8]. *C. rotundus* is widely distributed in the Mediterranean basin areas[9]. In Tunisia, it is widespread in the northeast, center and south of the country[10]. *C. rotundus* is a traditional herbal medicine; the tuber part is one of the oldest known medicinal elements used for the treatment of dysmenorrhea and menstrual irregularities. Infusions of this herb have been used for pain, fever, diarrhea[11].

A number of pharmacological and biological activities including anti-inflammatory, antidiabetic, antidiarrhoeal, cytoprotective, antimutagenic antimicrobial, antioxidant, cytotoxic and apoptotic, anti-pyretic and analgesic activities have been reported for this plant[12,13].

Corresponding author: Nabiha Bouzouita
E-mail address: bouzouita.nabiha@gmail.com

Therefore, the objectives of the present study were to characterize the chemical composition of *C.rotundus*, to assess the biological activities of a methanol extract as an antioxidant and as an antimicrobial agent against standard microorganisms, and to investigate for the first time its inhibitory effect on three CYPs isoforms, namely CYP1A2, CYP3A4 and CYP2D6.

Experimental Section

Plant material and reagents

Cyperus rotundus tubers were collected from Kebili in the southwestern part of Tunisia. Samples were authenticated by the National Institute of Agronomic Research of Tunisia (INRAT).Vivid CYP blue screening kits were purchased from Invitrogen Corporation (Carlsbad, CA, USA). All the other reagents were of the highest purity available and were purchased from the Sigma-Aldrich Chemical Company.

Preparation of methanol extract

Fifty grams of *C. rotundus* tubers were crushed and extracted with 80% aqueous methanol (3 x 300 mL) by agitated maceration at room temperature for 72 h. The extracts obtained from three extractions were combined, filtered through a Whatman No.4 filter paper and concentrated under reduced pressure.

Total phenolic and flavonoid contents

The total phenolic content was assayed using the Folin-Ciocalteu reagent according to the method modified by Turkoglu et al. (2006)[14]. An aliquot (0.1mL) of a suitable diluted extract was added to 0.5 mL of the Folin-Ciocalteu reagent and 1mL of deionized water. The mixture was shaken and allowed to stand for 1 min, before adding 1.5 mL of 20% Na_2CO_3 solution. The absorbance at 760 nm was recorded after two hours.

The total flavonoid content was measured according to Al-Dabbas et al. (2006)[15], in which 1.5 mL of the extract was mixed with 1.5 mL of 2% $AlCl_3$ methanol solution. After 10 min, the absorbance was read at 367.5 nm.

Total phenolic and flavonoid contents are expressed in mg gallic acid equivalents per gram of extract (mg GAE/g) and in mg quercetin equivalents per gram of extract (mg QE/g), respectively. Analyses were performed in triplicate.

HPLC-DAD analysis

Chromatographic separation was carried out on a HPLC-DAD Agilent 1200 using a C18 column kept at 30°C. A gradient solvent comprised of 0.5% acetic acid in water (A) and methanol (B) was applied for a total running time of 40 min. The following proportions of solvent B were used for elution: 0.5-20 min, 0-85%; 20-30 min, 85%; and 30-40 min, 85-0%. The flow rate was 0.5 mL/min, and 5 µL of each solution at 100 mg/mL was injected and detected at two wavelengths (254 and 330 nm).

Antioxidant activity

DPPH radical scavenging assay: *C. rotundus* methanol extract was tested for the scavenging effect on DPPH radicals according to the method of Al-Dabbas et al. (2006)[15], in which 2 mL of extract solution in methanol was added to 2 mL of a 10^{-4}mol/L DPPH methanolic solution. Solutions were then kept in the dark at room temperature for 30 minutes. The scavenging activity on the DPPH was determined by measuring the absorbance at 517 nm until the reaction reached the steady state, using a UV-vis spectrophotometer. As a positive control, synthetic antioxidant BHT was used. All determinations were performed in triplicate. The DPPH radical scavenging activity was calculated using Equation 1:

Equation 1. % inhibition = [1- A_1/A_0] x 100

where A_1 and A_0 are the absorbance of the tested sample and the blank after incubation, respectively.

ABTS radical cation decolorization assay : Spectrophotometric analysis of $ABTS^{·+}$ scavenging activity was done according to a previously described method[16]. ABTS radical cations were produced by reacting 2 mM ABTS in distilled water with 70 mM potassium persulfate ($K_2S_2O_8$) stored in the dark at room temperature for 24 h. Then, 1 mL of the ABTS radical cation solution was added to 1 mL of *C. rotundus* extract at different concentrations. The absorbance was measured at 734 nm, 30 min after mixing the prepared solution. The percentage of radical scavenging was calculated for each concentration relative to a blank containing no scavenger using Equation2:

Equation 2. Inhibition (%) = [($A_0−A_1$)/A_0] × 100

where A_0 is the absorbance of the control and A_1 is the absorbance of the sample. The sample concentration providing 50% of the free radical inhibition (EC_{50}) was calculated by plotting inhibition percentage against sample concentration. EC_{50} was expressed as Equivalents Trolox (TEAC)

β-Carotene bleaching test: The antioxidant activity of *C. rotundus* extract was determined according to a slightly modified version of the β-carotene bleaching method described by Suja et al. (2005)[17]. Two milligrams of β-carotene were dissolved in 5 mL of chloroform and 0.5 mL β-carotene solution was mixed with 20 mg of purified linoleic acid and 200 mg of Tween-40 emulsifier in a round bottom flask. Then chloroform was removed in a rotary vacuum evaporator. The resulting mixture was diluted with 50 mL of oxygenated distilled water. To 4 mL of this emulsion, 0.2 mL of the test sample in ethanol (0.2 mg/mL) was added. BHT was used for comparative purposes. A solution with 0.2 mL of ethanol and 4 mL of the above emulsion was used as control. A mixture prepared as above without β-carotene served as the blank, and was used to zero the spectrophotometer. The tubes were covered with aluminum foil and maintained at 50°C in a water bath. Absorbance of the emulsion at 470 nm was taken at baseline (t = 0 min) and after every 15 min. Measurement of absorbance continued until the color of β-carotene disappeared in the control reaction (t = 120 min). All determinations were performed in triplicate. Antioxidant activity coefficient (AAC) was calculated according to the following Equation 3:

Equation 3. AAC = [$A_{A\,(120)}$-$A_{C\,(120)}$ / $A_{C\,(0)}$-$A_{C\,(120)}$] × 1000

where $A_{A\,(120)}$ is the absorbance of antioxidant at 120 min, $A_{c\,(120)}$ is the absorbance of the control at 120 min and $A_{C\,(0)}$ is the absorbance of the control at 0 min.

Antibacterial activity

Disc diffusion method: The *in vitro* antibacterial activity of the tested sample was carried out by the disc diffusion method against two Gram positive bacteria (*Staphylococcus aureus* ATCC 25923 and *Streptococcus A*ATCC 11 700) and four Gram negative bacteria (*Pseudomonas aeruginosa* ATCC 9027, *Salmonella enteritidis* ATCC 14 028, *Escherichia coli* ATCC 25922 and *Klebsiella pneumoniae* ATCC 13833). In this test, nutrient agar (NA) was used as the culture medium[18].The NA plates were prepared by pouring 15 mL of molten medium into sterile Petriplates. The plates were allowed to solidify for 5 min and 100μL of the inoculum suspension was swabbed uniformly; the inoculum was allowed to dry for 5 min. Different concentrations of extract dissolved in dimethyl sulfoxide (DMSO) (20, 50, 75 and 100 mg/L) were loaded on 6 mm sterile individual discs. The loaded discs were placed on the surface of the medium and the compound was allowed to diffuse for 5 min. The plates were then incubated at 37°C for 24 h. The negative control was prepared using the respective solvent. At the end of incubation, inhibition zones that had formed around the discs were measured with a transparent ruler in millimeters. The test was performed in triplicate.

Submerged culture method: The MIC and MBC were determined using the submerged culture method[19] with some modifications. The activities of six strains of microorganism were determined as follows: equal volumes of each bacterial strain culture, containing approximately 1 10^6CFU/mL, were applied onto flasks containing 20 mL of nutrient broth and the methanol extract at concentrations ranging from 20 to 5000 mg/L. These serially diluted cultures were then incubated at 37°C for 18 h, then 1 ml of the culture medium was removed from each broth assay flask and sub-

cultured in fresh nutrient agar. After incubation at 37°C for 24 h, the MIC was determined as the lowest concentration at which the microorganism did not demonstrate visible growth, and the least concentration showing no visible growth on sub-culture was taken as the MBC.

Cytochrome P450 enzyme inhibition

The inhibitory effect of the methanol extract was determined using 96-well microtiter plates (Sumilon 96F)[20], based on reading the fluorescence of 3-cyano-7-hydroxycoumarin produced by CYP hydroxylase activities at an excitation wavelength of 460 nm and an emission wavelength of 409 nm. CYP inhibitory activity was assayed using the vivid CYP Blue Substrate. 10μM 7-benzyloxymethyloxy-3-cyanocoumarin (BOMCC) was the substrate used for CYP3A4, and 7-ethyloxymethyloxy-3-cyanocoumarin (EOMCC) was the substrate used for CYP1A2 (3μM) and CYP2D6 (10μM). The positive control was erythromycin for CYP3A4, safrole for CYP1A2 and cimetidine for CYP2D6. In the first step, serial dilutions of the methanol extract were performed by distilled water. In the second step, 40 μL of each sample was added to 50 μL of a CYP/NADPH-CYP reductase mixture in a 96-well plate (5nM of CYP1A2, 10nM of CYP2D6 and 5nM of CYP3A4). The obtained solution was then pre-incubated for 20 min at room temperature. The reaction was then initiated by adding 10 μL of the substrate/NADP+ mixture. The plate was incubated for 15 min. Finally, the reaction was ended by the addition of 10 μL of each stop solution (10 μM of ketoconazole for CYP3A4, 30 μM α-naphthoflavone for CYP1A2 and 15 μM quinidine for CYP2D6). Activity was measured as the rate of fluorescent metabolite production over the course of the reaction. The IC_{50} values were calculated by linear interpolation. The test was performed in triplicate.

Statistical analysis

Results are expressed as mean ± SD of three independent experiments. Statistical significant differences were considered at $p < 0.05$ using one-way analysis of variance (ANOVA) with Statgraphics Centurion.

Results and Discussion

Chemical composition

The total phenol content of *C. rotundus* extract, reported as mg gallic acid equivalents/g dried extract, was estimated to be equivalent to 83.6 ± 5.42 mg EGA/g; this value is comparable to that found by Ardestani and Yazdanparast, (2007)[9] which was 78.55 mg EGA/g in a dried extract. The flavonoid content was equal to 32.65 ± 3.5 mg EQ/g. The analysis of methanol extract composition by HPLC-DAD allowed us to characterize its phenolic composition.

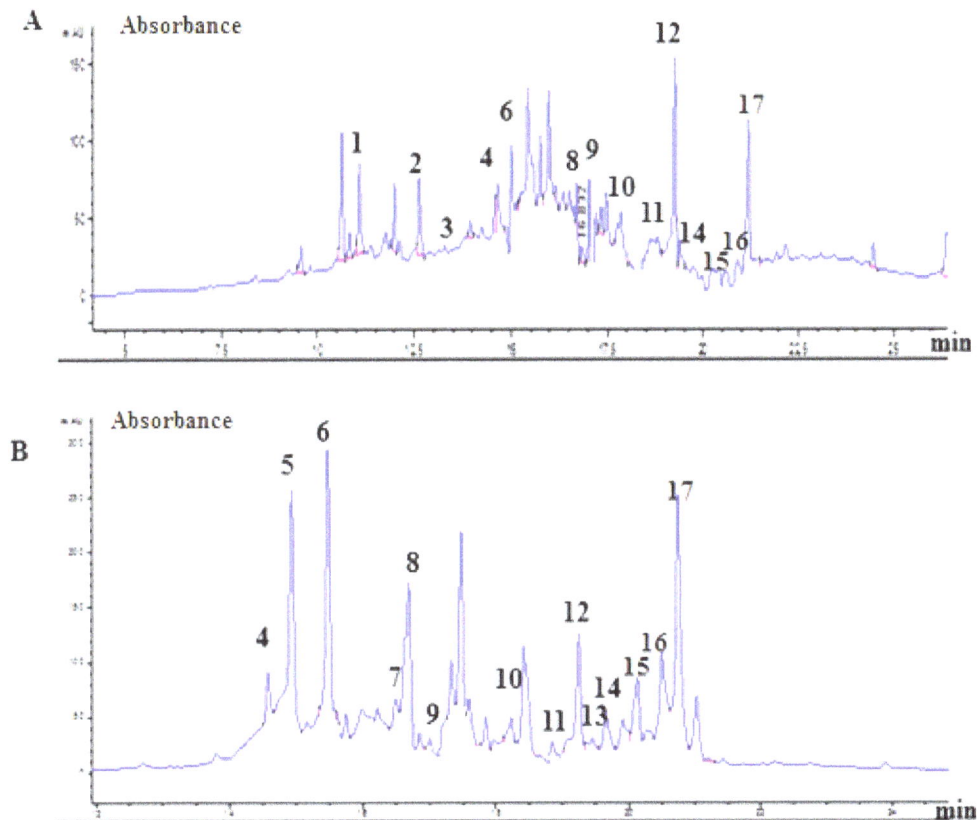

Figure 1. HPLC-DAD chromatograms of *Cyperus rotundus* methanol extract at 254 nm (A) and 330 nm (B)

The chemical composition of *Cyperus* extract (Fig.1) was characterized by the presence of phenolic acids (chlorogenic (1), caffeic (2), ferulic (3), *p*-coumaric (4), syringic (5), cinapic (6), salicylic (7), ellagic (8) and trans-cinnamic (9)) and flavonoids (myricetin (10), quercetin (11), luteolin (12), hesperetin (13), genistein (14), kaempferol (15), apigenin (16) and rhamnetin (17)) which can be classified into several groups (hydroxycinnamic acids, hydroxbenzoic acids, flavanols, flavanones and flavones).

Previous RP-HPLC analysis of *C. rotundus* revealed the presence of phenolic compounds such as gallic acid, caffeic acid, salicylic acid,*p*-coumaric acid, luteolin, quercetin, kaempferol and epicatechin[21,22].

The crude extract of *C.rotundus* was used to assess the antioxidant and antimicrobial activities and the inhibitory effect against cytochrome P450. The HPLC/DAD analysis of the plant material allowed us to identify the individual phenolic compounds present in the studied material that could participate in the biological activities of the extract. However, the use of the crude extract to assess the different biological activities could be more informative than using individual phenolic compounds. The extract may be more beneficial than isolated constituents, since a bioactive individual component can change its properties in the presence of other compounds present in an extract[23].

Antioxidant activity

The antioxidant activity of the *C.rotundus* methanol extract was assessed by radical scavenging assays using DPPH and ABTS radicals and the β-carotene bleaching method (Figure 2).

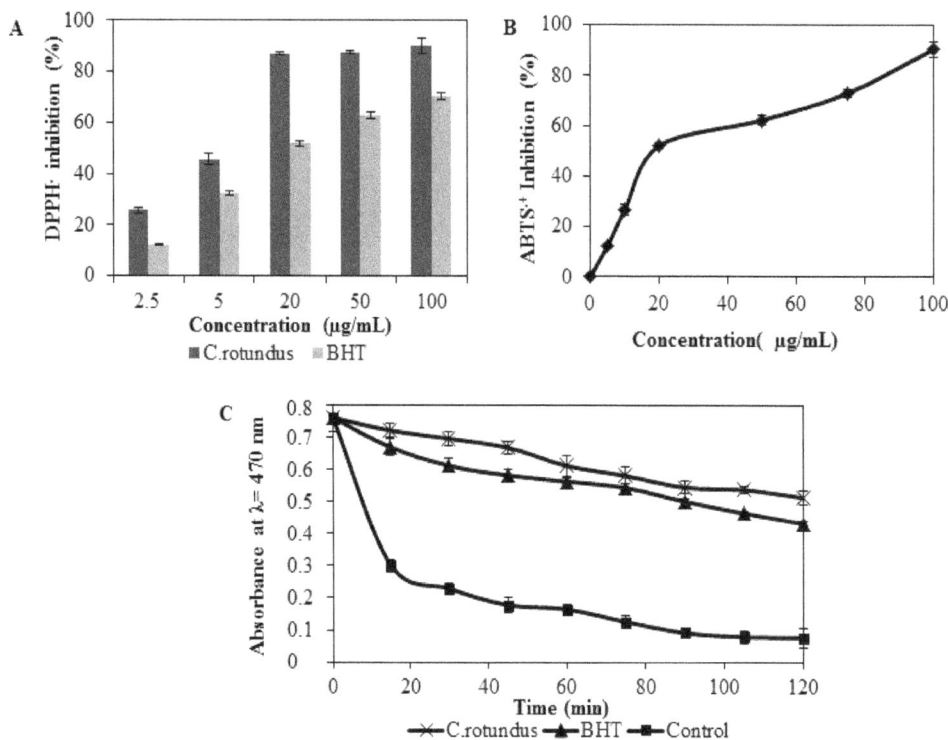

Figure 2. Evaluation of antioxidantactivity of *Cyperusrotundus*methanolextract by DPPH· test (A), ABTS·+ test (B) and β-carotenebleachingmethod (C)

Free radical scavenging tests

As shown in Fig.2A and Fig.2B, the *C. rotundus* extract (CRE) exhibited potent radical scavenging ability in a dose-dependent manner, pointing to its antioxidant activity. The antioxidant ability of CRE to scavenge purple colored DPPH· And blue-green colored ABTS·+radicals was compared to that of BHT and Trolox, respectively. CRE showed free radical scavenging activity with high potency. The DPPH radical scavenging activity provided an EC_{50} value of 5.76 ± 0.83 μg/mL (lower than that of BHT, 18.4± 1.3μg/mL).The EC_{50} value for the ABTS·+ radical was 18.8 ± 0.6μg/mL which represented 12072.87 μM ETrolox/g of the extract as the TEAC value.

β-carotene bleaching test

As shown in Figure 2C, there was a considerable decrease in absorbance in the control sample. This reduction was due to the accumulation of malondialdehyde compounds from linoleic acid oxidation, which is not stable. In the presence of CRE and BHT, the absorbance is more stable, which can be explained by their antioxidant activity based on their capacity to reduce malondialdehyde formation. Further oxidation causes malondialdehyde to be converted to secondary products such as alcohols and acids that cannot be detected. The antioxidant activity of CRE was found to be higher than that of BHT with an AAC of 670± 4.2.

The present results are in agreement with those from previous research by Kilani et al. (2011)[13], Nagulendran et al. (2007)[21] and Singh et al. (2012)[24] who demonstrated the antioxidant activity of *Cyperus* extracts. The antioxidant activity of CRE could be related to its phenolic composition. There seems to be a good correlation between its phenolic content and the antioxidant activity.

The presence of some phenolic acids such as chlorogenic and caffeic acids as well as the flavonoids quercetin and luteolin have been shown to have important antioxidant activity[25].

Antibacterial activity

The crude extract was screened for its antibacterial activity against *S. aureus, Strept. A, E. coli, S. enteritidis, P. aeruginosa,* and *K. pneumoniae* using the disc diffusion and submerged culture methods. The choice of these microorganisms was made due to the fact that some of them are causative agents of human intestinal infection.

Table 1. Inhibition diameters of *Cyperus rotundus* methanol extract against six bacterial strains

Bacterial strain	Inhibition Diameter (mm)				
	Antibiotic	Extract Concentration mg/L			
	50μg/mL	20	50	75	100
Gram+ bacteria					
Staphylococcus aureus ATCC 25923	15.7±1.2	6.0±0.5	7.0±0.4	11.0±1.0	13.0±1.2
Streptococcus A ATCC 11 700	14.2±0.8	4.0±0.2	5.0±0.7	9.0±1.1	11.0±0.9
Gram- bacteria					
Pseudomonas aeruginosa ATCC 9027	2.1±0.5	0	0	3.0±0.8	5.0±0.3
Salmonella enteritidis ATCC 14 028	14.7±1.4	3.0±0.1	5.0±0.3	7.0±0.7	9.0±1.2
Esherichia coli ATCC 25922	12.3±0.5	2.0±0.5	3.0±0.2	6.0±0.4	8.0±0.0
Klebsiella pneumoniae ATCC 13 833	12.6±1.5	0	2.4±0.9	5.0±0.6	7.0±0.3

Means ± SD (n = 3), Internal diameter = 6mm, Inhibition diameter = external diameter-internal diameter, significant difference (p<0.05)

The inhibition diameters (mm) of the extract on the selected microorganisms are shown in Table 1. The results indicated greater inhibition at the highest concentration. This is in agreement with a previous report[26] that the mode of action of plant extracts is concentration-dependent. It was found that Gram positive bacteria were more sensitive than Gram negative bacteria. *S.aureus* presented the highest inhibition zone of (13 mm) at 100 mg/L while *P.aeruginosa* was the most resistant bacterium with only a 5 mm inhibition zone at the same concentration.

Several studies have demonstrated the higher sensitivity of Gram positive bacteria compared to Gram negative which can be attributed to the difference in the outer layers of Gram negative and Gram positive bacteria. Gram negative bacteria, regardless of the cell membrane, have an additional layer of the outer membrane, which consists of phospholipids, lipopolysaccharide and proteins; the membrane is impermeable to most molecules. Nevertheless, the presence of pores in this layer will allow the free diffusion of molecules with a molecular weight below 600 Da[27].

Table 2. Minimum inhibitory concentration (MIC) and minimum bactericidal concentration (MBC) of *Cyperus rotundus* methanol extract

Strains	MIC (mg/L)	MBC (mg/L)
Staphylococcus aureus ATCC 25923	20	200
Streptococcus A ATCC 11 700	50	250
Pseudomonas aeruginosa ATCC 9027	250	>5000
Salmonella enteritidis ATCC 14 028	75	1500
Esherichia coli ATCC 25922	150	2000
Klebsiella pneumoniae ATCC 13 833	200	3000

This result was confirmed by the submerged culture method as shown in Table 2. The MIC and the MBC showed that *S. aureus* was the most sensitive bacterium with MIC and MBC values of 20 and 200 mg/L respectively. The antimicrobial activity of several *C. rotundus* extracts was previously investigated on Gram positive and Gram negative bacteria; the ethyl acetate and methanol extracts showed significant inhibitory activity against *S. enteritidis*, *S. aureus* and *E. Faecalis*[13,24]. The antimicrobial capacity of CRE is attributed to its phenolic compounds which are well-known for their antimicrobial activity[28].

The antimicrobial properties of phenolic compounds have been reported many times. Cowan (1999)[29] showed that quercetin is one of the most active principal antimicrobial agents in plant extracts. Caffeic acid, chlorogenic acid and gallic acid have been shown to exhibit some antimicrobial properties[29,30].

Cytochrome P450 enzyme inhibition

The inhibitory effect of CRE was investigated against three CYP enzymes .The results show that CYP inhibition was dose-dependent (Data not shown). The inhibition percentages of CYP1A2, CYP3A4 and CYP2D6 activities reached 85.36% for 0.2 mg/mL, 83.25% for 0.05 mg/mL and 67.28% for 0.025 mg/mL of the extract, respectively.

The EC_{50} values are presented in Table 3. The EC_{50} value is negatively related to inhibitory CYP activity. A lower EC_{50} value indicates greater inhibition of CYP activity by the tested sample. The lowest EC_{50} value of the *C.rotundus* methanol extract (11.13μg/mL) was obtained with CYP2D6, which was the most inhibited enzyme, followed by CYP3A4 and CYP1A2 with EC_{50} values of 13.44and 59.48μg/mL, respectively. The EC_{50} values of CRE were higher than those of the reference compounds (significant difference, p<0.05).

Table 3. IC_{50} Values of *Cyperus rotundus* methanol extract for cytochrome P450 enzyme CYP1A2, CYP3A4 and CYP2D6 inhibition activity

Samples	IC_{50} (µg/mL) CYP		
	1A2	2D6	3A4
C.rotundus extract	59,48± 0,02	11,13± 0,04	13,44± 0,01
Safrole[a]	0,64 ± 0,10	-	-
Cimetidin[b]	-	5,19 ± 1,74	-
Erythromycin[c]	-	-	0,62 ± 0,10

[a]positive contol of CYP1A2
[b]positive contol of CYP2D6
[c]positive contol of CYP3A4

Our results are comparable to those of Usia et al. (2006)[31] who demonstrated the inhibitory capacities of cytochrome P450 CYP3A4 and CYP2D6 by several medicinal plant extracts.

The CYP inhibition activity observed for CRE could be attributed to the presence of phenolic compounds such as phenolic acids and flavonoids (flavones, flavanones and flavonols). Previous research has indicated the strong inhibitory activity of the latter compounds on cytochrome P450[32]. Flavonols such as quercetin and kaempferol that possess a 3-OH group have inhibitory activity against CYP2D6, CYP1A2 and CYP3A4. Flavonoids with more phenolic hydroxyl (-OH) groups strongly inhibit CYP3A4 activity[33].

It is important to note that the additive and synergistic effects of phytochemicals in plants are responsible for their potent bioactive properties[34]. This, by implication, means that both minor and major phenolic compounds can contribute to the antioxidant, antimicrobial and the cytochrome P450 enzyme inhibition activities exhibited by the studied plant material.

Conclusion

The results of the present investigation reveal that the use of *C.rotundus* extract as an antioxidant nutraceutical may reduce the oxidation and microbial infections. It may also be an effective regulator of cytochrome P450 enzymes, partly due to the protective action provided by its phenolic compounds.

References

1- C. Wendakoon, P. Calderon, D. Gagnon, **2012**, J. Med. Act. Plants, 1 (2), 60-68.

2- F. P. Karakaş, A. Yildirim, A. Türker, 2012, Turk. J. Biol., 36, 641-652.

3- P.A. Cox, M.J. Balick, **1994**, Sci. Am., 270, 82−87.

4- P. Saha et al., **2011**, Int. J. Res. Pharm. Sci., 2(1), 52-59.

5- E. Scio et al., **2012**, In Phytochemicals as Nutraceuticals - Global Approaches to Their Role in Nutrition and Health ; ed. by V. Rao; InTech: Croatia, **2012**, pp. 21-42

6- S. Badal, M. Shields, R. Delgoda, In Enzyme Inhibition and Bioapplications; ed. by R.R. Sharma; InTech: Croatia, **2012**, pp.39-56).

7- R.D. Bruno, V.C.O. Njar, **2007**, Bioorg. Med. Chem., 15, 5047-5060.

8- O.A. Lawal, A.O. Oyedeji, **2009**, Molecules, 14, 2909-2917.

9- A. Ardestani, R. Yazdanparast, **2007**, Int. J. Bio. Macromol., 41, 572-578.

10- A.Cuénod, G. Pottier-Alapetite, A. Labbe, Flore analytiqueetsynoptiquede la Tunisie: Cryptogamesvasculaires, Gymnospermes et monocotyledons ; Imprimerie S.E.F.A.N: Tunis, **1954**, pp. 160-165.

11- S.J. Uddin, K. Mondal, J.A. Shilpi, M.T. Rahman, **2006**, Fitoterapia, 77, 134-136.

12- K. H. Kumar, S. Razack, I. Nallamuthu, F. Khanum, **2014**, Indus. Crop. Prod., 52, 815- 826.

13- J. S. Kilani et al., **2011**, S. Afr. J. Botany, 77(3), 767-776.

14- A. Turkoglu, M.E. Duru, N. Mercan, I. Kivrak, K. Gezer, **2007**, Food Chem., 101, 267-273.

15- M.M. Al-Dabbas, K. Kanefumi, D.X. Hou, M. Fujii, **2006**, J. Ethnopharmacol., 108, 287-293

16- R. Re, N. Pellegrini, A. Proteggente, A. Pannala, M. Yang, C. Rice-Evans, **1999**, Free Rad. Biol. Med., 26 (9/1), 1231-1237

17- K.P. Suja, A. Jayalekshmy, C. Arumughan, **2005**, Food Chem., 91, 213-219.

18- M.S. Shihabudeen, H.H. Priscilla, D.K. Thirumurugan, **2010**, Int. J. Pharma Sci. Res., 1(10), 430-434.

19- P. I. Alade, O. N. Irobi, **1993**, J. Ethnopharmacol., 39, 17-174.

20- Y.T. Huang, J.I. Onose, N. Abe, K. Yoshikawa, **2009**, Biosci. Biotechnol. Biochem., 73(4), 855-860.

21- K.R. Nagulendran, S. Velavan, R. Mahesh, V.H. Begum, **2007**, E-J. Chem., 4 (3), 440-449.

22- Z. Zhou, C. Fu, **2013**, Chem. Nat. Compd., 48 (6), 963-965.

23- C. G. Barnabas, S. Nagarajan, **1988**, Fitoterapia, 3, 508-510.

24- N. Singh, B.R. Pandey., M. Bhalla, M. Gilca, **2012**, Indian J. Nat. Prod. Resour., 3(4), 467-476.

25- U. Özgen et al. **2011**, Rec. Nat. Prod., 5(1), 12-21.

26- P.M. Furneri, A.Marino, A. Saija, N.Uccella, G. Bisignano, **2002**, Int. J. Antimicrob. Agents, 20, 293-296.

27- H. Falleh et al., **2008**, C. R. Biol., 331, 371-379.

28- J.P. Rauha et al., **2000**, Int. J. Food Microbiol., 56, 3-12.

29- M.M. Cowan, **1999**, Clin. Microbiol. Rev., 12, 564-582.

30- S. Chirumbolo, 2011, J. Med. Microbiol., 60 (10), 1562-1563.

31- T. Usia, H. Iwata, A. Hiratsuka, T. Watabe, S. Kadota, Y. Tezuka, **2006**, Phytomed., 13, 67-73.

32- L.C.Cheng, L.A. Li, **2012**, Toxicol.Appl. Pharmacol., 258, 343-350.

33- P.C. Ho, D.J. Saville, **2001**, J. Pharm. Pharm. Sci., 4(3), 217-227.

34- R. H. Liu, **2003**, Am. J. Clin. Nut. 78, 517-520.

Permissions

List of Contributors

Nawal Benzidia, Salem Bakkas and Layachi Khamliche
Laboratory of Organic Chemistry, Bioorganic and Environment

Anas Salhi
Laboratory of Water and Environment Faculty of Science, University Chouaïb Doukkali, El Jadida, Morocco

S.M.M. Akram
Department of Chemistry, Islamia College of Science & Commerce, Srinagar India
Department of Chemistry, University of Kashmir, India

Aijaz Ahmad Tak
Department of Chemistry, Islamia College of Science & Commerce, Srinagar India

Peerzada and G. Mustafa
Department of Chemistry, University of Kashmir, India

Javid A. Parray
Centre of Research for Development, University of Kashmir, India

Alioune Fall, Massène Sène and Mohamed Gaye
Laboratrie de cheimie de Coordination Organique (LCCO),Departement
De Chimie,Faculte des Sciences et Techniques: Université Cheikh Anta Diop de Dakar, Sénégal

Insa Seck and Matar Seck
Département de Chimie, Faculté de Medecine, de Pharmacie et d'Odonto-stomatologie, Université Cheikh Anta Diop, Dakar, Sénégal

Generosa Gómez and Yagamare Fall
Departamento de Química Orgánica, Facultad de Química and Instituto de Investigación Biomedica (IBI), University of Vigo, Campus Lagoas de Marcosende, 36310 Vigo, Spain

Sarra Elgharbi, Karima Horchani-Naifer and Mokhtar Férid
Laboratory of Physical Chemistry of Mineral Materials and their Applications, National Research Center in Materials Sciences, Technopole Borj Cedria B.P. 73-8027 Soliman, Tunisia

Nabil Fetteh
Research Center CPG Métlaoui 2134 Gafsa Tunisia

Karima Benhamed and Leila Boukli-Hacene
Laboratoire de Chimie Inorganique et Environnement, BP 119, Université de Tlemcen, Algeria

Yahia Harek
Laboratoire de Chimie Analytique et Electrochimie BP119 Université de Tlemcen Algeria

Rym Dhouib Sahnoun, Kamel Chaari and Jamel Bouaziz
Laboratory of Industrial Chemistry, National School of Engineering, University of Sfax, BP 1173, 3038 Sfax, Tunisia

Said Bakkali, Abdelillah Benabida and Mohammed Cherkaoui
Laboratoire de Matériaux d'Electrochimie et d'Environnement, Faculté des Sciences, Université Ibn Tofaïl, 14000 Kenitra, Maroc

Otávio Augusto Chaves, Edgar Schaeffer, Carlos Maurício R. Sant'Anna, Dari Cesarin-Sobrinho and Aurelio Baird Buarque Ferreira
Departamento de Química ICE Universidade Federal Rural do Rio de Janeiro, Rodovia BR-465, Km 7,Seropedica/RJ, Brazil

José Carlos Netto-Ferreira
Departamento de Química ICE Universidade Federal Rural do Rio de Janeiro Rodovia BR-465, Km7,Seropédica/RJ, Brazil
Instituto Nacional de Metrologia, Qualidade e Tecnologia (INMETRO), Divisão de_Metrologia Química,Duque de Caxias/RJ, Brazil

Mohammad Asif
Department of Pharmacy, GRD (PG) Institute of Management and Technology, Dehradun, (Uttarakhand), 248009, India

Thiago Alessandre da Silva, Luiz Pereira Ramos, Sônia Faria Zawadzki and Ronilson Vasconcelos Barbosa
Laboratory of Synthetic Polymers: Federal University of Paraná, Department of Chemistry CEP – 81.531-980 Brazil

Ridha Djellabi and Mohamed Fouzi Ghorab
Laboratory of Water Treatment and Valorization of Industrial Wastes, Chemistry department, Faculty of Sciences, Badji-Mokhtar University, BP12 2300, Annaba, Algeria

Claudia Letizia Bianchi
Università degli Studi di Milano, Dip. Chimica & INSTM-UdR Milano, Milano, Italy

Giuseppina Cerrato and Sara Morandi
Università degli Studi di Torino, Dipartimento di Chimica & NIS Interdepartmental Centre & Consorzio INSTM-UdR Torino, Italy

Zina Barhoumi and Noureddine Amdouni
Unité Physico-Chimie Des Matériaux Solides, Département de Chimie, Facultédes Sciences de Tunis, Campus Universitaire, 2092 El Manar Tunisie

Jean François Stumbé
Laboratoire de Photochimie et d'Ingénierie Macromoléculaires, Université de Haute Alsace, 3rue A.werner 68093 Mulhouse cedex, France

Amalendu Pal
Department of Chemistry, Kurukshetra University, Kurukshetra 136 119, India

Nabila Slimani Alaoui, Anas El Laghdach, Mostafa Stitou and Aniss Bakkali
Laboratory of Water, Studies and Environmental Analysis, Department of chemistry, Faculty of Sciences, University Abdelmalek Essaâdi, B.P. 2121, Mhannech II, 93002 Tetouan, Morocco

Zoila Gándara, Pedro-Lois Suárez, Generosa Gómez and Yagamare Fall
Departamento de Química Orgánica, Facultad de Química and Instituto de Investigación Biomedica (IBI),University of Vigo, Campus Lagoas de Marcosende, 36310 Vigo, Spain

Alioune Fall, Massène Sène, Ousmane Diouf and Mohamed Gaye
Laboratoire de Chimie de Coordination Organique (LCCO), Département de Chimie, Faculté des Sciences et Techniques: Université Cheikh Anta Diop de Dakar, Sénégal

Subhabrata Chaudhury, Shukun Li and William A. Donaldson
Department of Chemistry, Marquette University, Milwaukee, WI 53201-1881 USA

Maha Ameur and Hassen Amri
Selective Organic Synthesis & Biological Activity, Faculty of Science, University of Tunis El Manar, 2092 Tunis, Tunisia

Sarra Sibous, Touriya Ghailane, Serrar Houda, Rachida Ghailane, Said Boukhris and Abdelaziz Souizi
Laboratory of Organic, Organometallic and Theoretical Chemistry, University of IbnTofail, B.P. 133, 14000 Kenitra, Morocco

Mohammed Saad Mutlaq Al-ghamdi and Hossein Bayahia
Department of Chemistry, Faculty of Science, Albaha University, Albaha, 65431, Kingdom of Saudi Arabia

P. Murali Krishna
Department of Chemistry, Ramaiah Institute of Technology, Bengaluru-560 054, Karnataka, India

K. Hussain Reddy
Department of Chemistry, Sri Krishnadevaraya University, Anantapur-515 003, AP, India

Imane Hamdou, Mohamed Essahli and Abdeslam Lamiri
University Hassan 1, Laboratory of applied chemistry and environment, faculty of science and technology, Settat-Morocco

Pawel Leszczyński and Andrzej Sporzyński
Faculty of Chemistry, Warsaw University of Technology, Noakowskiego 3, 00-664 Warsaw, Poland

Nusrat Jalbani, Shahid Bhutto, Suhail Rahujo and Farooq Ahmed
Pakistan Council Scientific and Industrial Research Laboratories Complex, Karachi, Pakistan

Marwa Sahraoui, Asma Abderrahmen, Rim Mlika and Hafedh Ben Ouada
Laboratoire des Interfaces et Matériaux Avancés, Université Monastir, Faculté des Sciences de Monastir, Département de Physique, Avenue de l'environnement 5000 Monastir, Tunisia

Abdelhafidh Gharbi
Laboratoire de la Matière molle, Faculté des Sciences de Tunis, Tunisia

Ismahen Essaidi, Nabiha Bouzouita, Hayet Ben Haj Koubaier and Ahmed Snoussi
Ecole Supérieure des Industries Alimentaires, 58, Avenue Alain Savary; 1003 Tunis, Tunisia

Faculté des Sciences de Tunis ElManar, Laboratoire de Chimie Organique Structurale : Synthèse Chimique et Analyse Physico-chimique, Tunis, Tunisia

Zeineb Brahmi and Naoki Abe
Department of Nutritional Science, Faculty of Applied Bio-Science, Tokyo University of Agriculture, 1-1-1 Sakuragaoka, Setagaya-Ku, Tokyo 156-8502, Japan

Hervé Casabianca
Centre National de Recherche Scientifique, Service Central d'Analyse, 5 rue de la Doua 69100, Villeurbanne, France

Index

www.ingramcontent.com/pod-product-compliance
Lightning Source LLC
Chambersburg PA
CBHW080631200326
41458CB00013B/4590